The Geek Atlas

128 Places Where Science & Technology Come Alive

John Graham-Cumming

O'REILLY®

Beijing · Cambridge · Farnham · Köln · Sebastopol · Tokyo

The Geek Atlas: 128 Places Where Science & Technology Come Alive
by John Graham-Cumming

Published by O'Reilly Media, Inc., 1005 Gravenstein Highway North, Sebastopol, CA 95472.

O'Reilly books may be purchased for educational, business, or sales promotional use. Online editions are also available for most titles (*my.safaribooksonline.com*). For more information, contact our corporate/institutional sales department: (800) 998-9938 or *corporate@oreilly.com*.

Editor: Julie Steele
Production Editor: Rachel Monaghan
Production Services:
Newgen North America
Copyeditor: Emily Quill
Proofreader: Rachel Monaghan
Indexer: Julie Hawks
Cover Designer: Monica Kamsvaag
Interior Designer: Ron Bilodeau
Illustrator: Robert Romano

Printing History:

May 2009: First Edition.

ISBN: 978-0-596-52320-6

[LSI] [2013-11-01]

Contents

Spain

Switzerland

Taiwan

UK

California

Connecticut

District of Columbia

Florida

Hawaii

Idaho

Illinois

Massachusetts

Maryland

Michigan

Virginia

Washington

West Virginia

Miscellaneous

000

Introduction

A Mind Forever Voyaging

For my 13th birthday my parents took me around Dove Cottage, former home of the English poet William Wordsworth. I remember only one thing from that visit—an underground stream that kept one of the rooms cool, making it ideal storage for food. There in Wordsworth's house was a little bit of science waiting to be discovered.

Unfortunately, finding great scientific places to visit isn't as easy as finding the homes of long-dead poets, painters, or writers. Call any tourist office around the world and ask about scientific, mathematical, or technological attractions, and you'll be greeted with either a long silence or a short list of the obvious famous science museums. This is a pity, because if there's one thing that makes science stand apart, it's the willingness of scientists to freely share what they do. And the world is full of wonderful sites, museums, and seemingly random places that make the geek heart pound a little harder. Many of them are even free of charge.

This book's 128 places is a personal list of sites to visit where science, mathematics, or technology happened or is happening. You won't find tedious little third-rate museums, or plaques stuck to the wall stating that "Professor X slept here" among the selections. Every place has real scientific, mathematical, or technological interest.

Not all of the places feature man-made inventions or discoveries. There are also natural phenomena such as the moving Magnetic North Pole (Chapter 128) and the Aurora Borealis (Chapter 81). And there are a few graves of famous scientists, but rest assured that those graves have equations on them.

Each place has its own chapter, and each chapter is split into three parts: a general introduction to the place with an emphasis on its scientific, mathematical, or technological significance; a related technical subject covered in more detail; and practical visiting information. The book can be used as a true travel guide (and I hope you have the opportunity to visit some of the places), but it is also for the armchair traveler, whom I hope is inspired to put the book down and learn more about the science, mathematics, and technology covered.

In the technical descriptions, I've tried to simplify the science without dumbing it down to the point of using analogies and metaphors instead of actually describing the ideas. So as you flip through the book, you'll see the sorts of pictures you'd find in a travel guide, but also a lot of diagrams and equations. (Any reader who doesn't want to deal with the equations can safely read the first part of each chapter.)

There's also quite a bit of abstract mathematics covering topics like set theory, transfinite numbers, Fermat's Last Theorem, prime numbers, group theory, and more. To non-mathematicians these topics may be off-putting, but I suggest you stick with them. I promise that when you understand Cantor's diagonalization argument (page 180) showing the existence of different magnitudes of infinity, a whole world of pure mathematical beauty will be revealed. (Yes, I admit it—I'm a mathematics nerd.)

One thing that I've been asked by reviewers again and again is to recommend one single must-see place. Picking one place is next to impossible—there's just so much great science out there—but I will admit to shedding a tear every time I see the Difference Engine at the Science Museum in London (Chapter 77). It's mathematics in motion and arithmetic in action.

A disappointing trend with science museums today is a tendency to emphasize the "Wow!" factor without really explaining the underlying science. If it appears that I've overlooked your favorite museum, ask yourself whether it has any of the following annoying attributes: a short name ending with an exclamation mark, a logo featuring pastel colors or a cuddly cartoon mascot, or an IMAX theater. Believe me, there's more real "Wow!" in understanding Foucault's measurement of the speed of light (page 60) than in any movie, no matter how big it is.

A number of places do not appear in this book because tours have ended or been severely curtailed on account of "security concerns." This applies almost exclusively to the United States, where some interesting places for scientific tourism are now considered too sensitive for the general public. Other places have restricted access to U.S. citizens only (which is ironic, given the contribution of non-Americans to U.S. scientific research); some of these have been included in the hope that citizenship restrictions will be lifted in the coming years.

If your favorite scientist is missing, it may simply be because there's no single place that sums up his or her work. For example, I was hoping to include rocketry pioneer Robert Goddard, but none of the possible places is really suitable—the Goddard Rocket Launching Site in Massachusetts is just a historic marker in the middle of a golf course; the Goddard Space Flight Center in Maryland is great, but there are only so many NASA sites this book could hold; and the Roswell Museum in New Mexico bills itself more as an art museum than anything else. Nevertheless, of the three, the Roswell Museum is the best—it does, after all, have most of Goddard's equipment on display.

Other places almost made the cut, but weren't quite up to scratch. I wanted to include the Florence Nightingale Museum in London, because Nightingale's contribution to using graphics to illustrate quantitative information is largely overlooked. She was the first woman to become a member of the Royal Statistical Society, and her diagrams of the causes of death at a military field hospital are on display at the museum. Unfortunately, the display is a poor reproduction that provides no real information about what the visitor is seeing or about Nightingale's contribution to the use of graphics (including pie charts).

If you're still asking yourself why your favorite place isn't covered, please visit the book's accompanying website and tell me about it. I'm probably not aware of it and would love to go and visit. Who knows, there might be enough good ideas for a second edition!

—jgc
May 2009

P.S. As it turns out, I probably should have paid more attention at Dove Cottage. Years later I discovered that Wordsworth was an admirer of Newton, and wrote the following lines inspired by Newton's statue at Trinity College, Cambridge:

> The antechapel where the statue stood
>
> Of Newton with his prism and silent face,
>
> The marble index of a Mind for ever
>
> Voyaging through strange seas of Thought, alone.

Conventions Used in This Book

Throughout this book I've used metric units. Since the book covers a mixture of countries, such as the U.S., which uses its own mixture of Imperial and customary units; France, which uses metric; and the U.K., which uses a hodgepodge of Imperial and metric, I decided to take the scientific option and go metric. This also avoids problems like the difference in volume between a U.S. gallon and an Imperial gallon.

Metric units may feel a little peculiar to American readers since I speak about distance between places in the U.S. in kilometers. To make up for that, despite being British, I use U.S. spelling throughout.

For place names, I've opted for the format *Place Name, Nearest Village/City/Town, Country* for everywhere but the U.S. Within the U.S., the format is *Place Name, Nearest City/Town, State*. I use two-letter U.S. state abbreviations. Non-American readers who are unsure which of MI, MO, and MS is Missouri can check in with the U.S. Postal Service at *http://www.usps.com/ncsc/lookups/usps_abbreviations.html*.

Each place in the book has a Practical Information section that typically gives the address of the associated website if there is one, and directions if not. In addition, every place has its latitude and longitude listed, ready for use with a GPS navigation device or an online mapping service.

A quick visual reference at the start of each chapter gives some basic practical information; the following four icons are used:

FREE Entry to this place is free.

Refreshments are available on site.

This place is suitable for children.

This place can be visited in any kind of weather.

But please remember that things change. A previously free attraction may start charging a fee, a restaurant may close, or access to the site may be restricted. Check with the place before visiting; if things have changed, please let me know so that future editions of this book can be updated.

We'd Like to Hear from You

We have verified the information in this book to the best of our ability, but you may find that features have changed or that we may have made a mistake or two (shocking and hard to believe). Please let us know about any errors you find, as well as your suggestions for future editions by writing to:

O'Reilly Media, Inc.
1005 Gravenstein Highway North
Sebastopol, CA 95472
800-998-9938 (in the United States or Canada)
707-829-0515 (international or local)
707-829-0104 (fax)

We have a web page for this book where we list examples and any plans for future editions. You can access this information at:

http://www.oreilly.com/catalog/9780596523206

You can also send messages electronically. To be put on the mailing list or request a catalog, send an email to:

info@oreilly.com

To comment on the book, send an email to:

bookquestions@oreilly.com

For more information about our books, conferences, Resource Centers, and the O'Reilly Network, see our website at:

http://www.oreilly.com

Safari® Books Online

When you see a Safari® Books Online icon on the cover of your favorite technology book, that means the book is available online through the O'Reilly Network Safari Bookshelf.

Safari offers a solution that's better than e-books. It's a virtual library that lets you easily search thousands of top tech books, cut and paste code samples, download chapters, and find quick answers when you need the most accurate, current information. Try it for free at *http://my.safaribooksonline.com*.

001

Parkes Radio Telescope, Parkes, Australia

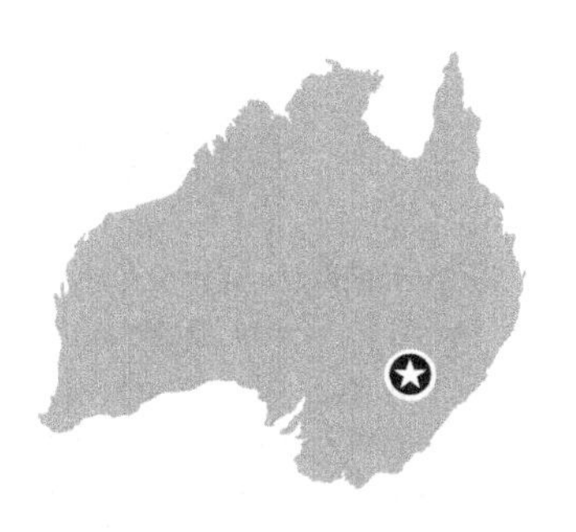

32° 59′ 59.8″ S, 148° 15′ 44.3″ E

The Dish

If you find yourself in the land down under (or just happen to be lucky enough to live there), then there's one landmark that shouldn't be missed, because it's the star of a movie that is all about science. The film is *The Dish*, and its star is the Parkes Radio Telescope (Figure 1-1).

Figure 1-1. The Dish; courtesy of Alex Cheal (alexcheal)

Halfway between Melbourne and Brisbane, and 20 kilometers north of the small town of Parkes, stands the gracefully curved dish that since 1961 has been listening to the southern sky for radio transmissions.

Pulsars (The Original Little Green Men)

In July 1967, a team at Cambridge University was working on a radio telescope to listen to signals from quasars (radio sources coming from somewhere in the sky). But they discovered something strange—a pulsating radio signal from the heavens that lasted 40 milliseconds and repeated regularly every 1.337 seconds. At first they thought it must be noise from some earthly source, but quickly realized it was not and named the source LGM-1 (Little Green Men-1).

What they had discovered was a pulsar (or pulsating star), but having never heard a radio source of such regularity coming from the sky, they speculated that it might be a message from a distant civilization, or a navigational beacon for some unknown beings traveling the universe.

Today, LGM-1 is known as the pulsar CP1919, and is one of more than 1,000 known pulsars. And pulsars have been discovered that emit more than just radio waves—they also eject regular bursts of X-rays and gamma rays.

Pulsars are created by rapidly rotating neutron stars. A neutron star is created when a star undergoes a supernova, running out of energy and suddenly collapsing in on itself because of its own gravity. This results in an incredibly dense and compact corpse of a star: a neutron star.

Neutron stars are typically less than 20 kilometers across (the distance from the town of Parkes to the telescope), but contain more mass than the Sun. The interior of a neutron star is made up of neutrons, because the incredible pressure in the star has forced protons and electrons together, eliminating any charge. Deep inside the neutron star, the pressure is so great that there's probably a soup of even more fundamental particles such as quarks.

Neutron stars are prevented from completely collapsing by the Pauli Exclusion Principle, which says that no two neutrons (in fact, no two identical fermions: see page 118) can occupy the same place at the same time.

It is believed that as the neutron star rotates, its magnetic field interacts with charged particles, leaving its surface to generate electromagnetic radiation. The radiation may be in the radio spectrum or in the form of gamma or X-rays. It leaves the star in a continuous beam emanating from the star's magnetic north and south poles. This is illustrated in Figure 1-2.

Figure 1-2. A pulsar with its magnetic field and beams of radiation

Since the star's rotational poles and magnetic poles are offset, the beam isn't continuously pointed at the Earth, and only sweeps the Earth (appearing to us as a pulse) once per rotation.

In truth, much about pulsars is still not understood, even after 40 years of listening to them. Perhaps the Cambridge University team's LGM-1 designation was correct after all, and Parkes has been missing extraterrestrial "g'days" since it opened in1961.

It was the Parkes Radio Telescope's role in picking up transmissions from the Apollo 11 moon landing that made it a movie star. When Neil Armstrong and Buzz Aldrin landed on the moon on July 20, 1969, their transmission of telemetry and television pictures was initially sent to the Goldstone Observatory in the Mojave Desert in California. But problems developed, and NASA switched to getting the signals from the Honeysuckle Creek receiver near Canberra in Australia. Shortly thereafter, they switched to Parkes and found the signal so good that they stuck with the Dish for the rest of the transmission.

The Dish has been involved in a number of other space missions, including Voyager 2 (when it passed close to Uranus and Neptune and returned images of the planets); the Giotto probe that flew close to Halley's Comet in 1986; and the Galileo probe, which photographed Jupiter in 1997.

The Dish has a small visitor center and two small cinemas showing films explaining radio astronomy and the solar system. While the visitor center is free, there's a small fee for the cinemas. Since the Dish is located far from civilization, there's a café serving drinks and meals, and also free picnic facilities and barbeque equipment.

Unfortunately for visitors, the Dish itself is in constant use and is not open for tours. But the observatory occasionally hosts open days when the general public can climb up into the Dish's rotating mounting, and afterward attend a talk by one of the telescope's scientists. In the past, the open days have been rounded out by a screening of *The Dish* under the starlit sky.

As you approach the Dish, switch off anything with a radio in it (such as a mobile phone)—the Dish is listening for very faint radio signals from across the cosmos, so it doesn't need to listen to you nattering away. Since Parkes is a bit remote, consider making the trip to coincide with a major local event: the Parkes Elvis Festival (held annually in the second week of January) is not to be missed.

But the 64-meter-wide Dish's main job is radio astronomy, with a special emphasis on pulsars (see sidebar). Over the years, the Dish has been upgraded to make it more and more sensitive to the incredibly faint signals that reach the Earth's surface.

Practical Information

Information about the Parkes Radio Telescope and other Australian observatories is available from *http://outreach.atnf.csiro.au/*.

002

Zentralfriedhof, Vienna, Austria

48° 8′ 58″ N, 16° 26′ 28″ E

A Scientist Among the Composers

If you need an excuse to visit the beautiful Austrian capital, then use the Zentralfriedhof (Central Cemetery) as your reason. Although the cemetery may not be one of the most famous attractions in Austria, it is the final resting place of many celebrated Austrians (and others), including Beethoven, Brahms, Schubert, four Strausses, and a host of other artists and politicians. But the grave that's waiting for scientific visitors is the one with a fundamental equation of thermodynamics written upon it.

That grave belongs to Ludwig Boltzmann, the Austrian physicist who created statistical mechanics (which helps to explain how the fundamental properties of atoms, such as mass or charge, determine the properties of matter) and showed that the laws of mechanics at an atomic level could explain the second law of thermodynamics (roughly that heat cannot flow from a cool body to a hotter body) via Boltzmann's Equation (see Equation 2-1).

$$S = k \log W$$

Equation 2-1. Boltzmann's Equation

Boltzmann lived during the 19th century (he died just after the turn of the 20th) and firmly believed that matter was composed of atoms and molecules. Despite the fact that Dalton (see Chapter 55) had described atomic weights in 1808, there was still debate about the existence of atoms. But Boltzmann used what others considered to be an unproven theory to apply, to great effect, probability theory to the physical world through statistical mechanics.

Along with James Clerk Maxwell (see Chapter 35) and Josiah Willard Gibbs, Boltzmann was one of the most important physicists of the 19th century. His grave (Figure 2-1) is a testament to his importance, with his famous equation cut into the stone and featuring an imposing bust of the scientist. He is buried alongside members of his family.

Figure 2-1. Boltzmann's grave; courtesy of Martin Röll (martinroell)

The cemetery itself is enormous—2.4 square kilometers in size with around three million people buried there, making it one of the largest cemeteries in Europe. One area contains the tombs of notables, of which Boltzmann is the only scientist.

While you are in Vienna, there's also a small museum in Freud's former home that is worth a visit.

Practical Information

The number 71 tram has multiple stops at the Zentralfriedhof. Boltzmann's grave is in section 14C of the cemetery, which is closest to the Zentralfriedhof Tor 2 stop.

Statistical Mechanics and Entropy

The relationship between the macrostates (such as volume, temperature, and pressure) and the microstates (the location, mass, and velocity of individual atoms) of a material is fundamental to statistical mechanics, and Boltzmann laid its foundations. The macrostates are easily measured; the microstates are not.

Externally, a bottle full of air might be described by a small number of macrostates—its exact volume, temperature, and pressure could be measured, for example. But inside the air, the molecules are moving around and bumping into each other, and for any fixed macrostates the microstates are constantly changing. Nevertheless, there's a relationship between the micro and macro.

Boltzmann's entropy can be thought of as a measure of the degree of chaos inside the bottle, or a measure of the number of different ways the molecules of air can arrange themselves to achieve the same volume, pressure, and temperature.

Imagine a pack of 52 playing cards dropped on the floor. You can think of them as having one macrostate: the number of playing cards that are face up. That macrostate, like volume, temperature, and pressure, can be easily measured. But for any specific number of face-up cards, there are many different combinations of individual face-up cards: that is, for any macrostate there are many possible microstates (each card has its own microstate specifying whether it is up or down).

If all the cards are face up, then there's little chaos: each card's microstate is known. The same applies if all are face down. The most chaos occurs at the midpoint of these two extremes: when 26 cards are face up and 26 cards are face down. Then there are 495,918,532,948,104 possible microstates.

Boltzmann's famous equation (as interpreted by Max Planck) uses the number of possible microstates, W, for any given set of macrostates. The constant k is known as Boltzmann's constant. The resulting value, S, the entropy, is actually a macrostate just like volume, temperature, or pressure and can be calculated for an ideal gas (for more on ideal gases, see page 211).

For any given volume, temperature, and pressure, the entropy, S, is a measure of how uncertain we are about the internal state of the gas.

Boltzmann's equation is engraved on his tomb because it is the fundamental link between the microscopic world of moving, colliding atoms and the macroscopic world of temperatures, pressures, and volumes.

003

Atomium, Brussels, Belgium

50° 53′ 41″ N, 4° 20′ 28″ E

An Iron Crystal

The Atomium building in Brussels was built for the International Exhibition in 1958. It represents the crystal structure of iron (actually just one of the allotropes of iron; see sidebar) and is constructed from steel with an aluminum skin. Like the Eiffel Tower before it, the Atomium was intended to be a temporary structure, but survived because of its popularity.

The Atomium consists of 9 spheres representing iron atoms, connected by 20 tubes representing the bonds between the iron atoms, forming a cube with iron atoms at the vertexes and a single iron atom at the center. The cube structure sits balanced on one vertex for aesthetic reasons, and is supported by extra pillars connected to spheres near the ground. The entire structure is over 100 meters tall.

Between 2004 and 2006, the Atomium was extensively renovated. The corroded aluminum was removed and replaced with stainless steel (which is made from a different allotrope of iron than the allotrope represented by the Atomium itself).

There's a lot going on inside the Atomium. In the sphere at the base, there's an exhibition covering the 1950s and the International Exhibition of 1958. One of the other spheres contains an exhibition space that houses temporary exhibits. The topmost sphere offers a restaurant and a panoramic view over Brussels. There's even a sphere used exclusively by school trips—the children get to sleep inside.

If you're wondering why there's no photograph of the Atomium here to show you just how cool it is, it's because you can't photograph it and publish the picture. Through Belgian copyright law, the Atomium insists that it owns the copyright of photographs even if they're taken by a third party, and that a large fee must be paid for any reproduction. Happily, we have the Web. To see this incredible structure, do a Google Image search for "Atomium".

Iron Allotropes

The iron crystal represented by the Atomium is actually just one of three possible structures (or allotropes) of iron: alpha, beta, and delta. Allotropy (forming different structures from a single element) is not limited to iron; other elements (such as carbon and oxygen) also have allotropes.

When molten iron starts to cool down, it first forms the delta allotrope, which has an iron atom at each vertex of a cube and a single iron atom in the middle of the cube. Each atom connects to four other atoms (their adjacent neighbors along the sides of the cube and the central atom). This structure is called a body-centered cubic, and is the structure represented by the Atomium (see Figure 3-1).

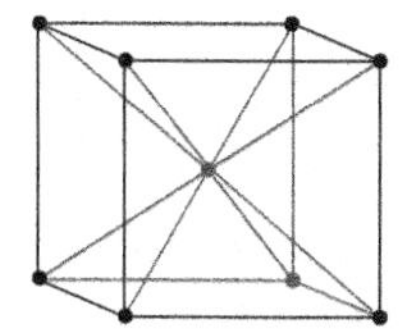

Figure 3-1. Body-centered cubic

The Delta Allotrope of Iron

As iron cools further (below 1394°C), it then forms the gamma allotrope, which has a different structure. Gamma iron is a face-centered cubic (see Figure 3-2); it has a cubic structure with atoms at the vertexes, but also a single atom in the center of each of the cube's faces. Gamma iron is used in the production of stainless steel, where it is alloyed with chromium. (The chromium forms an invisible oxide layer on the outside of the stainless steel, which protects it from corrosion.)

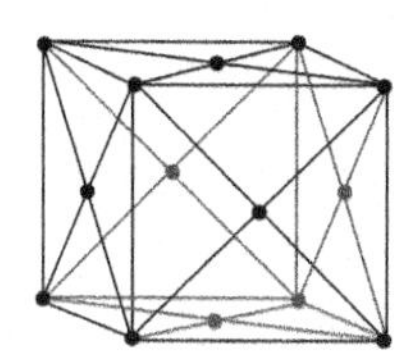

Figure 3-2. Face-centered cubic

When below 912°C, iron forms its final allotrope, the alpha form. This has the same structure as the delta allotrope, but with the important property that once it falls below 770°C it becomes ferromagnetic (770°C is the so-called *Curie point*, above which a ferromagnetic material loses its magnetism). The alpha allotrope is used in cast iron and steel making.

Carbon Allotropes

Probably the most interesting allotropes are those made from carbon. Carbon has many different allotropes including diamonds, graphite, and carbon nanotubes.

Diamonds form a complex structure that consists of tetrahedrons of four carbon atoms joined together to fit inside a cube. This structure is known as a diamond cubic (see Figure 3-3).

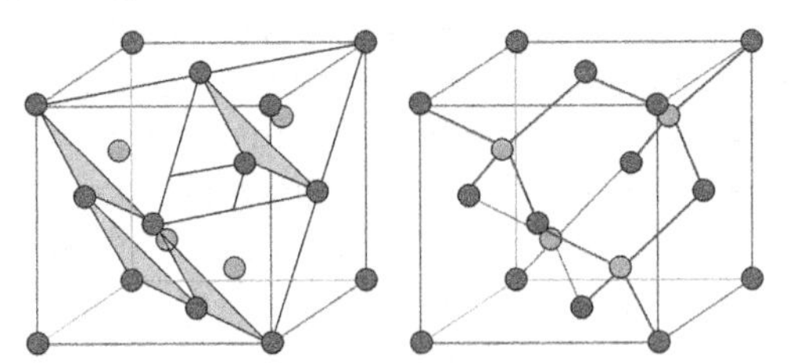

Figure 3-3. Diamond structure

If you don't have diamonds at home, you've probably got another carbon allotrope: graphite, which is found in pencil leads. Graphite is almost the polar opposite of diamonds—it's soft, black, and opaque. The carbon in graphite consists of hexagonal units that link together into flat sheets (Figure 3-4). The sheets lie on top of each other and in the presence of air they can slip around, making graphite a useful lubricant.

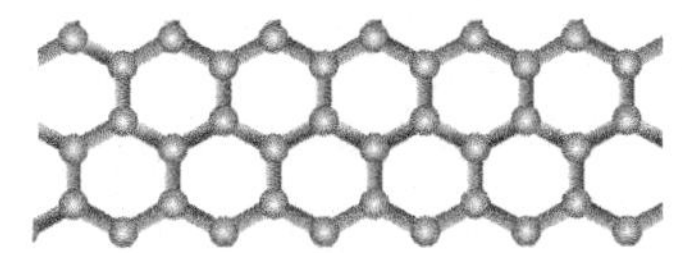

Figure 3-4. Graphite structure

Yet another allotrope of carbon forms tubes (known as nanotubes). A nanotube consists of a sheet of carbon atoms in the hexagonal structure seen in graphite, but that has been rolled up to form a tube (Figure 3-5).

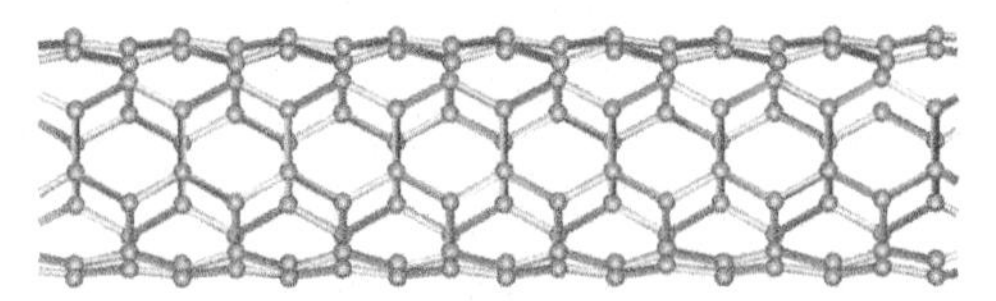

Figure 3-5. Carbon nanotube

Nanotubes are extremely strong and stiff, are very good conductors of heat, and can form either semiconductors or good conductors. But they are a relatively recent discovery, and have only lately become available commercially. They are likely to be as important to the 21st century as iron was to the 19th.

Although the interior of the Atomium is interesting, there's no real science to be found there, so you can easily avoid the entrance fee and view the structure from the outside. After all, copyright law doesn't apply to images on the retina.

Practical Information

Information about the Atomium and details on how to visit it are available at *http://www.atomium.be/* (click "EN" for English).

004

Baddeck, Nova Scotia, Canada

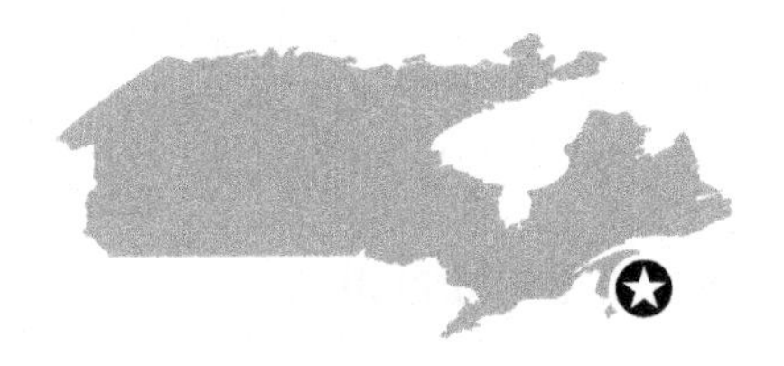

46° 6′ 0″ N, 60° 45′ 15″ W

Alexander Graham Bell's Summer Home

Following in the footsteps of Alexander Graham Bell isn't easy: he moved frequently and worked on a wide variety of inventions. He was born in 1847 at 16 South Charlotte Street in Edinburgh, Scotland, and was home-schooled there until high school. He moved to England as a young man and helped his father teach deaf people to speak. He emigrated to Canada with his family in 1870. He spent years in the United States, mostly in the Boston area, and became a U.S. citizen; his parents remained in Canada in Brantford, Ontario, and Bell visited frequently and had a workshop there.

But the best place to understand Bell's life and work is in Nova Scotia, where Bell lived from 1889 until his death in 1922.

Bell is known, of course, for the invention of the telephone, but he started out trying to invent a method of sending multiple telegraph signals down the same wire. He worked on his harmonic telegraph, which would send multiple signals—each having its own pitch—down the same wire at the same time, while continuing to teach deaf students in Boston.

In secret, because he feared that people would steal his ideas, Bell and his assistant, Thomas Watson, were also working on sending speech by wire. On March 10, 1876, Bell drew a diagram of a working telephone in his notebook (Figure 4-1).

Part of the text reads:

> Mr Watson was stationed in one room with the Receiving Instrument. He pressed one ear closely against S and closed his other ear with his hand. The Transmitting Instrument was placed in another room and the doors of both rooms were closed.
>
> I then shouted into M the following sentence: "Mr Watson—Come here—I want to see you." To my delight he came and declared that he had heard and understood what I said.

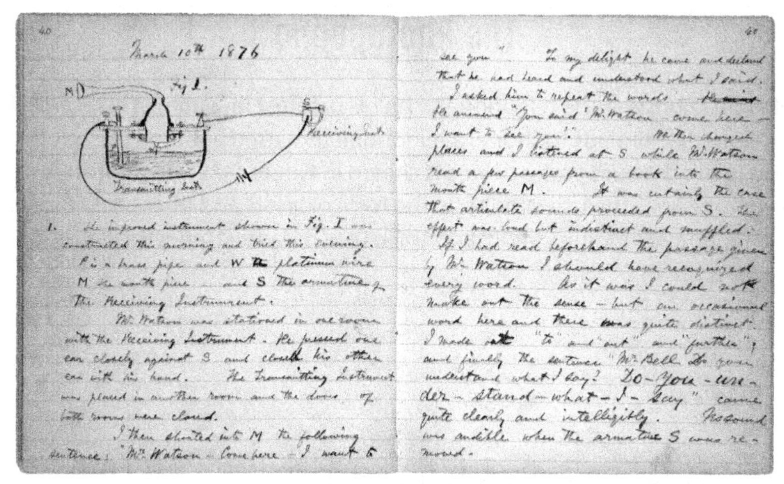

March 10th 1876

Fig. 1.

Receiving Inst.

Transmitting Inst.

1\. The improved instrument shown in Fig. I was constructed this morning and tried this evening. P is a brass pipe and W the platinum wire M the mouth piece and S the armature of the Receiving Instrument.

Mr. Watson was stationed in one room with the Receiving Instrument. He pressed one ear closely against S and closed his other ear with his hand. The Transmitting Instrument was placed in another room and the doors of both rooms were closed.

I then shouted into M the following sentence: "Mr. Watson – Come here – I want to see you." To my delight he came and declared that he had heard and understood what I said.

I asked him to repeat the words. He answered "You said 'Mr. Watson – come here – I want to see you.'" We then changed places and I listened at S while Mr. Watson read a few passages from a book into the mouth piece M. It was certainly the case that articulate sounds proceeded from S. The effect was loud but indistinct and muffled. If I had read beforehand the passage given by Mr. Watson I should have recognized every word. As it was I could not make out the sense – but an occasional word here and there was quite distinct. I made out "to" and "out" and "further"; and finally the sentence "Mr. Bell Do you understand what I say? Do-you-un-der-stand-what-I-say" came quite clearly and intelligibly. No sound was audible when the armature S was removed.

Figure 4-1. Bell's telephone: March 10, 1876

They changed places, with Watson shouting into the microphone. Bell's notes continue:

> and finally the sentence "Mr Bell. Do you understand what I say? DO-YOU-UNDERSTAND-WHAT-I-SAY" came quite clearly and intelligibly.

Bell quickly patented the telephone and established the Bell Telephone Company. With money from his invention, Bell was able to continue researching other ideas. He married and built a large house on Cape Breton Island. The house, near the village of Baddeck, is still owned by the Bell family.

The nearby Alexander Graham Bell National Historic Site features a museum of Bell's life and work that showcases his teaching of the deaf, his invention of the telephone, and, among his many interests, his fascination with hydrofoils. In 1919, his HD-4 hydrofoil achieved the world-record speed of almost 71 mph rising above the water. A reconstruction of the HD-4 is on display at the museum.

Bell also helped run the Aerial Experiment Association from Baddeck. The AEA was founded in 1907 with the help of engine expert Glenn Curtiss (see Chapter 112 for information on the Glenn H. Curtiss Museum). One of the AEA's aircraft, the June Bug, won the first aeronautic prize for a 1-kilometer flight.

Practical Information

Parks Canada has information about visiting the Alexander Graham Bell National Historic Site at *http://www.pc.gc.ca/lhn-nhs/ns/grahambell/index_e.asp*.

The Photophone

Bell didn't consider the telephone to be his greatest invention. He reserved that honor for the photophone, a telephone that used light to transmit the voice (Bell's drawing of the device is shown in Figure 4-2). The photophone's design presages the use of light (in the form of lasers and fiber optic cables) for telephone communication, and the use of parabolic reflectors as receivers (see Chapter 48).

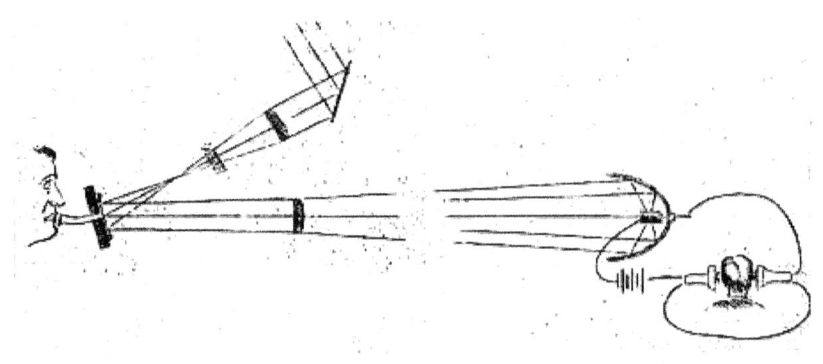

Figure 4-2. The photophone

Bell had established the Volta Laboratory (which would later become the famous Bell Labs) in Washington, DC. There, on June 3, 1880, Bell transmitted voice across the street using his photophone. He was so sure that the photophone was his greatest invention that he even deposited an example of the device sealed in tin boxes at the Smithsonian Institution to ensure that he would be recognized as its inventor. The boxes were finally opened in 1937.

The photophone worked by focusing sunlight onto a mirror. The mirror vibrated with the speaker's voice and reflected light onto a parabolic mirror at the receiver. Positioned at the focus of the parabolic mirror was a selenium cell. Selenium's resistance changes with the amount of light falling on it, so Bell used the changing resistance to reproduce the sound transmitted by the light.

005

Mendel Museum of Genetics, Brno, Czech Republic

49° 11′ 27.46″ N, 16° 35′ 34.85″ E

Ten Years of Observing Peas

In 1865, an Austrian monk living in Brno in what is now the Czech Republic wrote and then published a paper that is now recognized as the foundation of modern genetics. In his paper, *Experiments in Plant Hybridization*, Gregor Mendel summarized the result of over a decade of experimentation with 30,000 *Pisum sativum* plants (otherwise known as the humble garden pea).

Mendel described how seven key traits were passed from generation to generation of plants. He carefully observed seed shape and color, flower color, pod shape and color, and size and shape of the plant's stem.

Through careful experimentation of cross-breeding different strains of peas, he discovered that some traits appeared to be dominant and others were recessive. He posited that each pea plant contained two copies of each trait, and that certain forms of an individual trait could overpower (or dominate) others. He showed that violet-colored flowers dominated white-colored flowers, for example, and went on to identify the dominant forms of each of the seven traits studied. Only if both copies of the trait were the recessive version would that version actually be expressed (resulting in, say, a white flower).

This led him to develop two laws of inheritance (see sidebar), which have stood the test of time. With our modern knowledge of chromosomes and DNA, the mechanism underlying Mendelian inheritance is now understood. But Mendel knew none of that; he simply theorized from the data he had in front of him. The key to Mendel's success was that he carefully documented what he saw, and made sure to begin with plants that produced exactly one of the traits he was observing (for example, plants that only had offspring of the same flower color) so that he knew he was starting from a pure, known point.

At the same time that Darwin (see Chapter 6) was describing the process of evolution, Mendel was unlocking the mechanism that made evolution work.

Mendel's Laws

Mendel's two laws are known as the Law of Segregation and the Law of Independent Assortment, and between them they describe the rules of inheritance.

Mendel's Law of Segregation says that in passing traits down to offspring, only one gene from each parent is selected (we now know that this happens because only one half of each parent's DNA is used in creating the child). The Law of Independent Assortment says that traits are passed down independently (for example, there is no connection between flower color and the shape of seeds).

Mendel proposed that each trait in the pea plant (such as flower color) was actually dictated by two characteristic units (which he called factors). Today, these factors are called genes. By experimentation, Mendel discovered that some genes are dominant and some recessive. In pea plants, the gene for violet flowers is dominant and the gene for white flowers is recessive.

Labeling the violet gene as V and the white gene as w, a pea plant could contain any combination of the two (VV, Vw, wV, or ww), but if one of them was V the flowers would appear violet. (The combination of genes is called the genotype, and the physical expression of the genes is called the phenotype.) Because the V gene is dominant, only plants with the ww genotype would have a white-flower phenotype.

In his experiments, Mendel started out with two groups of pea plants that produced homogeneous children for a specific trait (such as all white-flowered children, or all violet-flowered children). This ensured that both genes for that trait were identical in the parent plants. That is, his white-flowered plants had the ww genotype, and the violet-flowered plants had the VV genotype.

He then cross-bred these two groups. In observing two generations of plants (depicted in Figure 5-1) from these pure parents, he noticed a ratio of 1:3 (for every white-flowered child, there were three violet-flowered children), and from his knowledge of combinatorics he was able to deduce the dominant/recessive split.

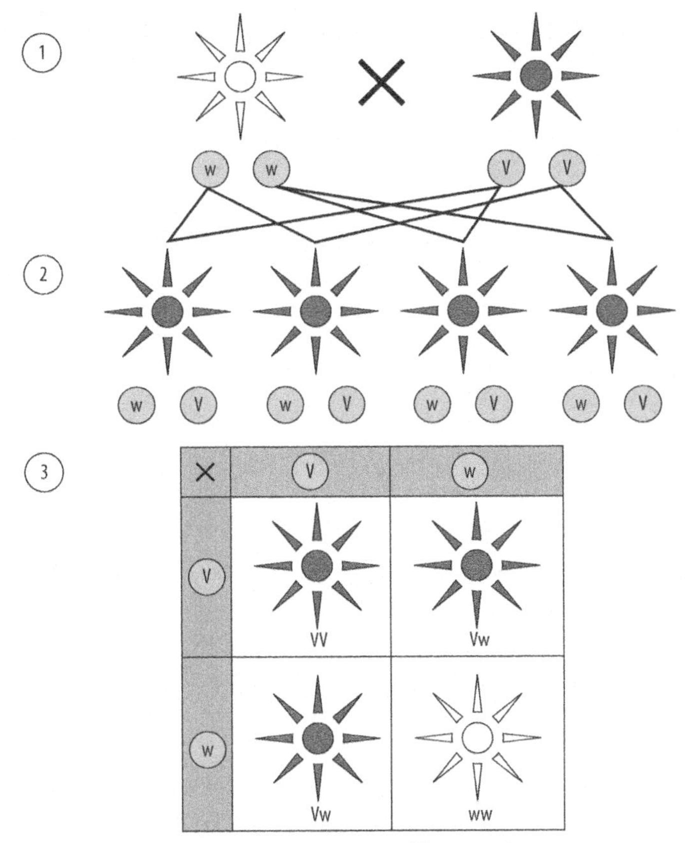

Figure 5-1. Inheritance of flower color

He also mixed traits to see if any were connected, and he was also able to determine the lack of connections simply by looking at the ratios. He did all this based on his own observations, without any knowledge of the underlying mechanism. Yet, he was correct (at least for the types of traits he observed).

Traits that follow Mendel's Laws are now known as Mendelian traits, and the process described above is Mendelian inheritance. In humans, things like blood type, the type of earwax you produce, and whether you have a cleft chin are all traits that follow Mendel's Laws.

Some traits don't have a single gene controlling them—the individual genes follow Mendel's Laws, but the expressions of them don't. In humans, eye color is determined by at least two genes (one controlling brown versus blue, the other green versus hazel).

Today, the Mendel Museum of Genetics can be found at the Abbey of St. Thomas, the Augustinian monastery where Mendel lived and worked. The museum covers everything from Mendel's experiments through the unravelling of the secrets of DNA to the present day. On the floor leading to the museum is a trail of DNA letters.

Outside there are the foundations of the greenhouses that Mendel used, and a restored garden filled with pea plants illustrating the fact that violet is the dominant flower color. There's also a collection of beehives; Mendel was busy with many other projects: he kept bees, recorded weather patterns, studied mathematics, and taught at a local school. Eventually he became Abbott of the monastery.

The monastery is in picturesque Old Brno, close to the city center; it's possible to visit it and the rest of the city in a day. Vienna, Prague, and Bratislava are all a two-hour drive away.

If you are staying in Brno (or elsewhere), then it's essential to sample Czech beer. Close to the monastery is the Starobrno brewery, which has been making beer since Mendel's time. Don't drink too much, though: the museum holds regular lectures on genetics by scholars from around the world (check their website for a calendar), and you'll need to have a clear head to be able to follow along.

Practical Information

Visitor information about the Mendel Museum is available in English at *http://www.mendelmuseum.muni.cz/en/*. Information is provided in English throughout the museum, and guided tours in English are also possible.

006

Galápagos Islands, Ecuador

0° 40′ 0″ S, 90° 33′ 0″ W

The Second Voyage of the Beagle

The Galápagos Islands are an archipelago of volcanic islands about 1,000 kilometers off the coast of Ecuador in the Pacific Ocean. They were visited in 1835 by the British survey ship HMS *Beagle*, which sailed from 1831 to 1836 around South America and on to Australia before returning to Britain, gathering information along the way about safe landing places and navigable rivers. The most famous passenger aboard was the 22-year-old Charles Darwin.

Darwin spent most of the voyage ashore, surveying the geology of the land and collecting specimens of local fauna, flora, and fossils. As the voyage progressed, he kept a journal, and copies of the journal and his specimens were sent back to Britain. By the time Darwin returned home, he was a minor scientific celebrity.

Unbeknownst to his shipmates, the ideas that would become his famous theory were forming in Darwin's mind during the voyage. A year after his return, Darwin sketched his "tree of life" diagram in a notebook, and went on to work out the theory of natural selection.

On the voyage, Darwin spent over a month surveying the Galápagos Islands. Because the islands were far from the nearest land, and because there were so many of them, they made an ideal location for observing different varieties of the same species. Darwin noted that tortoises, mockingbirds, and finches were present on different islands, but differed from island to island.

The finches were a key clue to the theory of natural selection, although Darwin thought that the birds he had collected on different islands were unrelated. It was only upon his return to London that it became clear that these were 12 species of finch dissimilar from any other finches in the world. He reasoned that the finches had evolved specific beak sizes and shapes because of different sources of food on the various islands, writing:

> Seeing this gradation and diversity of structure in one small, intimately related group of birds, one might really fancy that from an original paucity of birds in this archipelago, one species had been taken and modified for different ends.

Artificial Speciation

Geographic isolation is one of the major ways in which a single species can split into two. If a species, such as Darwin's finches, becomes separated into two groups isolated by some geographical feature (such as mountains or oceans), it's possible for allopatric speciation to occur—the two groups evolve in different ways in accordance with their environment.

Going beyond the evidence of existing species, experiments have been performed to test the theory of allopatric speciation, deliberately splitting a population in two and exposing the two groups to different environments. After a number of generations, the evolution of the two groups has often diverged enough that they are no longer interested in mating with one another. Scientists often use fruit flies for artificial speciation experiments.

Fruit flies are a favorite tool of biologists because they have a short lifespan, are easily cultured in laboratories, and are readily available. The 1933 Nobel Prize in Medicine was awarded to the American scientist Thomas Hunt Morgan for the discovery that chromosomes (long, single pieces of DNA) carried the genes responsible for the inheritance of traits. He used the fruit fly *Drosophila melanogaster*, and its many mutations, to confirm his theory.

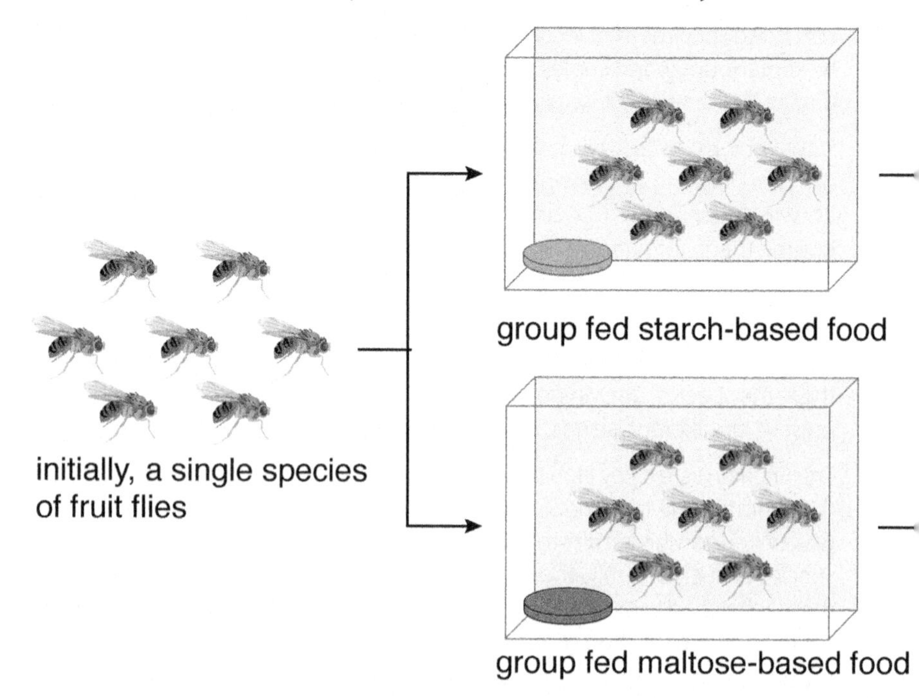

Figure 6-2. Dodd's artificial speciation experiment

In 1989, Diane Dodd (of Yale University's Department of Biology) reported on an experiment using *Drosophila pseudoobscura* (a species of fruit fly). This experiment involved taking a population of fruit flies and splitting it into two isolated groups (see Figure 6-1). One group was fed a diet of maltose; the other was fed a diet of starch. The experiment was repeated with an additional three populations, each split into the same two groups with the same maltose or starch diet.

The separate groups were allowed to feed, reproduce, and die for about one year. The flies were then bred for a single generation, all feeding on the same mixture of cornmeal, molasses, and agar. They were then given the opportunity to mate with one another in a series of tests that mixed originally maltose-fed and starch-fed males with samples of either maltose-fed or starch-fed females.

The experiment showed that the two populations had become behaviorally isolated—they would only mate with flies that had come from a population fed with the same food. Thus, isolation because of geography, which could lead to different food supplies being available, could result in populations that would no longer breed together.

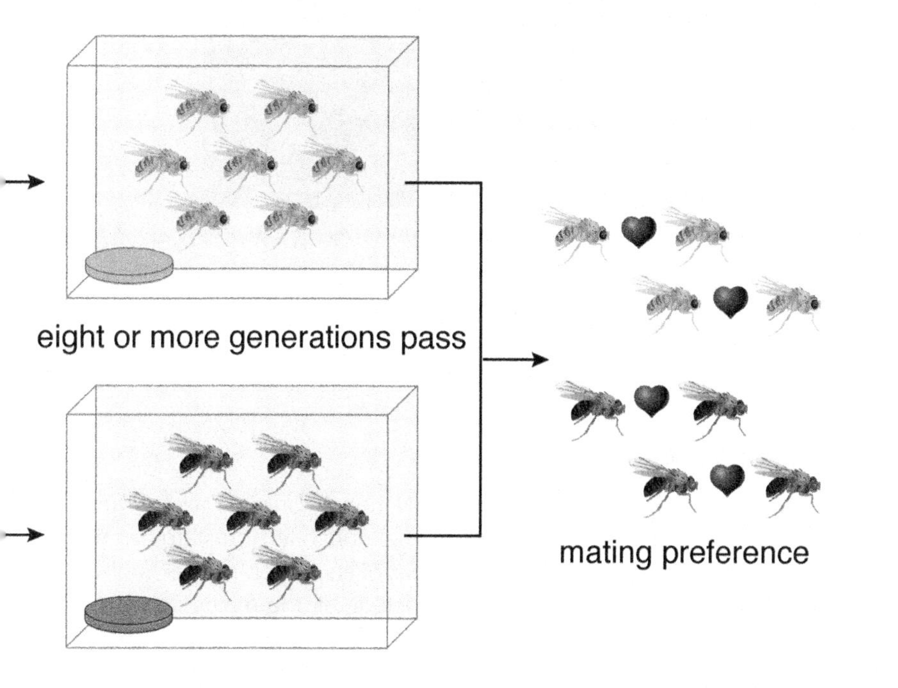

Today, almost all of the Galápagos Islands are a national park, and the surrounding sea is a marine sanctuary.

By far the best way to see the islands is to book a boat tour that hops from island to island (especially important to get a feel for the different habitats that helped create the different species) and has sleeping accommodations on board. There are hotels on the islands, but staying in one defeats the purpose of the visit—it's essential to get out and about to see what Darwin saw. The islands are also a great place for diving and snorkeling.

The largest settlement on the islands is on Isla Santa Cruz; here you can visit the Charles Darwin Research Station, where many of the islands' species of tortoise are looked after and studied. One tortoise, nicknamed Lonesome George (see Figure 6-2), is thought to be up to 90 years old. Unfortunately, he is the last known Pinta giant tortoise (he weighs 88 kilograms and is a meter across) and has failed to mate with younger females of similar species. Also on Isla Santa Cruz is the El Chato Tortoise Reserve, where it's possible to see many giant tortoises in one place.

*Figure 6-1. **Lonesome George; courtesy Oliver Lee (o spot)***

Any Galápagos tour should include a visit to Isla Fernandina for its colony of marine iguanas, Isla Bartolomé for its barren landscape and the hike up the now-dormant volcano, and Isla Espanola for the variety of wildlife (including a large colony of sea lions).

Practical Information

General information about the Galápagos National Park is available from *http://www.galapagospark.org/*. Many tour operators offer tours around the islands. Ecoventura (*http://www.ecoventura.com/*) offers carbon-neutral, English-speaking tours that last seven nights and include all the major sights and snorkeling.

Information about the Charles Darwin Research Station can be found at *http://www.darwinfoundation.org/*.

007

Airbus, Toulouse, France

43° 39′ 14.32″ N, 1° 21′ 45.11″ E

The A380

Passengers flying into the small airport in Toulouse are often surprised by the large number of jet aircraft parked near the main runway. These include many Airbus A380s painted in corrosion-resistant green paint, and the peculiar A300-600ST (better known as the Beluga and one of the most voluminous aircraft in the world).

None of these aircraft are ferrying passengers from Toulouse; they are awaiting a final paint job, final tests, or delivery to an airline. Toulouse is the headquarters of Airbus, and its runway is where newly built jets take to the sky for the first time. It was from the Toulouse runway that the Concorde first flew in 1969, and the double-decker Airbus A380 had its first test flight there in 2005.

Close to Toulouse airport are the Airbus factories in which the A380s undergo final assembly along with almost every other aircraft that Airbus manufactures. (The small Airbus A318, A319, and A321 aircraft are assembled in Germany.)

One way to see these aircraft is to visit Toulouse airport and sit upstairs in the viewing area overlooking the runway. Wait long enough and you're sure to see a newly built aircraft making a test flight, or the enormous Beluga transporter arriving with parts from another Airbus site. The Beluga (Figure 7-1) is big enough to carry segments of the International Space Station or entire helicopters, and is used by Airbus to transport the entire fuselage of other Airbus aircraft from place to place.

Figure 7-1. An Airbus Beluga at Toulouse airport; courtesy of DigitalAirlines.com

For a clearer view, you can also book a tour that starts with a video presentation of the first A380 takeoff. Airbus had video cameras all over the aircraft for the occasion, and watching the expressions on the pilots' faces is priceless. French speakers will probably catch some colorful language from the pilots when things don't go quite as intended!

The tour continues with a bus trip around the Airbus site to view the various factories and the delivery center where aircraft are weighed before being accepted by customers. The highlight is entering the A380 factory, where the assembly of A380s can be viewed from a specially constructed viewing area.

To complete the trip, it's possible to visit two Concordes. If you book ahead, you can climb aboard the very first production Concorde, which flew from 1973 to 1985 (and was the French president's official plane). You can also board a more recent Concorde, which flew with Air France until 2003.

Tours end at the Airbus visitors shop, which has a surprisingly good selection of memorabilia and scale aircraft models.

Practical Information

Tours of the Airbus site are organized by a separate company; for details, see *http://www.taxiway.fr/* (click the British flag for information in English). It is absolutely essential to book well ahead for these tours, as they fill up quickly.

Composite Materials

One of the key features of the new A380 aircraft is its use of composite materials in place of metals to reduce the aircraft's weight; 25% of the A380 airframe is made of composite materials, saving a total of 1.5 tonnes. In addition, 22% of the A380 is built from carbon, glass, or quartz-fiber reinforced plastic; the remaining 3% is GLARE (glass-fiber aluminum laminate).

Structural materials fall into four important categories: metals, polymers, ceramics, and composites. Metals are the most obvious materials and cover everything from bronze and iron through steels, exotic metals like titanium and zirconium, and alloys of two or more metals.

Polymers cover everything from wood and fibers to glues, rubber, Bakelite, nylon, acrylic, and all the plastics. Ceramics include stone, glass, cement, bricks, and pottery.

Composite materials have as long a history as all the others, starting with mixtures of straw and mud to make primitive bricks through laminates of different woods (such as plywood). Their history in aircraft manufacturing is also long; even the massive Spruce Goose used laminated wood for its skin (see Chapter 117).

Since the 1960s, more exotic composites have become available. These are commonly used for boat hulls, skis, tennis rackets, and most recently for large parts of aircraft, including the A380. Composite materials used for aircraft consist of a fibrous material made from carbon, glass, or quartz (known as the reinforcement) and a second material that holds the fibers in place (called the matrix).

For example, carbon-fiber reinforced plastic is made from sheets of carbon-fiber that are bonded together using an epoxy matrix. The carbon-fiber sheets can be placed in a mold to create the required shape, and the epoxy is added and set to create the final composite component.

GLARE is made from very thin sheets of aluminum sandwiched together with sheets of glass-fiber reinforced plastic. The entire bundle is held together by the same matrix (typically an epoxy) used to make the glass-fiber sheets.

008

The Arago Medallions, Paris, France

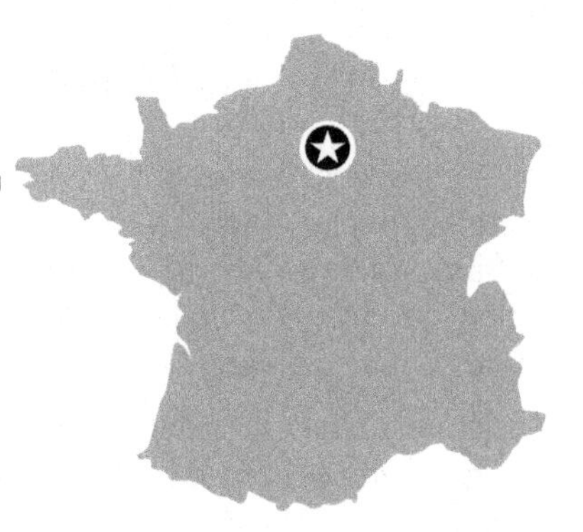

48° 50′ 4.23″ N, 2° 20′ 11.41″ E

François Arago and the Paris Meridian

François Arago was a French mathematician, physicist, and astronomer (and briefly the Prime Minister of France in 1848) who studied optics, magnetism, electricity, and astronomy. He strongly supported the then-controversial theory that light was made up of waves; he demonstrated that a rotating metal disc affects the motion of a magnetic needle suspended above it (for more on magnetism, see Chapter 75); and he showed that light moves more slowly in dense media. He was also a popular orator, and gave public lectures on astronomy for over 30 years.

But Arago is best known today as one of the directors of l'Observatoire de Paris (the Paris Observatory), where he lived and worked. The Observatory was established in 1667 with the support of King Louis XIV of France. Initially, the Observatory's mission was to improve the instruments and maps needed for marine navigation. On Midsummer's Day in 1667, the outline of the Observatory was traced on the ground, with measurements being made to determine the location for the Paris meridian. The Paris meridian (and hence the French definition of 0° of longitude) runs through the middle of the Observatory, and the entire building and site are aligned north-south along it.

One of Arago's early tasks (while he was just secretary of the Observatory) was to help extend the Paris meridian. The length of the meridian was also used to calculate the length of a meter. The French Academy of Sciences had earlier decreed that one meter was to be one ten-millionth of the length of the Paris meridian from the North Pole to the Equator (one quarter of the way around the Earth). Arago was sent on a mission to continue the meridian line southwards into Spain and all the way to Formentera (the smallest and most southerly of the Balearic Islands). Measurement of the distance from Dunkerque and Barcelona (both along the Paris meridian) was used to determine the length of the meter to within 5 millimeters.

The Paris meridian was replaced by the Prime Meridian (which runs through Greenwich, UK; see Chapter 49) at the International Meridian Conference in 1884, but the French continued to use their own meridian as a reference point until 1911. To find the Paris meridian today, look for 2° 20′ 14″ E on your GPS. Even better than looking at your GPS, you can follow a set of 135 bronze medallions set into streets, pavements, and courtyards throughout Paris.

These Arago medallions were the work of a Dutch artist, Jan Dibbets, who was commissioned to create a memorial to François Arago in 1994. The medallions run along the Paris meridian in central Paris and take in some of the city's best-known monuments.

A trip along the path of the Arago medallions should begin near the Observatory itself. Just to the south of the Observatory garden on Place de l'Ile de Sein is a monument to Arago, and a single Arago medallion can be found at its base. Crossing the road toward the Observatory reveals another medallion set into the pavement. If the Observatory garden is open, it's possible to walk inside and see the meridian line set into the ground; if not, skirt around the side following Rue du Faubourg Saint-Jacques and turn left onto Rue Cassini. This brings you to the entrance to the Observatory, where you'll find more medallions and an atomic clock to set your watch to very accurate French time.

A gentle walk up Avenue de l'Observatoire approximately follows the meridian north and affords a wonderful view of the Basilique du Sacré-Cœur in the distance. You'll reach le Jardin de Luxembourg (a large park containing the French Senate), where you'll find more medallions and a view westward to the Eiffel Tower (see Chapter 18). To the east of the park is le Panthéon (see Chapter 13).

Exit the park on the north side and walk up Rue Garancière to the church Saint-Sulpice: the meridian passes close to it. Carry on up Rue Mabillon to Boulevard Saint Germain, where another Arago medallion awaits.

You can follow the Arago medallions north until you reach the River Seine at the Quai de Conti. Then cross the river via the romantic Pont des Arts and you are at le Musée du Louvre. The meridian line crosses the museum's courtyard (with more medallions to be found outside and inside the museum) and continues on to Le Palais Royal. Further north, the line comes close to l'Opera, the medallions become more spaced out, and a little further on is the Place Pigalle and le Moulin Rouge.

Some of the medallions are missing (most notably the medallion closest to the pyramid inside the Louvre) and some have been replaced with a different design commemorating la Meridienne Verte (The Green Meridian). The Green Meridian was a project in the year 2000 to plant a line of trees (and hold a picnic) along the line of the Paris meridian throughout France. At least emotionally, the French haven't yet ceded to Greenwich.

True North and Longitude

Meridians are imaginary lines that follow the Earth's surface north-south, cutting through the North and South Poles. Longitude is measured as the angle from the Prime Meridian (see Chapter 49), but any north-south line is a meridian of the Earth. The direction of the North Pole (which runs along any meridian line) is known as True North. It's possible to find the meridian line for any point on Earth, and thus the direction of True North, with patience, a long stick, and a piece of string.

On a sunny day, place a stick vertically in flat ground and follow the shadow it casts. Start in the morning and mark the position of the end of the shadow at regular intervals (Figure 8-1). The shadow will move as the Earth rotates and the Sun's position in the sky changes.

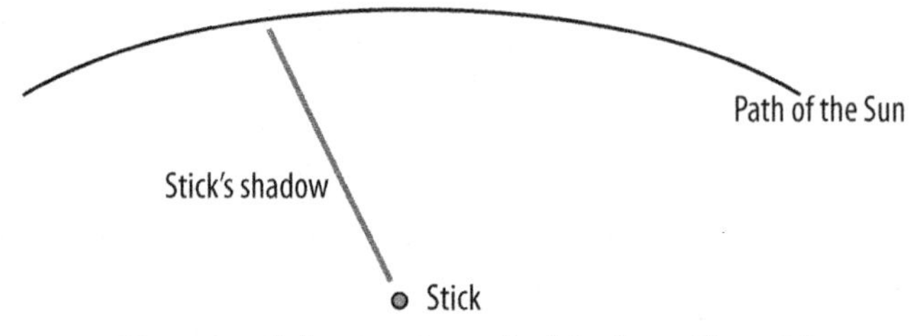

Figure 8-1. Following the path of the Sun with a stick

When the shadow is at its shortest, it is pointing to True North (if you're in the northern hemisphere; in the southern hemisphere it will point to True South) and the shadow is tracing out the meridian line.

Since it's hard to tell exactly when the shortest shadow occurs (it happens at noon local time, but you'd need a clock calibrated to your exact longitude to know when precisely that is), you might want to take two measurements to find the meridian line.

Once the Sun's path has been traced out, use a piece of string attached to the stick to draw an arc of a circle that cuts through the path of the Sun in two places (see Figure 8-2). Try to make these as far apart as possible, using the biggest circle you can.

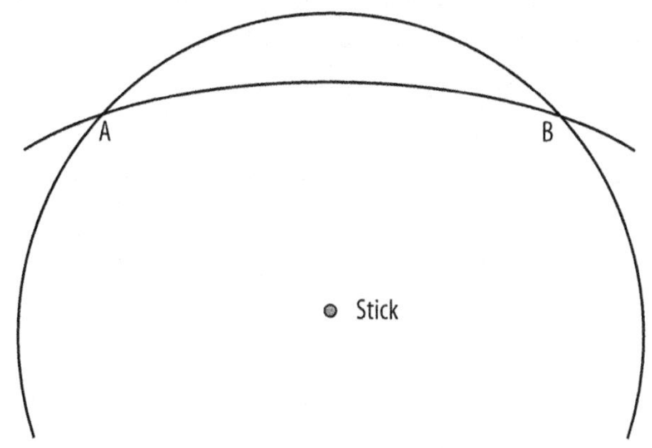

Figure 8-2. A circle cutting through the path of the Sun

The line from A to B (the two points where the circle cuts through the path of the Sun) is perpendicular to the meridian, and the meridian is at its center. So to find the meridian and True North, bisect the line AB by drawing a pair of circles: one centered on A and the other centered on B (see Figure 8-3). For simplicity, use the same string with the same length: the two circles will intersect at the stick and at another point, labeled C in the figure.

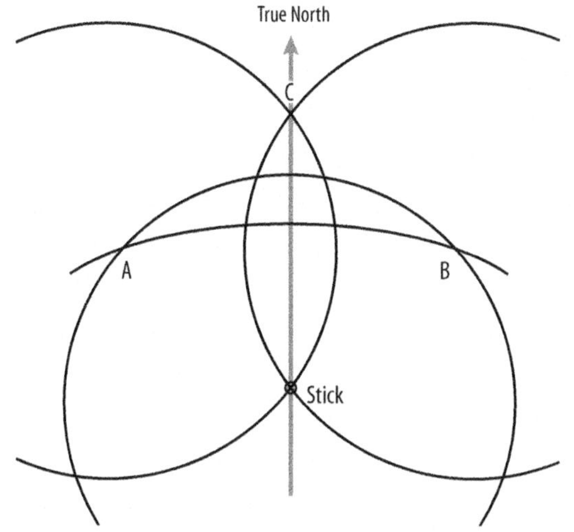

Figure 8-3. Creating the perpendicular bisector to the line AB

The line from the stick to point C is the meridian, with True North in the direction from the stick to C.

Practical Information

The Paris Observatory has an excellent site with details of its work; for the English version, see *http://www.obspm.fr/presentation.en.shtml*.

The Observatory itself is open on the first Saturday of the month, and guided tours are available on Tuesdays and Thursdays. The guided tour takes in the Cassini Room, which runs along the Paris meridian, and a map of France on the floor shows the meridian as it crosses the country.

009

Beaumont-de-Lomagne, France

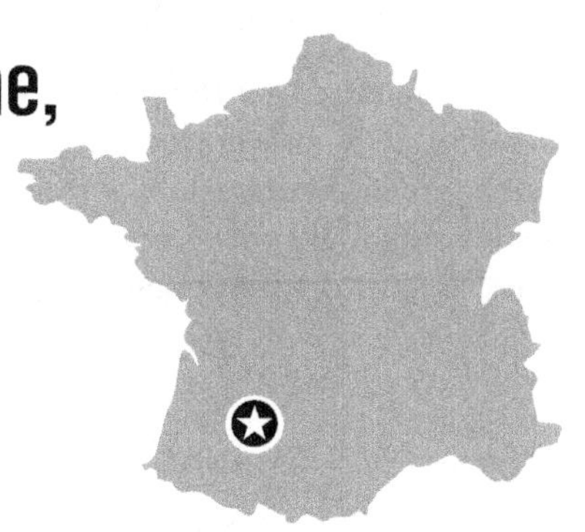

43° 52′ 58.63″ N, 0° 59′ 15.40″ E

Pierre de Fermat

The birthplaces of famous scientists and mathematicians are often disappointing: at best there may be a small plaque giving the briefest biography and dates of birth and death. Happily, mathematician Pierre de Fermat's birthplace at Beaumont-de-Lomagne is an exception. His former home is a delightful small museum full of hands-on mathematics.

Founded in 1276, Beaumont-de-Lomagne is a *bastide* (a fortified town) about 58 kilometers northwest of Toulouse. In the middle sits La Halle, a covered square typical of many towns in the region. It was originally built in the 14th century, was replaced in the 17th, and has been maintained ever since. On market days, the tiled roof protects a lively market selling local produce—the surrounding area is an important agricultural part of France, and the market sells everything from fruit to foie gras, and the town's speciality: garlic.

Just off the marketplace on Rue Pierre de Fermat is L'Hôtel Pierre de Fermat—Fermat's home. The 15th-century stone building is adjacent to the tourist office (where the staff speaks some English) and contains the Fermat Museum. If the museum isn't open, ask at the tourist office for the key. Entry is free of charge.

The museum traces Fermat's life and work (in French), but of greatest interest are the mathematical puzzles and games. Pieces of wood allow you to try your hand at proving the Pythagorean Theorem geometrically by correctly placing the squares and triangles.

You can also find magic squares and triangles where numbers must be placed so that the rows, columns, and diagonals all add up to the same number (a version of Sudoku). The museum also has a large collection of 3D puzzles, and an intriguing game involving a piece of string and a board covered in nails where the string must be threaded around the nails, touching each one, following a graph painted on the board. The museum has plenty of stuff for mathematicians of all ages.

Fermat's Last Theorem

Fermat's famous Last Theorem, so called because it was the last one of his theorems to be proven, looks relatively simple to anyone familiar with the Pythagorean Theorem. In that theorem, the square of the hypotenuse of a right-angled triangle is equal to the sum of the squares of the other two sides, as illustrated in Equation 9-1.

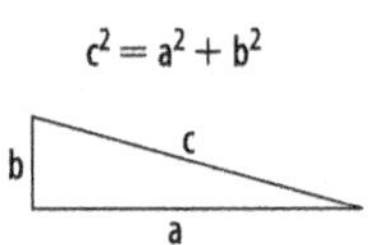

Equation 9-1. The Pythagorean Theorem

One version of the Pythagorean Theorem restricts a, b, and c to whole numbers (i.e., they can't be fractions or numbers like π). Equations that work only on whole numbers are called Diophantine equations (after the Greek mathematician Diophantus). Because of the Pythagorean Theorem, the Diophantine equation $c^2 = a^2 + b^2$ has many solutions (such as the familiar 3-4-5 triangle).

Fermat's Last Theorem (Equation 9-2) is a generalization. It says that there are no solutions for the Diophantine equation $c^n = a^n + b^n$ where n is bigger than 2.

$$c^n = a^n + b^n$$
$$n > 2$$

Equation 9-2. Fermat's Last Theorem

Part of the theorem's fame comes from the fact that Fermat claimed to have a proof. He wrote in the margin of a copy of Diophantus's book *Arithmetica* (which is all about Diophantine equations):

> It is impossible to separate a cube into two cubes, or a fourth power into two fourth powers, or in general, any power higher than the second into two like powers. I have discovered a truly marvellous proof of this, which this margin is too narrow to contain.

Fermat's Last Theorem was finally proven 350 years later, in 1994, by British mathematician Andrews Wiles. The proof is very complex, and relates Fermat's simple equation to a mathematical object called an elliptic curve.

A local non-profit association, Association Fermat-Lomagne, maintains the museum and puts on an annual mathematical festival (usually held in October). The festival features more mathematical games, talks, puzzles, and conferences. The association also provides tours and workshops for all skill levels (with advance reservations).

The city of Toulouse, where Fermat worked as a lawyer, is an hour from Beaumont-de-Lomagne and has many good hotels and restaurants. It's also the home of Airbus (see Chapter 7) and has a large aerospace industry—don't be surprised to see an Ariane rocket close to the road when you're driving from Beaumount-de-Lomagne.

Practical Information

The town of Beaumont-de-Lomagne's website is at *http://www.beaumont-de-lomagne.fr/*, and the Association Fermat-Lomagne can be reached by email at *fermat.lomagne@wanadoo.fr*.

010

Château du Clos Lucé, Amboise, France

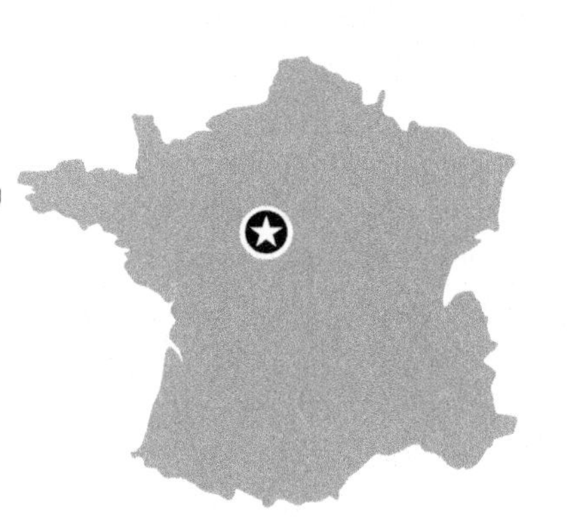

47° 24′ 36″ N, 0° 59′ 29″ E

The Final Home of Leonardo da Vinci

Southwest of Paris, the fertile Loire Valley was home to the kings of France and their associated nobility throughout the Renaissance. The valley contains hundreds of châteaux, many of which are open to the public, built by kings and noblemen starting in the 10th century. In one of these châteaux, Leonardo da Vinci lived and worked for the last three years of his life.

He came to the Château du Clos Lucé in 1516 at the invitation of King François I. The King lent da Vinci the château and gave him a pension on which to live. Today, the château has been restored to its Renaissance state and contains a museum of da Vinci's inventions.

It also contains da Vinci's bedroom (where he lived and died) and his work room. There's even an underground passage that is said to lead to the King's nearby château and was used by King François I when visiting.

When da Vinci arrived in France, he brought with him three precious paintings: *The Virgin and Child with St. Anne*, *St. John the Baptist*, and the *Mona Lisa*. All three are now in the Louvre in Paris. While at Clos Lucé, da Vinci continued painting, illustrating, and inventing, and worked on architectural projects for the King and irrigation systems between the Loire and Saône rivers.

Today, the gardens of the château contain transparent reproductions of many of da Vinci's greatest paintings and sketches, including the Vitruvian Man, depicting a naked man inside a circle and square, his legs and arms outstretched into two different positions showing the proportions of the body.

Also in the gardens is a landscaped area designed to bring to life da Vinci's botanical drawings, his study of geological features, and his knowledge of water. The garden contains a double-decked wooden bridge invented by da Vinci.

English-speaking visitors can take an audio tour, and the château guide book is available in English.

da Vinci's 40 Inventions

Château de Clos Luce is much more than just da Vinci's former home: it houses a collection of 40 reproductions of his inventions, including machine guns, a tank, a cannon with adjustable firing angle, the first car, a design for a helicopter, a flying machine, a parachute, a glider, paddle boats, a swing bridge, and the double-hulled boat.

The exhibitions of his inventions are split between the château itself, where small models are on display, a separate hall containing an audiovisual presentation of da Vinci's life and work, and the garden.

Of particular interest in the château is da Vinci's clockwork car. The car (which looks more like a wheelbarrow) is powered by a pair of large springs. The hall contains a full-sized model, with a 12-meter wing span, of da Vinci's flying machine.

The gardens contain large models that the visitor can interact with. These include da Vinci's design for a helicopter (Figure 10-1), paddle boats, a water wheel, and a tank. There's even a model of his pyramidal parachute.

Figure 10-1. da Vinci's helicopter; courtesy of Betsy Devine

Practical Information

Full information about the château is available in English at *http://www.vinci-closluce.com*. A helpful guide to visiting the Loire Valley is available at *http://www.loirevalleytourism.com/*. A great way to visit the Loire region is by biking along the "La Loire à Vélo" path that follows the river, passing by many châteaux, including the Château du Clos Lucé.

011

Institut Pasteur, Paris, France

48° 50′ 24″ N, 2° 18′ 42″ E

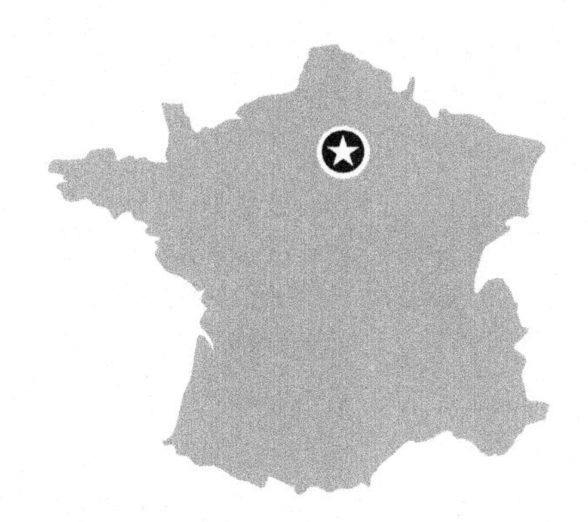

The Founder of Immunology

The French chemist and microbiologist Louis Pasteur is a household name because of the pasteurization of milk and other liquids. But Pasteur originally started out doing a doctorate on crystallography. He went on to disprove the theory of spontaneous generation (that bacteria and other living creatures such as maggots could appear from nowhere), and later showed how to create vaccinations using weakened forms of live diseases.

The Institut Pasteur is a private foundation that performs fundamental biological research. Pasteur is buried inside the institute, and the rooms he lived in during the last part of his life have been turned into a museum. Pasteur's home consisted of 10 rooms and two galleries with a grand staircase linking them, and it has been entirely restored to the state it was in when Pasteur was alive.

The museum is partly historical and partly scientific. The general living spaces show the comfortable life Pasteur and his wife enjoyed in the large apartment, and Pasteur's crypt is a Byzantine funeral chamber under the building. Of interest to scientists is an entire room dedicated to Pasteur's equipment and specimens.

Pasteur's crystallography work looked at tartaric acid ($C_4H_6O_6$), which occurs naturally in wine. Tartaric acid is chiral—it exists in two crystal forms that are mirror images of each other, yet cannot be superimposed (like human hands). One form rotates light to the left, the other to the right.

Pasteur went on to realize that the left-handed form of tartaric acid would aid fermentation, but the right-handed form would not. This led him to study fermentation, eventually demonstrating that it is caused by micro-organisms (called ferments) and that those organisms could be isolated and studied.

He then set out to discover where the ferments came from, and was able to show that they were present in the air or dust and were not spontaneously created. He did this with a simple experiment wherein a specially made bottle with a long curving neck was filled with meat broth. The neck allowed air to pass, but prevented dust from entering, and Pasteur observed that the broth did not ferment. (Some of these bottles remain, their contents still untouched, 150 years later.)

Further, he showed that if care was taken to isolate a specimen from micro-organisms, it would not ferment or rot. He then went on to show that some micro-organisms needed air to survive, and others did not (or even were harmed by the presence of oxygen).

These discoveries led him to study diseases, and he isolated the bacterium staphylococcus, streptococcus, and pneumococcus (see Figure 11-1).

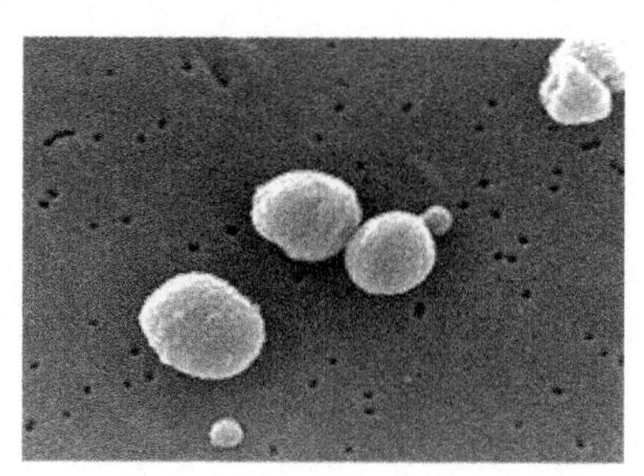

Figure 11-1. Pneumococcus (Image credit: CDC/Janice Carr)

At the time, the idea of vaccination against smallpox (using cowpox) was well known, having been popularized by Edward Jenner (see Chapter 44) in the late 1700s. Pasteur took this a step further by showing how to attenuate live viruses so that they could be used for vaccination. He successfully reduced the virulence of cholera, anthrax, and rabies and used them as vaccines.

Practical Information

Information for the Institut Pasteur is at *http://www.pasteur.fr/english.html*, and guided tours are available. There is printed information in English, and the room containing Pasteur's scientific instruments has a special audio commentary in English.

The museum is easy to reach: there's a stop named Pasteur on the Paris Metro lines 6 and 12.

Pasteurization, Flash Pasteurization, and UHT

Once Pasteur had realized that fermentation and diseases were caused by micro-organisms, he also realized that it was possible to kill the micro-organisms (or significantly reduce their number) by heating. Pasteur showed that heating wine to 55°C prevented wine diseases and stopped fermentation.

The basic process of pasteurization is still used today. Milk can undergo one of three processes to remove pathogens: pasteurization, flash pasteurization, or UHT.

Basic pasteurization involves heating milk to 63°C for 30 minutes in a large vat. This is hot enough and long enough to kill or attenuate Mycobacterium tuberculosis (the bacteria responsible for tuberculosis) and other bacteria that are relatively heat-resistant. At the same time, the overall quality of the milk is not spoiled.

Flash pasteurization (which is also used for fruit juices) is sometimes called High Temperature Short Time, and involves heating milk to 72°C for 15 seconds. This also preserves the quality of the milk and kills off harmful bacteria. It is performed by forcing the milk through small heated pipes, or between hot metal plates, to obtain the necessary rapid rise in temperature.

Neither of these processes sterilizes the milk. There are still some micro-organisms remaining, although they have been greatly reduced in number and will not cause disease in humans. Because micro-organisms are still present, the milk must be kept refrigerated to maintain its freshness.

UHT treatment, on the other hand, raises the temperature of the milk to 138°C for one or two seconds and does almost entirely sterilize it. UHT milk (sometimes called shelf milk) is packed in sterile containers and can last for months unrefrigerated. However, the taste of the milk is altered, giving it a cooked flavor not present with pasteurization.

Recently, pasteurization of milk has been replaced by the use of filters. Since bacteria are typically a few micrometers long, it's possible to remove them using a filter with smaller holes (of around 1 micrometer). At the same time, the constituent parts of milk are even smaller (proteins can be as small as 1 nanometer) and can pass through the filter.

Filtration of milk brings pasteurization full circle—when Pasteur was investigating the existence of micro-organisms, he used filters in his broth experiments to keep micro-organisms out. In addition, his research drove the creation of ceramic water filters, since they were then known to protect against disease.

012

The Jacquard Museum, Roubaix, France

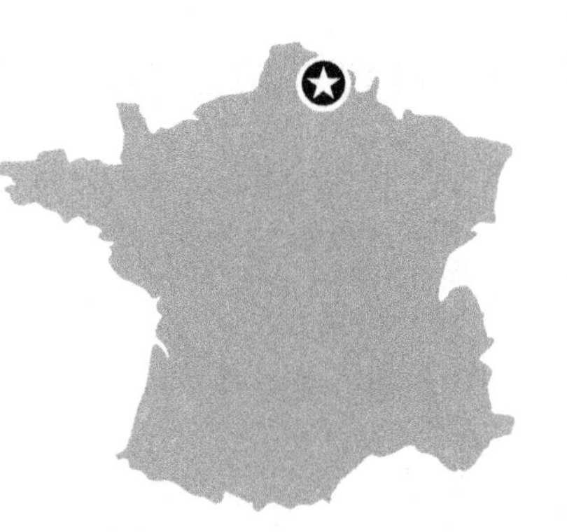

50° 40′ 59.05″ N, 3° 11′ 42.43″ E

The Punched Card

The history of computing can be partially traced to a 19th-century French inventor named Joseph Marie Jacquard, who was interested in improving weaving and ended up being a major influence on the Industrial Revolution. His Jacquard loom was able to produce intricate woven patterns by reading a string of punched cards containing the pattern to be created, mechanizing a previously labor-intensive task.

The punched cards controlled which threads fell over or under each other, thus creating a pattern in the woven cloth. Prior to Jacquard's automation of the loom, only simple patterns were possible because the positioning of threads was done by hand, or was at best partially automated.

A Jacquard loom was able—and still is, since the technology is alive today, with computers having replaced the cards—to produce an intricate pattern and repeat it by reading punched cards in a loop (see Figure 12-1).

Figure 12-1. Punched cards feeding into a Jacquard loom; courtesy of Justin Cormack

The place to see Jacquard's technology in action is the Jacquard Museum in Roubaix, France. The museum has a collection of 15 working punched-card and electronic looms. Tours of the museum are available in English and can explain and demonstrate the entire weaving process, from the creation of a design to turning the design into punched cards and operating the looms.

The looms on display date from the 18th, 19th, and 20th centuries. The museum also offers workshops on the weaving process and actively restores looms to working order. There's something for (almost) all the senses in the museum: the sound of the looms operating, the smell of the fabric and oil, hands-on exhibits, and the sight of these massive machines in operation.

After Jacquard, the punched card went on to be used in other areas. The inventor Herman Hollerith created a tabulating machine capable of reading punched census cards and producing statistical information by reading the card's holes using a simple electrical circuit. Hollerith's company handled the 1890 U.S. census, and later merged with another company to become IBM. There's a Hollerith tabulator on display at the Computer History Museum (seeChapter 86).

The Jacquard loom also inspired Charles Babbage (see Chapter 77), who envisaged using punched cards to store programs for his Analytical Engine, an entirely mechanical computer. For roughly the first 70 years of the 20th century, punched cards were used as the input mechanism for computers (see Figure 12-2).

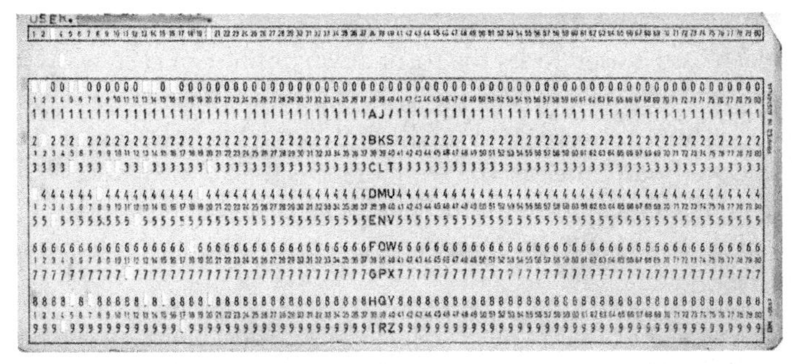

Figure 12-2. A 20th-century computer punched card; courtesy of Peter Renshaw (bootload)

Practical Information

The Jacquard Museum's website is at *http://madefla.50g.com/* and is available in both English and French. If you can't make it to Roubaix, an alternative is to visit a modern factory using Jacquard weaving techniques. In Northern Ireland, the Thomas Ferguson Irish Linen company in Banbridge, County Down, offers a factory tour that covers all aspects of the weaving process and explains the Jacquard loom itself. See *http://www.fergusonsirishlinen.com/*.

The Jacquard Weaving Process

The simplest sort of weaving has no pattern at all. Two sets of threads, one perpendicular to the other, are interleaved in a uniform pattern. One set of threads, called the warp, runs in parallel lengthways on a piece of cloth, and the other threads, called the weft, are woven above and below the warp threads (see Figure 12-3).

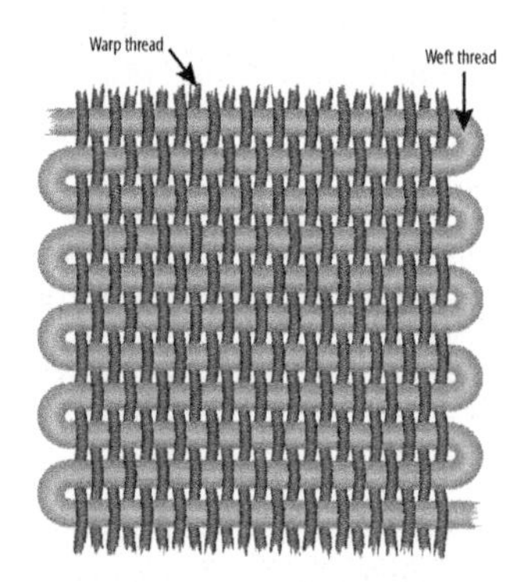

Figure 12-3. A plain weave

To weave the cloth, the warp threads are held stretched out, and every other warp thread is lifted to create a space. The weft thread is then drawn across the gap. Then the process is reversed: the opposite set of warp threads is lifted, and the weft thread is once again drawn across the gap.

Making a pattern involves lifting selected warp threads so that the weft thread shows above or below the warp. Before the Jacquard loom, only simple patterns were possible because lifting the appropriate warp threads was slow and difficult.

Jacquard's loom associates a single hole in a punched card with a single warp thread. Each card is used to set the up or down position of each warp thread for a single weaving of a weft thread. With the punched card in position in the machine, hooks are able to drop down through the holes and pull up their associated warp threads. The weft thread is then drawn across the gap, creating a single line of the pattern. The machine then moves onto the next card. Since patterns are likely to repeat, the machine works through a set of cards in a loop.

Creating the cards involves drawing the pattern to be created on squared paper and treating the pattern as a set of pixels. Each row corresponds to a single punched card, which is punched to create the appropriate pattern of lifted warp threads, with each thread controlling the appearance of a single pixel (where the warp and the weft intersect). The cards are then strung together and run through the loom in order.

The parallels with modern computing are clear: the output of the loom is a pixelated image drawn line by line (just like a standard computer display), and the input is a "program" written on punched cards. Ada Lovelace, who worked with Charles Babbage on his Analytical Engine, wrote:

> We may say most aptly that the Analytical Engine weaves algebraical patterns just as the Jacquard loom weaves flowers and leaves.

013

Le Panthéon, Paris, France

48° 50′ 46″ N, 2° 20′ 45″ E

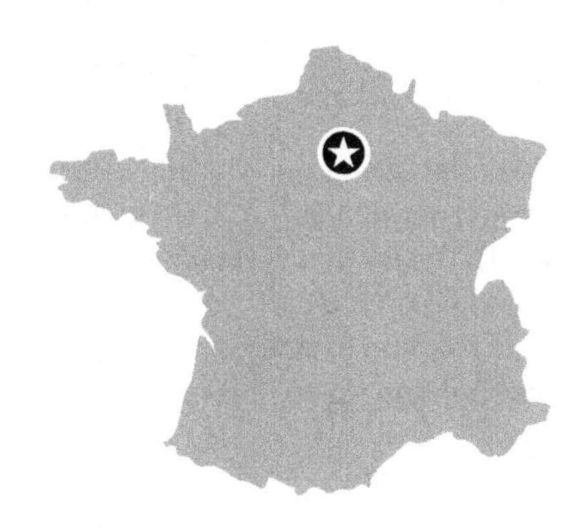

To the Great Men the Grateful Homeland

In the Latin Quarter of Paris, atop the Montagne Sainte-Geneviève, sits the Pantheon (*le Panthéon*), an immense burial place for the great and good men of France (and one woman). The Pantheon's facade is modeled on the Pantheon in Rome, with massive columns and the addition of a dome topped by a Christian cross testifying to its history as a church. The Pantheon was completed in 1789, just in time for the French Revolution, and it became the mausoleum that it is today by decree of the Revolutionary government.

The interior of the Pantheon is lavishly decorated with mosaics, statues, paintings, and frescoes. The view from the dome is equally impressive, with all of Paris visible. The Eiffel Tower, Notre Dame, and Le Jardin de Luxembourg (with the French Senate building) are close by, and the Louvre and Jardin des Tuileries are within easy walking distance. La Basilique du Sacré-Cœur is visible in the distance.

The Pantheon's position as a place of scientific interest relates to the list of famous scientists buried there, and most importantly to the 1851 demonstration by Léon Foucault of his pendulum, showing that the Earth was rotating. The pendulum still swings in the nave, and a video presentation is available in English of the working of the pendulum and its restoration in 1995.

Although it was known in 1851 that the Earth rotated about its axis (in 1543, Copernicus had proposed that the Earth rotated about the Sun, that the Earth itself rotated about its axis, and that the axis's angle accounted for the seasons), but Foucault's Pendulum was a simple demonstration of the Earth's rotation.

Foucault's Pendulum consisted of a 67-meter-long wire attached to the dome of the Pantheon with a 28-kilogram iron ball at the end of the wire. The pendulum was set in motion swinging back and forth in a straight line (with no left-to-right wobble) by tying it up with a piece of string and then burning through the string using a candle. When the pendulum was released, it swung back and forth in a line, but the line did not stay fixed relative to the floor.

Foucault's Pendulum

To understand how Foucault's Pendulum works, it's best to start by visualizing a pendulum that is set in motion directly above the North Pole (Figure 13-1). The Earth is twisting relative to the swinging pendulum, so if you are standing on the Earth, the pendulum will appear to undergo one complete 360° rotation in a day, since the Earth rotates once completely about its axis through the North Pole every 24 hours.

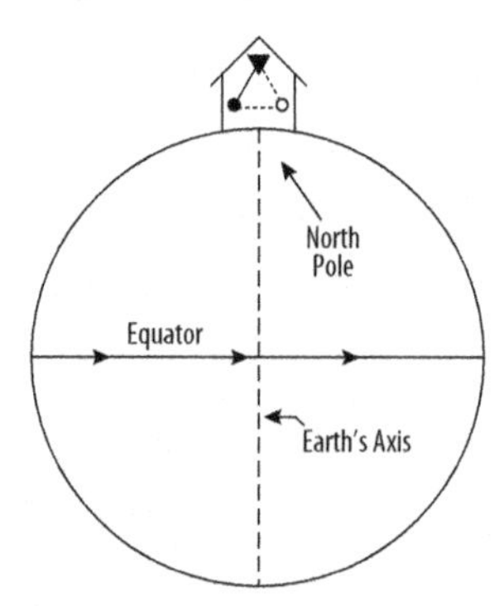

Figure 13-1. A Foucault's Pendulum at the North Pole

You can also try imagining a Foucault's Pendulum on the Equator (Figure 13-2). It doesn't rotate at all relative to the ground because the pendulum's fixed point moves with the Earth's rotation and there's no twisting effect. Instead, the ground is sliding around an axis that it is parallel to (in this case it's the Earth's axis).

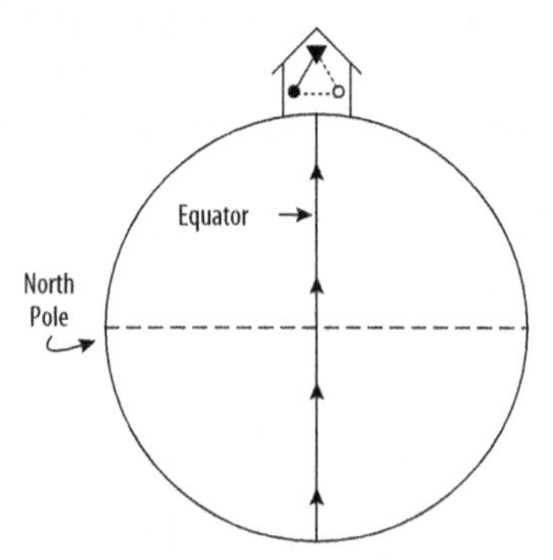

Figure 13-2. A Foucault's Pendulum at the Equator

So what happens at some latitude between the Equator and the pole? Here the ground twists more than at the Equator (where it doesn't twist at all) but less than at either pole (Figure 13-3). The latitude is just the angle relative to the Equator, and a Foucault's Pendulum at a latitude of Θ° (the North Pole is at 90°, the Equator at 0°) will undergo both movements: a twist and a slide.

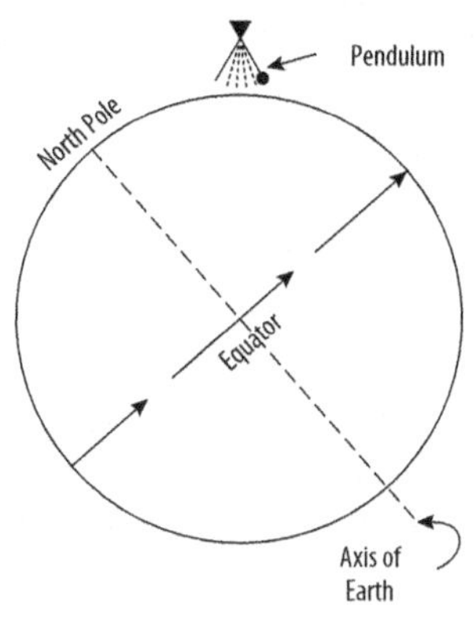

Figure 13-3. Foucault's Pendulum at an arbitrary latitude

The pendulum follows the sliding motion of the ground and moves along with it, but the twisting of the ground is proportional to the latitude. In fact, for a latitude of $\Theta°$, the pendulum will appear to rotate through an angle of $360° \times \sin \Theta$ in one day.

When Foucault demonstrated his pendulum in Paris, it swung through approximately 270°, which corresponds to the latitude of the Pantheon: that is to say, 48.85° N: $360° \times \sin 48.85 = 271.1°$. See Figure 13-4 for a photograph of Foucault's Pendulum in the Pantheon.

In the Northern Hemisphere, a Foucault's Pendulum will rotate clockwise, and in the Southern Hemisphere it will rotate counterclockwise.

Figure 13-4. Foucault's Pendulum in the Pantheon; courtesy of Robert Morton

As the Earth rotated, the pendulum continued to swing unaffected by the rotation (because it was free to swing from its suspension point), but the floor moved relative to the pendulum. After observing the pendulum for some time, this movement becomes apparent, leading to the conclusion that the Earth must have moved. Since the motion is predictable, such a pendulum can be used as a clock, and the pendulum in the Pantheon has a scale marked around it that can be used to tell the time by reading where it is swinging.

The Pantheon was also used for radio experiments between its dome and the Eiffel Tower. In 1898, the first radio communication was made between the two, and in 1901 the Eiffel Tower became a long-distance radio station for the French military.

Buried in the Pantheon are Pierre and Marie Curie (Marie Curie was, in fact, Polish, but became a naturalized French citizen), who are honored for their work with radiation and for the discovery of polonium and radium.

Also present are the remains of Louis Braille, the inventor of the Braille system commonly used by the blind for reading and writing.

Other famous French scientists buried in the Pantheon include Marquis de Condorcet (mathematician), Gaspard Monge (mathematician), Jean Perrin (Nobel Prize–winning physicist), Paul Langevin (physicist), Paul Painlevé (mathematician), Marcellin Berthelot (chemist), Lazare Carnot (mathematician), and Joseph-Louis Lagrange (mathematician and astronomer).

Practical Information

The Pantheon is easily reached on the Paris Metro: exit at the station Cardinal Lemoine, Place Monge, or Maubert-Mutualité, or take the suburban RER train to Luxembourg. Alternatively, you can incorporate the Pantheon on a walk through Paris following the old Paris meridian (see Chapter 8); the Pantheon is to the east of the meridian, just a short walk up Rue Soufflot.

014

Millau Viaduct, Millau, France

44° 4′ 46″ N, 3° 1′ 20″ E

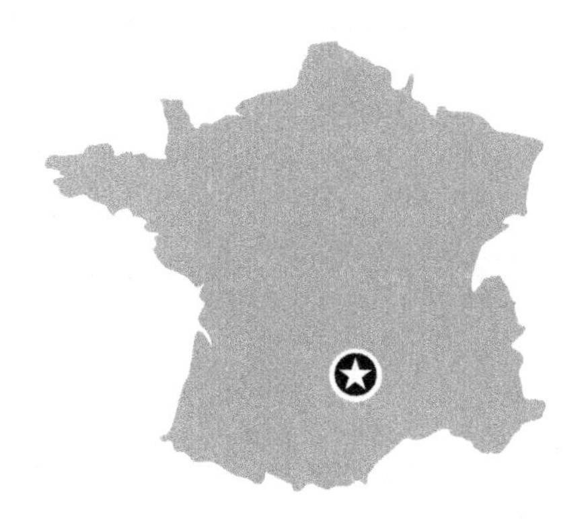

The World's Highest Road Bridge

The Millau Viaduct is over 2 kilometers long and stands 270 meters above the River Tarn in southwestern France. It is the highest road bridge in the world—its highest pylon is 343 meters tall (slightly taller than the Eiffel Tower, Chapter 18, and slightly shorter than the Empire State Building). And it's striking to see, because, despite its immense size, it seems almost fragile.

The viaduct is part of an important autoroute link that allows Parisiens fleeing the city in the summer to arrive at the warm beaches of southern France without having to descend into the Tarn river valley at Millau and climb back out again. Before the viaduct was built, Millau was a major point of congestion, as drivers had to leave the autoroute and drive along local roads. The cost of the bridge's construction was 400 million euros, but it's a small price to pay for not having thousands of overheated, frustrated Parisiens race through your town in July and August.

As well as being a piece of high-tech construction, the viaduct itself is full of high-tech monitoring equipment. It contains a large Ethernet network that links together equipment in each of the seven pylons. The equipment includes accelerometers (which are used to measure any oscillation of the viaduct to the nearest millimeter), thermometers, inclinometers, and anemometers to measure movement and the environment.

The tallest pylon also has fiber optic extensometers that measure any stretching of the viaduct materials and are capable of measuring changes of 1 micrometer. All the other pylons have electrical extensometers. Data from these sensors is transmitted in real time to a monitoring station situated at one end of the viaduct.

It's easy to get a great view of the viaduct because there's a special rest stop for that very purpose (the Aire du Viaduc de Millau), which can be accessed by taking exit 45 from the A75 autoroute. A short walk from the parking lot takes you close to one end of the viaduct for a spectacular view down its length.

Also at the rest stop is an exhibition explaining the construction of the viaduct. The tourist office in Millau organizes guided tours, using a shuttle bus that takes visitors underneath the viaduct and close to the tallest of the pylons supporting it.

And, of course, you can simply drive over it by taking the A75 (and paying the toll).

If you follow the A75 north for about an hour, you'll come to another interesting rest stop—the Aire de Garabit. The Garabit Viaduct is a railway bridge that was completed in 1884 by Gustave Eiffel and is still in use today for local train service. The curved underside of the viaduct is a mathematical shape called a catenary (see page 408).

Practical Information

For more information about the Millau Viaduct, check out its website (in English) at *http://www.leviaducdemillau.com/english/*. The Millau tourist office's English website is at *http://www.ot-millau.fr/index_gb.htm*.

Cable-Stayed Bridge

The Millau Viaduct is an example of a cable-stayed bridge, as opposed to the more familiar suspension bridge (see page 100). In a suspension bridge (Figure 14-1), cables are connected to the ground at each end of the bridge and then strung over the pylons. The bridge deck is connected to the main, hanging, cables via vertical rods or ropes. The weight of the deck is transferred into tension in the main cables. As the cables run over the pylons, this force is transferred into a vertical compression of the pylons; at the ends of the cables, the tension force is balanced by the strength of the anchor points in the ground.

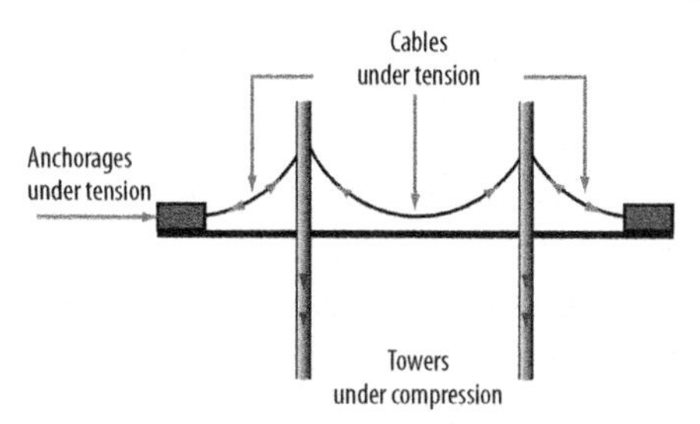

Figure 14-1. Suspension bridge

A cable-stayed bridge does not have anchor points, nor does it have cables that run the length of the bridge draped over the pylons. In a cable-stayed bridge, multiple cables fan out from each pylon, usually symmetrically, to support a section of bridge deck around the pylon.

A major difference from a suspension bridge is that the cables on a cable-stayed bridge are angled, and thus there's a force both vertically and horizontally on the bridge's deck. The deck has to be strong enough to resist the horizontal component of this force. In a suspension bridge, the deck is supported vertically and can be more slender. On the other hand, cable-stayed bridges are much stiffer than suspension bridges, which reduces any tendency of the deck to sway or oscillate.

The Millau Viaduct is an example of a harp cable-stayed bridge (Figure 14-2). The cables supporting the deck pass through the pylon at different heights (the cable nearest the pylon passes through the lowest point on the pylon).

Other cable-stayed bridges have a fan design (Figure 14-3), where the cables all pass through the same point at or near the top of the pylon.

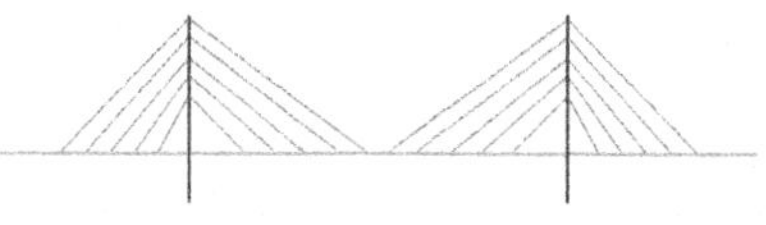

Figure 14-2. Harp cable-stayed bridge

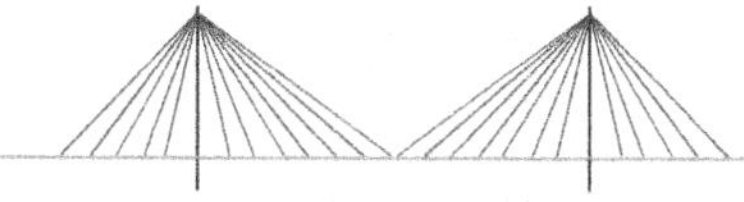

Figure 14-3. Fan cable-stayed bridge

015

Musée Curie, Paris, France

48° 50′ 36″ N, 2° 20′ 39″ E

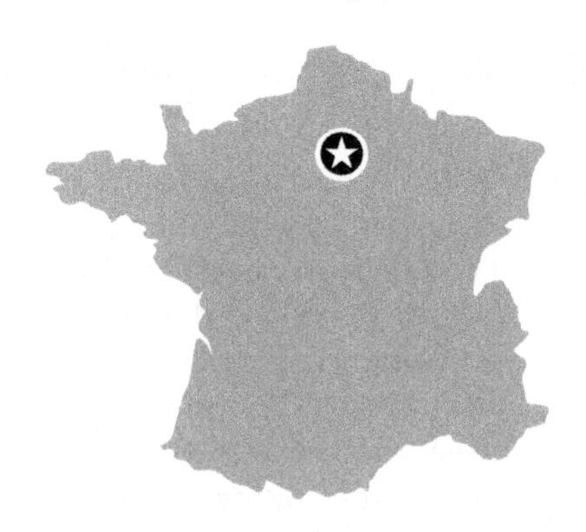

Radium

Visitors to the Musée Curie at the Institut Curie, where Marie Curie worked with radioactive materials with absolutely no safety equipment, will be happy to know that the building was decontaminated in the 1980s, making it safe to visit. The Curies were so unaware of the danger of radioactivity that Pierre Curie carried a sample of radium around in his pocket to show people, and Marie Curie had a glowing jar of radium salt as a night light.

The museum covers the life and work of two couples: Pierre and Marie Curie, and their daughter and her husband, Irène and Frédéric Joliot-Curie.

Pierre and Marie Curie discovered the radioactive elements polonium (named after Marie Curie's home country, Poland) and radium; they were also the first to use the word radioactive. The Joliot-Curies discovered "artificial radioactivity"—they were able to take a non-radioactive element like aluminum and make it radioactive by bombarding it with alpha particles emitted by polonium. All four of the scientists won Nobel Prizes.

The main part of this small museum consists of Marie Curie's office and her chemistry laboratory. Both have been restored to the state they were in at the time of her research. There's a good collection of the Curies' notes and equipment, including the apparatus used to detect radiation.

An amusing (and somewhat frightening) part of the museum details the craze of the 1920s and 1930s for using radioactivity in a variety of products. The museum has a "radium shop," which has a reproduction of a fountain producing radioactive water to drink, advertisements for radioactive wool (which was apparently especially good for babies), and a beauty powder containing radium and thorium.

Ionizing Radiation and Smoke Detectors

The Curies were able to detect radiation (and determine which elements were radioactive) by exploiting a technique developed by Marie Curie's doctoral supervisor Henri Becquerel. The radiation emitted by the uranium Becquerel was working with, and by Curie's radium, is *ionizing*—that is, the radiation has enough energy that it is capable of removing an electron from an atom. Removing an electron causes the atom to become charged, and the charge can be detected.

A radioactive source can ionize air, and a simple detector is a pair of metal plates connected to a DC source. One plate is negatively charged, and the other is positively charged. When radiation ionizes the air between the plates, the positively charged ionized atoms are attracted to the negatively charged plate, the free electrons knocked off by the radiation are attracted to the positively charged plate, and a measurable current flows through the plates.

The Curies' equipment had three parts (see Figure 15-1): piezoelectric quartz, Q; a quadrant electrometer, E; and an ionization chamber (the gap between plates A and B). The sample to be examined was placed on a plate of the ionization chamber, and that plate was connected to a battery. The other plate was connected to the electrometer, which was then connected to the piezoelectric quartz.

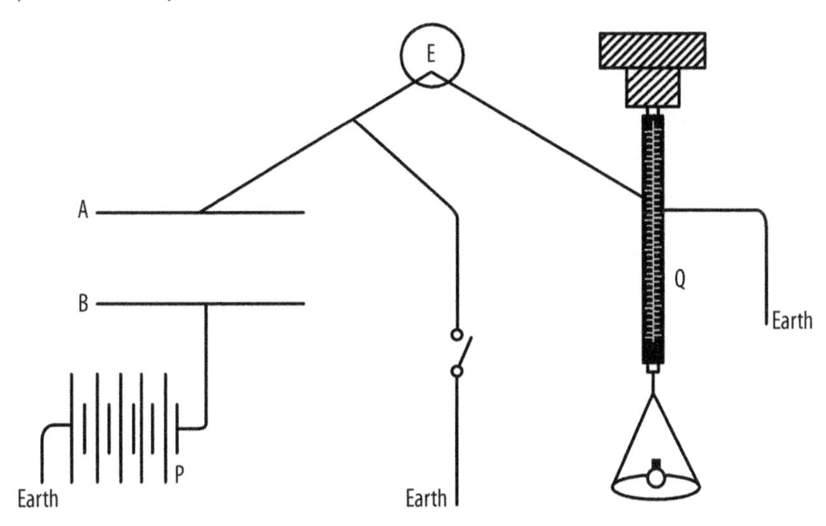

Figure 15-1. Piezoelectric quartz electrometer with ionizing chamber

The Curies would add weight to the quartz, changing the charge it created using the piezoelectric effect: certain materials, mostly crystals, will create an electric charge when stressed. By adding the right amount of weight, they were able to balance out the charge from the battery and ionized air. The electrometer served to indicate when the charges were balanced.

From the weight applied to the quartz, the charge could be calculated; from that, the strength of the radiation from the sample could be calculated in turn.

Most modern homes contain a small radioactive source and an ionization chamber—inside the household smoke detector. A smoke detector typically has a tiny piece of radioactive Americium 241 emitting alpha particles into an ionization chamber. The plates of the ionization chamber are charged by the smoke detector's battery.

When there's no fire, the ionized air creates a small current across the plates. But when smoke enters the chamber, it attaches to the ions and removes their charge. When this occurs, the current across the plates drops and the alarm will sound.

The museum also details the rapid discovery that radioactivity could be used to treat cancerous tumors. This led to the creation of the Radium Institute, where Curie worked on radiation and other scientists worked on the medical use of radioactivity. The Radium Institute became the Institut Curie in 1978.

Pierre Curie didn't die of radiation-induced sickness (he was hit by a carriage while crossing the street), but Marie Curie almost certainly did: she died of leukemia in 1934. Marie Curie is the only woman honored by a place in the Pantheon in Paris (see Chapter 13), where she is buried alongside her husband.

Practical Information

The museum is part of the Institut Curie, a French cancer research organization. Details are available (in English) from the institute's website at *http://www.curie.fr/*. The museum organizes tours in English once per week; the tours are free, and no reservation is needed.

016

Musée de l'Air et de l'Espace, Le Bourget, France

48° 56′ 50″ N, 2° 26′ 6″ E

"Concorde Lives On, She Is Only Sleeping"

There are many air and space museums in the world, but only one has two Concordes: the Musée de l'Air et de l'Espace outside Paris. And one of them is in spectacular condition because it was kept alive until 2007 by a team of volunteer mechanics and crew who maintained and flew it for Air France.

The centerpieces of the museum are Concorde 001 (the prototype Concorde that flew in 1969) and Concorde F-BTSD (known as Sierra Delta; see Figure 16-1), housed in a dedicated hall facing each other. The Sierra Delta was kept in good order by a team of volunteers called "Maintenance Concorde."

Figure 16-1. Concorde F-BTSD; courtesy of Sergio Colucci (scolucci)

Visitors lucky enough to be at the museum when the volunteers are working on the aircraft can see it with the power on and watch the famous movable nose cone being tested.

You can climb on board and peer into the cockpit through a glass screen. Inside the cockpit, a piece of graffiti left by the flight crew of one of the last flights reads (in French): "Concorde lives on, she is only sleeping."

The museum is also the oldest aeronautic museum in the world; it was founded in 1919 and contains an important collection of items. Aircraft from the Second World War are well represented, with a British Spitfire Mk 16, German Focker Wulf F190, U.S. Martin B-26 Marauder, French Dewoitine D.520, and a Soviet Yakovlev Yak-3.

There's a hall of prototype aircraft, which includes France's first jet aircraft (the SNCASO SO.6000 Triton), Leduc 010 (an experimental ramjet-powered aircraft), and the Nord 1500 Griffon (a ramjet/turbojet hybrid).

The Boeing 747 exhibit allows visitors to see the entire aircraft, from the cockpit to the passenger area and down into the baggage hold. Close to the 747 are Ariane 1 and Ariane 5 rockets.

Finally, a special exhibition honors French aviation pioneer Antoine de Saint Exupéry, and there's a hall dedicated to early flying machines, including gliders, balloons, and airships.

Practical Information

The museum's website is at *http://www.mae.org/*; visiting the museum is free of charge. Access to the museum is easiest by car, although it can also be reached by public transport from central Paris.

Supersonic Boom

One of the factors that led to Concorde's commercial demise was the sonic boom. The sonic boom generated by Concorde (and other supersonic aircraft) led to it being banned from flying supersonically over land.

When an aircraft flies below the speed of sound, it creates a disturbance in the air, which results in waves created at the aircraft's nose (Figure 16-2). These waves move away from the aircraft at the speed of sound (around 1,230 kph). Since the aircraft is flying below the speed of the waves, they move away from the aircraft in every direction.

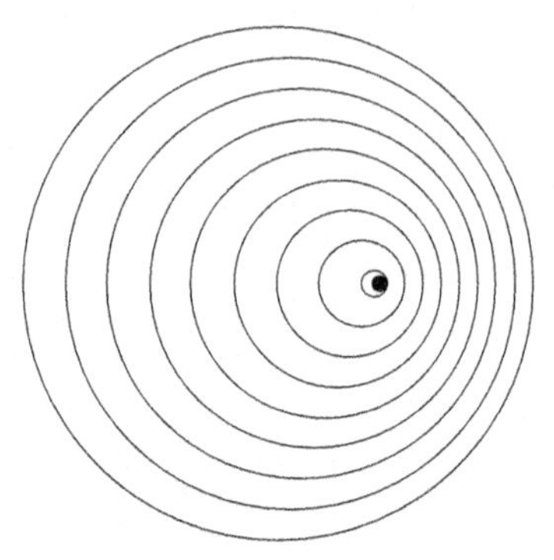

Figure 16-2. Pressure waves at subsonic speeds (aircraft flying to the right)

When flying at the speed of sound, the waves generated by the aircraft at the nose move at the same speed as the aircraft, and they bunch together creating a pressure front, or shock wave (Figure 16-3). This pressure front of coinciding waves of air is referred to as the sound barrier, and aircraft accelerating through the speed of sound experience a shock wave (which can cause control problems) when it reaches the critical speed.

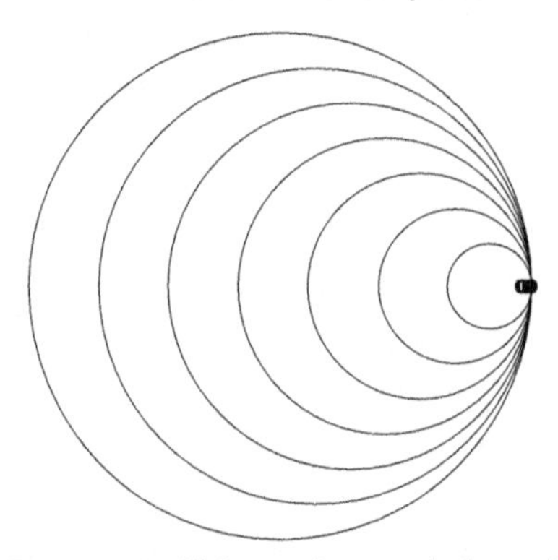

Figure 16-3. Flying at the speed of sound

Once through the sound barrier, the waves generated by the nose are not capable of keeping up with the aircraft, and they appear behind the nose. As the waves spread out, they form a cone shape referred to as the Mach cone (Figure 16-4). The faster the aircraft, the narrower the cone.

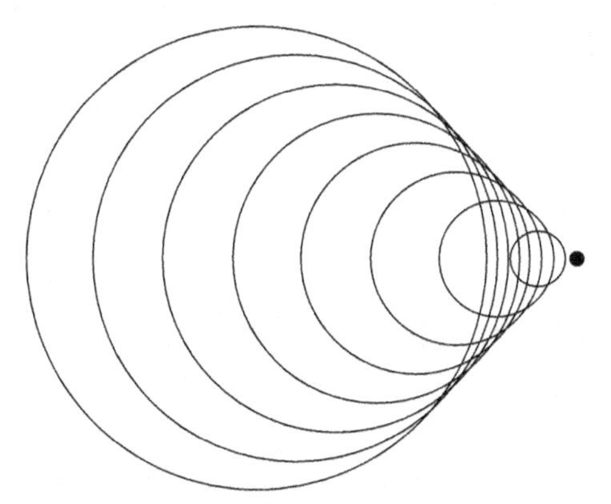

Figure 16-4. Flying supersonically

Inside the Mach cone, the air is traveling at subsonic speeds. For this reason, the wings of supersonic aircraft are swept backward so that they remain in subsonic air even when flying at supersonic speeds. The angle of the wing is determined by the speed of the aircraft and the resulting shape of the Mach cone.

The sonic boom arises because the waves are compressed together behind the aircraft, and its sound is multiplied by coinciding waves. When the Mach cone intersects the ground, the air pressure rapidly increases and then decreases with the shock wave, and the boom is heard.

In real aircraft there are two booms: one from the shock wave created by the aircraft's nose, and another from its tail.

017

Musée des Arts et Métiers, Paris, France

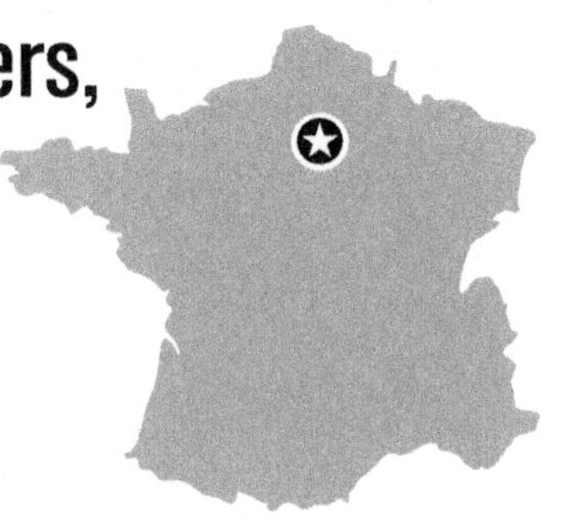

48° 51′ 58″ N, 2° 21′ 19″ E

Science Before and After the French Revolution

Britain and Germany may contest the title of best science museum with the London Science Museum and Munich Deutsches Museum (see Chapters 77 and 19, respectively), but France's Musée des Arts et Métiers (Museum of Arts and Trades) boasts of being the oldest (it was founded in 1794), and has a superb collection of devices dating from before 1750 to the present day.

The museum, situated in an 800-year-old priory, has a collection covering scientific instruments, materials, construction, communications, energy, mechanics, and transport.

The scientific instrument collection recounts the history of the creation of the metric system (with the meter, liter, and gram). The meter was to be one ten-millionth of the distance along the meridian from the North Pole to the Equator (running through Paris, of course), and when calculations were made in 1793 the meter was defined (with an error of 5 millimeters; see Chapter 8 for more on the meridian and meter). The gram was the weight of a cubic centimeter of water, and a liter was the volume of a cube with 10-centimeter sides.

Also in the scientific instrument collection is Foucault's device for measuring the speed of light. In 1862, Foucault measured the speed of light using simple apparatus to estimate the speed of light at 298,000 kps (he was off by less than 1%).

Pascal's calculating machines are also on display. His 1642 arithmetic machine performs addition by adding from right to left, carrying digits mechanically as the addition calculation continues. Despite its age, the Pascalina machine looks as though it might still work today.

The materials collection includes machines for spinning cotton, weaving (there's a Jacquard loom; see Chapter 12), and making paper, glass, porcelain, steel, and aluminum foil.

The communications collection has printing presses, Daguerre's cameras for making Daguerreotype photographs, and early telegraph equipment (including the entirely mechanical line-of-sight semaphore system used in France up until 1860 for long-distance communication). Cinema plays an important part in the collection, too, with the Lumière brothers' equipment.

Émile Baudot's telegraph equipment is also present in the collection. Baudot invented the predecessor of all computer codes used to represent characters. His Baudot code of 5 on-or-off bits allowed 32 different characters to be transmitted. He also invented time division multiplexing, whereby multiple telegraph signals could share the same line, transmitting up to five different messages at the same time.

Since much of the museum is in French, the best way for an English-speaking visitor to enjoy it is by renting a portable audio guide. The guide has over six hours of commentary and covers 175 of the major objects on display.

Once you are done exploring the museum, it's a short Paris Metro trip to Le Panthéon to see Foucault's Pendulum (see Chapter 13).

Practical Information

Information about the museum (in English and French) is available from *http://www.artsetmetiers.net/*. It's very easy to get to—the Paris Metro lines 3 and 11 both have a stop called Arts et Métiers.

Measuring the Speed of Light

Foucault's 1862 calculation of the speed of light relied on precision machinery made for that purpose, and a supply of air from an organ-maker's bellows. In his experiment, sunlight was focused into a beam and directed toward a plate onto which parallel transparent lines had been cut 0.1 millimeter apart, creating a horizontal scale. The light from this screen was split, with some of the light going directly into an objective lens where Foucault could view the scale. The rest of the light was directed toward a rotating mirror.

As the mirror turned, it reflected light against a series of mirrors. The light would bounce through the mirrors and back again onto the rotating mirror. From the rotating mirror, it bounced back toward the objective, where Foucault could compare its image of the scale with the image coming directly from the scale itself.

The key to measuring the speed of light was the insight that while the light was bouncing around Foucault's series of mirrors (see Figure 17-1), the rotating mirror was still turning. By the time the light returned to it, the rotating mirror had moved a little, and this movement meant that the image in the objective of the bounced light was shifted relative to the original scale. The amount of shift was proportional to the speed of light.

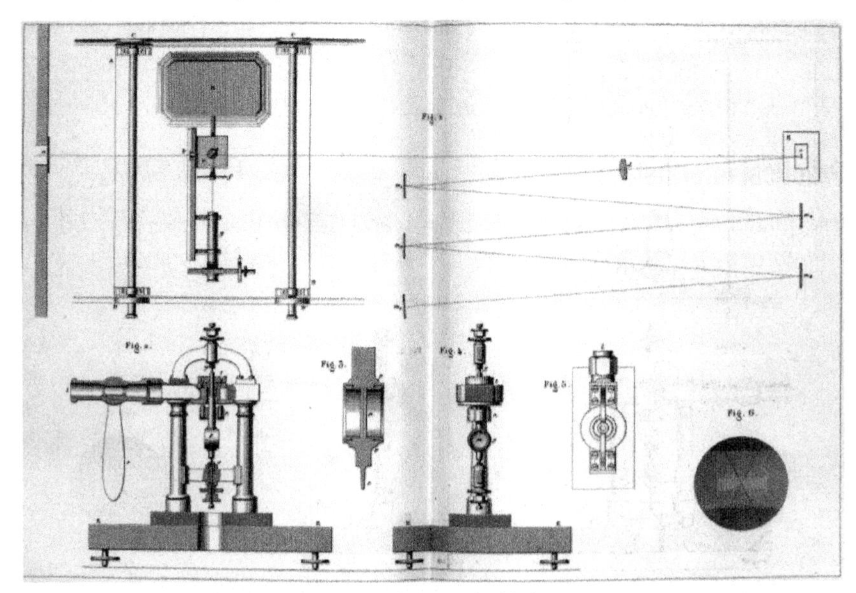

Figure 17-1. Foucault's speed of light equipment; courtesy of Service Interétablissements de Coopération Documentaire des universités de Strasbourg. Département du Patrimoine (34, boulevard de la Victoire 67000 Strasbourg, http://www-sicd.u-strasbg.fr/)

In Foucault's experiment, the mirror rotated at 400 revolutions per second; this was verified using a clockwork mechanism that rotated a slotted disk. When the slots in the disk appeared stationary when viewed in the light reflected from the rotating mirror, Foucault knew that it was rotating at precisely 400 revolutions per second.

The rotating mirror was powered by a small air turbine, which received air at a constant pressure from hand-pumped organ bellows. The bellows were filled with air, and the air pressure reaching the turbine was adjusted to achieve the precise rotation rate required.

The series of mirrors made the reflected light travel 20 meters before returning. Foucault then measured the difference between the two images of the scale. From this he knew the angle through which the mirror had turned while the light was being reflected through the series of mirrors. From that he was also able to calculate the length of time that the light had been bouncing around the 20 meters of mirrors, since he knew the rotation rate of the mirror very precisely.

Given the time and the distance, Foucault was then able to calculate the speed of light.

018

The Eiffel Tower, Paris, France

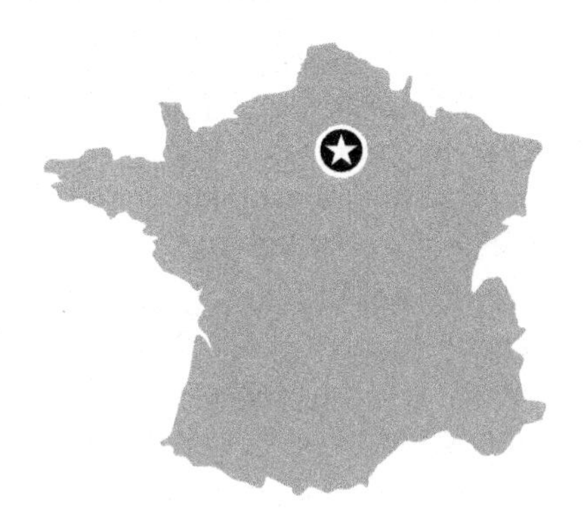

48° 51′ 29.88″ N, 2° 17′ 40.2″ E

The Great Men of Science

Most visitors go to the Eiffel Tower because it's an icon, or they go for the view from the top, or they wait for nightfall to watch the spectacular light show. But scientific travellers should go because the Eiffel Tower is filled with science and technology.

The tower was built between 1887 and 1889 by the French engineer Gustav Eiffel, and was the entrance to the 1889 World's Fair in Paris. It was designed as a temporary monument and was initially considered an ugly addition to the Paris skyline. It survived because of its scientific utility.

Almost the entire tower is constructed from puddle iron, which has a higher carbon content than wrought iron and therefore more tensile strength. Puddle iron is made by mixing the pig iron from a blast furnace (see page 282) with iron oxide (rust) and puddling it (stirring the molten mixture). Oxidation occurs inside the mixture, removing some carbon and other impurities. Puddling has the additional advantage that the composition of the resulting iron is controllable by adjusting the amount of iron oxide added.

The iron was formed into plates, which were riveted together with hot rivets on site. The rivets contracted as they cooled to tightly clamp the plates together. The tower currently stands 324 meters tall and contains 7,300 tonnes of iron. Visitors who make it to the top (just the long wait for the elevator can put many people off) are rewarded with a spectacular view of Paris, and a reproduction of Gustav Eiffel's office.

One of Eiffel's major concerns for the tower was its wind resistance. He had constructed many bridges (and the Statue of Liberty) before building the tower. The tower's graceful curving shape was designed to prevent it from being adversely affected by the wind (see sidebar). Eiffel himself said that it was the wind that determined the basic shape of the tower.

Around the base of the tower Gustav Eiffel had the names of 72 great French scientists and engineers inscribed just below the first balcony. Here you'll find Lagrange, Laplace, Lavoisier, Ampère, Navier, Gay-Lussac, Fizeau, Becquerel, Coriolis, Cauchy, Fresnel, Coulomb, Foucault, Arago, Poisson, Daguerre, Fourier, Carnot, and many more. The names are just visible from the ground, although a good pair of binoculars would help.

After the tower was constructed and had served its purpose, it might have been pulled down, but was largely saved by scientists. Eiffel used it for his own experiments, and meterologists found it useful for measuring pressure, humidity, and temperature at different heights. At the time, it was the tallest structure in the world.

In 1898, radio signals were sent from the Pantheon (see Chapter 13) to the Eiffel Tower, and in 1903 the tower was used by the French military for radio communication. In 1921, experimental television signals were sent from the top of the tower. Today the top is covered with radio antennas of various kinds.

A scientific day out in Paris can be had by following the Arago medallions (see Chapter 8), stopping off at the Pantheon, and finishing with the Eiffel Tower.

Practical Information

The Eiffel Tower has a website at *http://www.tour-eiffel.fr/*. The site is available in English.

The Shape of the Eiffel Tower

Gustav Eiffel wrote that "the curve exterior of the tower reproduces, at a determined scale, the same curve of the moments of the wind." In effect, Eiffel was saying that he had designed the tower with wind resistance as the guiding factor and that its shape was entirely dictated by the wind itself. It's perhaps not surprising, then, that the tower has such a graceful, almost natural, silhouette.

In 1885, Eiffel wrote a paper for the French Society of Civil Engineers in which he described the most significant part of the tower's design—he had eliminated any diagonal bars by ensuring that stress from the wind was transmitted exclusively down the exterior of the tower. This design dictated a specific curving shape.

In his diagram (Figure 18-1), Eiffel imagined the forces of the wind acting on the tower from one direction (P^{I}, P^{II}, P^{III}, P^{IV}). He then imagined any horizontal cut through the tower (such as MN in the figure) that passed through a pair of the walls. He explained that if the walls cut by the horizontal could be angled such that their imaginary intersection point (where the dotted lines at the top meet) were exactly where the resulting combined wind force was directed, then all the stress would run along the walls with a resulting force of zero across any diagonal. This simple result comes about because for a structure to be in equilibrium, all the moments passing through a point must sum to zero.

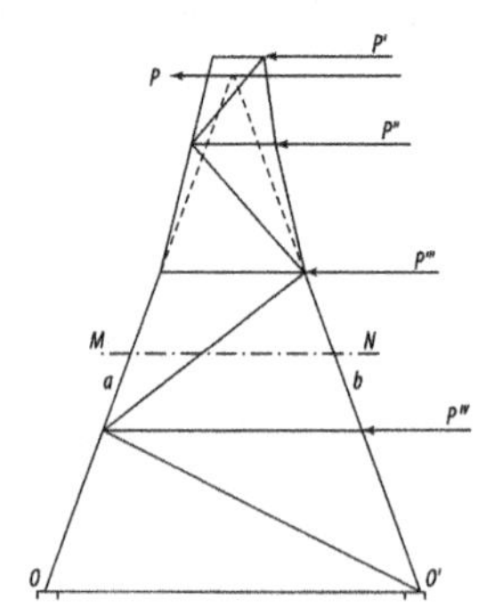

Figure 18-1. Eiffel's diagram

For further details of the mathematics behind the tower's shape, read the paper "Model Equations for the Eiffel Tower Profile: Historical Perspective and New Results" by P. Weidman and I. Pinelis. The paper shows that the shape of the tower is actually an exponential curve.

Of course the tower is not actually curved—it's made of a number of approximations to the actual curve dictated by the wind, and Eiffel built in a large safety factor. To this day, the Eiffel Tower stands without being bothered by the Parisian wind—the largest deflection of the tower was measured in 1999 when the top moved 13 centimeters.

019

Deutsches Museum, Munich, Germany

48° 7′ 48″ N, 11° 35′ 0″ E

The German Museum of Masterpieces of Natural Science and Technology

The Deutsches Museum is probably the largest science museum in the world. It opened in 1906 and sits in central Munich on an island in the middle of the river Isar. The main museum is complemented by two branch museums (both in Munich), one covering all aspects of flying (the "Flugwerft Schleißheim," housed in an old military airbase north of the city center) and the other covering ground transportation (the "Verkehrszentrum," situated close to the Theresienwiese where Oktoberfest takes place).

The main Deutsches Museum is simply enormous: it has 50,000 square meters of exhibition space, and around 28,000 objects are on display. On a sunny day, a visit to the museum should start in the main courtyard, where a human-powered sundial lets you tell the time: simply stand in the spot corresponding to the current month, and your shadow will tell you the time.

Entering the museum reveals its scale: one of the first exhibits you come to is a complete 19th-century fishing boat, the *Maria*, with a large slice through the side that enables visitors to see into the boat from the top deck to the keel. There's even a model rat in the bilges. Directly behind the boat is a steam-powered tug from the 1930s, and the rest of the room is filled with a variety of river craft including a coracle (a small walnut shell–shaped boat used mainly in Wales). The most famous exhibit in the maritime exhibition is the U1 submarine.

Another room displays a small part of the Deutsches Museum's collection of aircraft. High above your head hangs a unique view of part of a Lufthansa passenger jet. The jet has been sliced through from top to bottom, revealing a cross-section of the aircraft with the passenger seats, oxygen masks, windows, and overhead baggage compartments in the upper section. Below the passenger floor is the cargo hold filled with baggage containers.

The Diesel Cycle and the Planimeter

No German science museum would be complete without the diesel engine. The Deutsches Museum has a large collection of diesel engines, including the first built in 1897 by Rudolf Diesel. The American rights to build diesel engines were almost immediately bought by Adolphus Busch (one of the cofounders of the brewery Anheuser-Busch), who built a business making diesel engines in the U.S. The diesel engine's continuing popularity is due to its simplicity and high efficiency.

A basic diesel engine consists of a piston inside a cylinder. With the piston at the bottom of the cylinder, the cylinder is filled with air. The piston is forced up into the cylinder, compressing the air and causing it to heat up (Charles's law of thermodynamics says that the ratio of volume to temperature is constant: as the piston moves, the volume in the cylinder decreases, and the air heats up in response). Diesel fuel is injected into the top of the cylinder directly into the hot air in the form of small droplets. The air temperature (more than 700°C) is greater than the temperature at which the diesel will combust, and a small explosion occurs, forcing the piston down, generating power. Finally, the hot air and gases inside the cylinder are released from the cylinder.

This cycle of compressing air, burning diesel, forcing the piston down, and releasing the hot gases is known as the diesel cycle. The diesel cycle can be visualized by graphing the pressure and volume of the cylinder during the four parts of the cycle. The pressure in the cylinder is measured and the volume of the cylinder is known from the piston position.

The resulting PV (P for pressure, V for volume) diagram (Figure 19-1) can be used to calculate the power output of the engine as it operates.

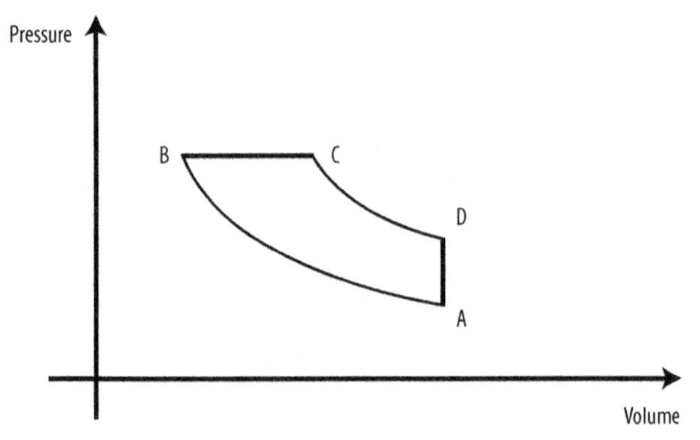

Figure 19-1. A PV diagram

In Figure 19-1, the line AB corresponds to the piston compressing the air in the cylinder, raising both the temperature and the pressure. The line BC sees the pressure remain constant with the volume increasing (as the piston is moved and the newly injected diesel is ignited). CD is the "power stroke" when the piston is forced down the cylinder by the burning diesel, thus creating power. The engine returns to its initial state via DA, where the piston is no longer moving and pressure is released as the remaining hot gases are released from the cylinder via an exhaust.

The work done by the engine can be calculated from the PV diagram by measuring the area enclosed by the graph. Work is actually obtained by integrating the pressure with respect to the volume: for an enclosed shape, the integral is simply the area. Before the advent of electronic systems (and in certain situations today), the area enclosed by the graph was measured using a mechanical device called a planimeter.

The planimeter was invented in 1854 by Swiss mathematician Jakob Amsler. It consists of a pair of rods joined together by a hinge. The end of one rod is fixed to the diagram, or table, and the end of the other rod is traced along the contour of the graph. The tracing end has a small wheel on it, which is used to measure the distance it travels (and hence the length of the contour). A mechanism (typically mounted at the hinge) takes the measurement from the wheel and displays the area.

Planimeters can be used to determine the area of any enclosed shape and are used by surveyors measuring on a map, architects measuring on plans, and engineers measuring the power output of engines expressed through PV diagrams.

The high efficiency of diesel engines at converting diesel to power makes them useful for everything from cars to ships. The museum's collection of diesel engines includes small engines as well as parts of massive marine diesel engines built to power German battleships.

Close by is a room dedicated to power machinery: machines for getting power from water, wind, or muscles (human or animal). The water exhibit includes water wheels from around the world and a variety of pumps and rams. The muscle-power exhibits show whims (where an animal walks in a circle turning a shaft) and tread-operated machines such as the dog tread wheel, where a dog runs inside a drum, creating power to keep a blacksmith's fire constantly fanned.

The Electric Power exhibit traces the history of electricity, displaying objects ranging from dynamos and generators to high-power switches, transformers, and overhead cables. The exhibit includes the generator used for the first transmission of three-phase power between Lauffen and Frankfurt, a distance of 175 kilometers, at the International Electro-Technical Exhibition in 1891. The highlight of the exhibit is the demonstration of high-voltage electricity. Three times a day, AC power at 300 kilovolts is demonstrated, and a 800-kilovolt lightning strike is created.

The museum's bridge-building room features a full-scale suspension bridge that you can walk over, and numerous exhibits about bridge building.

The computer section features the first programmable computer, the Zuse Z3. The computer was originally built in 1941, and subsequently lost when its inventor's Berlin apartment building was destroyed by Allied bombing. A replica of that machine was built in the early 1960s and is now on display in the museum. The computer's core consists of about 2,000 relays capable of performing mathematical calculations using binary; programs were fed into the machine using a punched tape. The Z3 had just 64 words of memory; each word had 22 bits. That's not even enough memory to store this sentence, but it was enough to perform statistical work for the Nazi government.

Even excluding the two branches, the Deutsches Museum is much too large to visit in a single day (or even in several days). In addition to the areas already discussed, there are exhibits covering Petroleum and Natural Gas, Metals, Welding and Soldering, Materials Testing, Machine Tools, Machine Components, Hydraulic Engineering, Energy Technologies, Physics, Optics, Electron Microscopy, Nuclear Physics, Musical Instruments, Chemistry, Pharmacology, Glass Blowing, Ceramics, Paper, Printing, Astronautics, Photography, Textiles, the Environment, Astronomy, Geodesy, Mathematics, Telecommunications, Agriculture, Chronometry, Weights and Measures, and Amateur Radio. There's also a Foucault's Pendulum (see Chapter 13), a model railway, a planetarium, and a reconstruction of a Spanish cave, complete with cave paintings.

For children, there's a dedicated space under the museum called Das Kinderreich (Children's Kingdom) aimed at ages 3 to 8. The Kinderreich is full of hands-on exhibits for children, including musical instruments, a working canal lock with toy boats and plenty of water, an Archimedes' screw, a Mercedes fire engine to clamber over, computers, and more. For children who manage to get wet, the staff is on hand with hairdryers.

For the really large exhibits, you need to visit the two branches. The Flugwerft Schleißheim features a collection of military aircraft and missiles, gliders, helicopters and VTOL (vertical take-off and landing) aircraft, engines, flight simulators, and hang gliders. The Verkehrszentrum has everything from Formula 1 race cars to trams, buses, and trains.

Practical Information

The Deutsches Museum is open every day (excluding certain German public holidays) from 9 a.m. to 5 p.m. The easiest way to get to the main museum is by taking the U1 or U2 underground train to Fraunhofer Strasse Information, in English, can be obtained from *http://www.deutsches-museum.de/en/information/*.

Note that the museum has an excellent photography policy, and allows the use of still and video cameras for private purposes.

020

Peenemünde Historical Technical Information Center, Peenemünde, Germany

54° 8′ 16″ N, 13° 46′ 8″ E

The Fieseler Fi 103 and the A-4

In the history of rocketry, places like the Kennedy Space Center (Chapter 94) or White Sands Missile Range (Chapter 106) garner all the attention, but the roots of rocket science can be traced to a spot three hours north of Berlin, on the German coast at Peenemünde. It was at Peenemünde during World War II that Werner von Braun and others worked to perfect rocket-based weapons and launched the first rocket into space.

Two major weapons came out of Peenemünde: the Fieseler Fi 103 (better known as the V-1, or doodlebug), and the A-4 (better known as the V-2).

The V-1 was a comparatively simple rocket fired from a sloping steam-powered launcher. It consisted of a pulse jet engine (see sidebar) and a simple guidance system that steered the V-1 until it was close to its target. The distance to the target was calculated with an anemometer in the nose, which drove a counter counting down to zero. The counter was set so that it would reach zero just before reaching the intended target, based on the prevailing winds and speed of the weapon. When the counter reached zero, the weapon's controls would jam, sending it into a steep dive in which the engine would cut out. Since the V-1 was otherwise very noisy, this sudden silence served as a warning to the intended victims that an explosion would occur close by.

But the real prize for Soviet and Allied forces looking for the technological spoils of war was the team that developed the V-2. The V-2 rocket first reached space (at a height of over 80 kilometers) in October 1942, and went on to be produced and fired by the thousands at the UK, Holland, France, and Belgium.

The V-2 had a rocket motor that mixed alcohol and liquid oxygen to create thrust, and a sophisticated guidance system that in some late models included radio signals to guide the rocket to its target.

At the end of World War II, Wernher von Braun and his team of over 100 men surrendered to the U.S. Army and were transported to the U.S. along with tons of their equipment and rockets. The team was installed at White Sands Missile Range and went on to test-fire more V-2s before moving to Huntsville, Alabama, where they designed the Saturn V rocket used to put Neil Armstrong and Buzz Aldrin on the Moon.

After World War II, Peenemünde became part of the German Democratic Republic (under the influence of the Soviet Union), and the Soviet army destroyed most of the facilities. After Germany's reunification in 1990, Peenemünde was turned into a monument to the work performed there, and to the slave laborers who toiled (and died) producing V-1s and V-2s.

Today the site features a museum that focuses on the research activities, the weapons, and the willing and unwilling participants. In addition, the museum explains Peenemünde's impact on rocketry and civilian space flight after World War II. There are reproductions of the V-1 and V-2 weapons, and exhibits of original equipment that survived the Soviet destruction of the site.

There are only a few remaining buildings at the site. One is the Nazi-era power station that remained in use until 1990; it now houses the museum. The other building is the factory where liquid oxygen was produced; it is now off limits on account of its poor condition.

To get a feel for the scope of the site, you can follow a special trail that covers 22 kilometers and takes in some of the remains of the extensive research and test facilities built in the 1940s. Because the Royal Air Force bombed Peenemünde in the single largest air raid of the entire Second World War, you definitely don't want to wander off the trail because of the risk of finding unexploded ordnance. Also off the trail is Test Stand VII, where the V-2 rockets were tested; only the foolhardy who ignore the warnings of danger make it to the spot where von Braun started the Space Race.

Practical Information

Information in English about the Peenemünde center can be found at *http://www.peenemuende.de/*.

Pulse Jet Engine

A reaction engine is an engine that shoots out a stream of hot gases, causing an equal and opposite reaction that drives the rocket forward. The engine used in the V-1 was a pulse jet, a simple reaction engine that had hardly any moving parts. In fact, one version of a pulse jet engine has no moving parts at all.

A pulse jet consists of a tube that is flared at one end to create an exhaust nozzle, and has an air intake at the other end. Near the end with the air intake, there's a chamber that contains a spark plug, and a valve covering the air intake opening that connects to the outside of the chamber (see Figure 20-1). Initially the chamber is filled with a mixture of fuel (such as propane gas) and air, and the spark plug is set off.

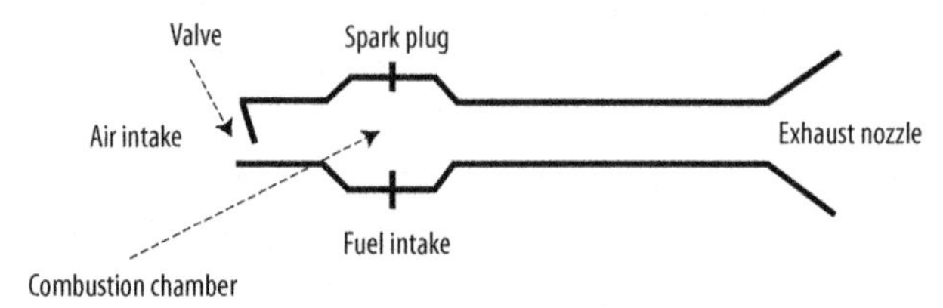

Figure 20-1. A pulse jet engine

The spark ignites the fuel, which explodes and expands rapidly to fill the chamber with hot gases. This expansion slams shut the valve, preventing the gases from leaving the chamber through the air intake. That leaves only the exhaust as a possible exit for the gases, and they expand rapidly down the exhaust and out to produce thrust.

As the gases leave the exhaust, the pressure in the chamber drops until the valve opens again under normal air pressure and the chamber fills up with air (at the same time fuel is either injected or sucked in). As this happens, some of the hot gases leaving the exhaust get sucked back into the chamber by the low pressure. These gases are hot enough to ignite the fuel and start the cycle again without needing the spark plug.

This cycle repeats itself, generating thrust and a characteristic loud, rhythmic buzzing sound (this is how the V-1 got its "doodlebug" nickname). This sound is one of the great problems with pulse jet engines—they are hardly stealthy (during the Second World War, engine noise from the V-1s could be heard from kilometers away) and are loud enough to damage hearing.

It's possible to eliminate the valve completely, resulting in an engine with no moving parts at all, by making a 180° bend in the pulse jet so that both the exhaust and the air intake point in the same direction (see Figure 20-2). When the fuel/air mixture is ignited, the gases expand both down the exhaust and through the short air intake.

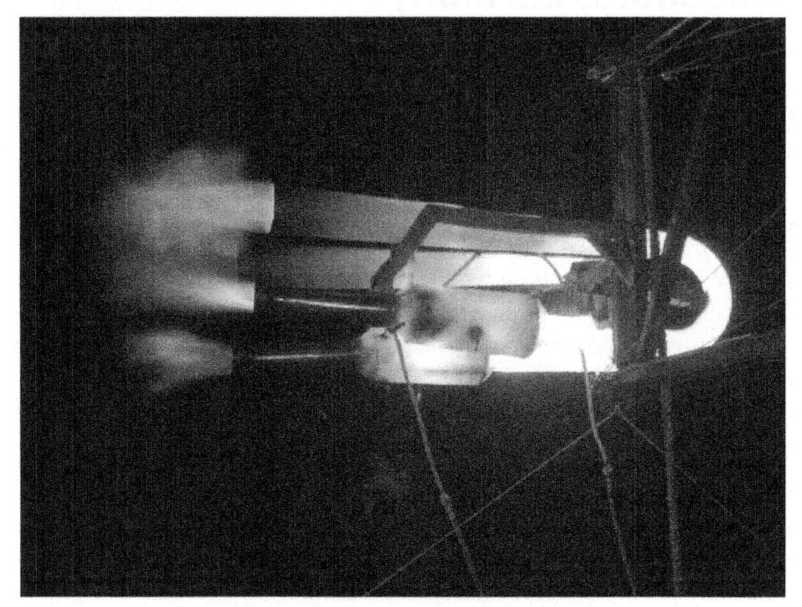

Figure 20-2. A pair of valveless pulse jets; courtesy of Eleanor Lovinsky, the Madagascar Institute

Air quickly rushes back in through the air intake because of the reduced pressure in the chamber and the fact that the air intake is much shorter than the exhaust. Shortly afterward, the hot gases return from the exhaust because of the low pressure. The hot gases ignite the new fuel/air mixture and the cycle continues.

It's also possible to further improve the performance of a pulse jet by adding augmenters. When the hot gases leave the exhaust (and air intake), they no longer provide any thrust. By adding a slanted nozzle behind the exhaust (and with a gap between it and the exhaust), the hot gases heat the surrounding air, which then expands through the nozzle and provides extra thrust.

021

Röntgen Museum, Remscheid, Germany

51° 11′ 37.90″ N, 7° 15′ 33.79″ E

The Discovery of the X-Ray

In 1895, Wilhelm Conrad Röntgen discovered a type of radiation known today as X-rays. Six years later, his discovery earned him the first Nobel Prize in Physics, in part because of the almost immediate adoption of X-rays in medicine. Within two months of his discovery, Röntgen presented a paper entitled "On a New Kind of Rays," in which he detailed his breakthrough and speculated that these new rays were perhaps similar in composition to light.

Röntgen used a sheet of paper coated with barium platinocyanide to detect X-rays. Under the influence of radiation, barium platinocyanide fluoresces; it had been used by others, such as the Curies (see Chapter 15), to detect the presence of radiation from radioactive substances. Röntgen discovered that a sheet of barium platinocyanide–coated paper would fluoresce even when it was meters from the X-ray source.

His X-ray source was a vacuum tube called a Crookes Tube, which had been invented by the British physicist William Crookes 20 years earlier. Crookes had used the tube (which consisted of an evacuated glass tube containing a pair of metal electrodes) to discover "cathode rays" (which we now call electrons), because the accelerated electrons created a green glow when they hit the glass at the end of the tube.

By encasing a Crookes Tube in black cardboard, Röntgen made it impossible for any light (including ultraviolet and infrared) to escape, and was then able to measure the absorption of X-rays by diverse materials such as paper, plywood, copper, lead, gold, silver, and his own hand. He even got his wife involved by taking an X-ray photograph of her hand, complete with a ring on the third finger (Figure 21-1).

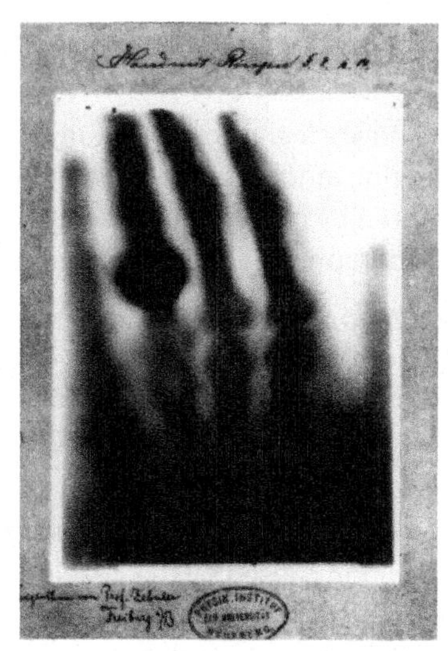

Figure 21-1. X-ray of Röntgen's wife's hand

Röntgen's life and discovery, and the impact of X-rays on medical science up to the present day, are commemorated in the Röntgen Museum in his hometown of Remscheid (near Düsseldorf). The museum has displays of original equipment used by Röntgen, including a collection of X-ray tubes, and his Nobel Prize medal.

The museum is not just about the past: there's also a collection of modern X-ray equipment that traces the development of X-ray technology for medical purposes. Some of the pieces are quite large, and despite limited space the collection includes general-purpose X-ray machines and portable X-ray equipment. There are machines for dental X-rays and for taking mammograms. The museum also has CAT scan machines that work by rotating an X-ray tube around part of the patient's body, producing an image of a slice of the body.

Since X-rays have important uses outside medicine, the museum also covers their use in X-ray crystallography (which was used to unravel the shape of DNA, see page 267), in non-destructive testing of devices by examining their interiors, in screening of people and luggage for security purposes, and in examining artistic works.

Practical Information

Details of the museum are at *http://www.roentgen-museum.de/*. The website is entirely in German, but the museum does have English-speaking staff, and a guide book printed in English. Remscheid is about a 40-minute drive from Düsseldorf.

Bremsstrahlung and K-Shell Emission

The use of X-rays is well known: fire a stream of X-rays at a human body, and they are absorbed in differing amounts by tissues and bones. The resulting pattern of X-rays can be seen by placing a photographic plate in their path; the X-rays affect the photographic material in a manner similar to light.

But the production of X-rays is a different matter. They are created in two different ways by electrons hitting a metal target at high speed. One method, Bremsstrahlung, relies on rapid deceleration of the electrons; the other, K-shell emission, on a collision between a pair of electrons.

A basic X-ray tube (Figure 21-2) consists of a heated filament, K, that gives off electrons by the thermionic effect (the filament is so hot that electrons are released, because the vibration caused by the heat is enough to overcome the forces keeping them in place). The electrons accelerate toward an anode, A, made from a tough metal such as tungsten.

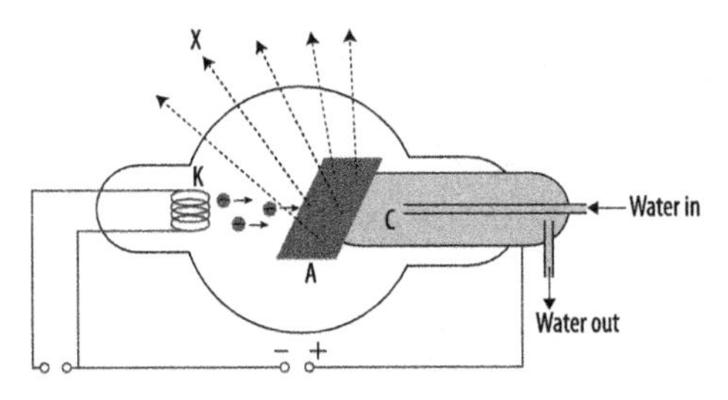

Figure 21-2. X-ray tube

When the electrons hit the metal anode, most of their energy goes into heating it up. For this reason, the anode is made of a metal that can withstand high temperatures and frequent cooling (for example, by pumping cooling water behind it). Some of the electrons cause Bremsstrahlung and K-shell emission of X-rays.

The electrons that cause Bremsstrahlung (German for "braking radiation") create X-rays when they are deflected by the electric field around the nucleus of atoms in the tungsten anode (see Figure 21-3). The electrons are accelerated toward the anode by its positive charge; if they get close enough to a tungsten nucleus, their path is bent and they lose speed (and energy). Since energy is conserved, the lost energy has to go somewhere, and is emitted as X-rays.

Some electrons actually hit other electrons surrounding the tungsten nuclei and knock them out of place. X-rays are emitted when the arriving electrons knock electrons from the innermost shell (the *K-shell*) of the tungsten nucleus. This causes electrons from a higher shell to fall into place in the K-shell.

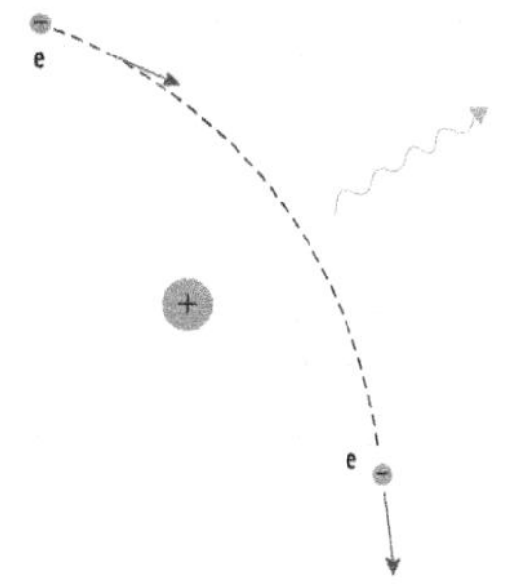

Figure 21-3. Bremsstrahlung

In doing so those electrons lose energy, and because once again energy must be conserved, radiation is emitted in the form of radiation known as characteristic X-rays. The exact frequency of the characteristic X-rays can be predicted from the atomic number of the element used for the anode—they are characteristic of a particular element (see, for example, Figure 21-4).

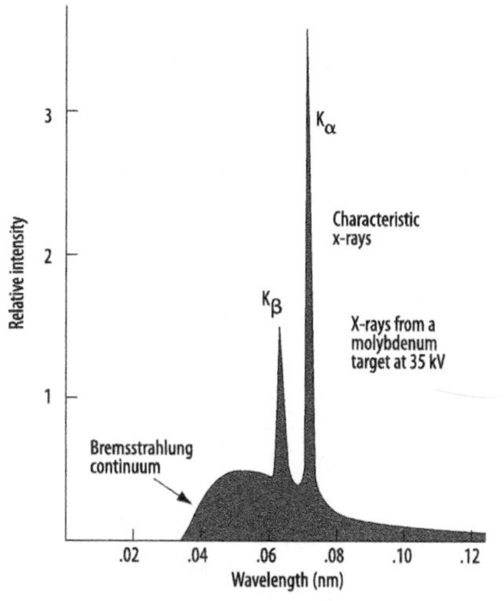

Figure 21-4. Wavelengths of X-rays from Bremsstrahlung and K-shell emission; courtesy of HyperPhysics

The X-rays emitted by K-shell emission are much stronger (have higher energy) than those produced from Bremsstrahlung, but they appear at only two distinct frequencies. Those frequencies are known as K_α (which corresponds to an electron filling a gap in the K-shell from the shell above) and K_β (which corresponds to an electron filling the gap from two shells above).

The X-rays produced by the Bremsstrahlung method appear over a continuum of wavelengths because the braking effect depends on how close an electron came to a tungsten nucleus.

022

Stadtfriedhof, Göttingen, Germany

51° 31′ 53.71″ N, 9°54′ 37.18″ E

An Elephant's Graveyard

The 1,000-year-old city of Göttingen in Germany is best known for its famous university. The Georg-August-Universität Göttingen was founded in 1734 and has welcomed some of the greatest scientists of Europe (as well as assorted kings and politicians including Otto von Bismarck). Sadly, Göttingen is rarely mentioned among the top European universities (Oxford, Cambridge, and others get that honor), but it has fostered some of the best scientific minds of the last 300 years.

The mathematicians Wilhelm Ackermann, Carl Friedrich Gauss, Richard Dedekind, Bernhard Riemann, Felix Klein, Johann Peter Gustav Lejeune Dirichlet, and David Hilbert all studied or taught at the university. The physicists associated with the university include Paul Dirac (Nobel Prize in 1933), Max Born (Nobel Prize in 1954), J. Robert Oppenheimer (see also page 416), Max Planck (Nobel Prize in 1918), Enrico Fermi (Nobel Prize in 1938), Werner Heisenberg (Nobel Prize in 1932), and Wolfgang Pauli (Nobel Prize in 1945). Chemists are also well represented: Adolf Butenandt (Nobel Prize in 1939), Otto Hahn (Nobel Prize in 1944), Walter Haworth (Nobel Prize in 1937), Gerhard Herzberg (Nobel Prize in 1971), Irving Langmuir (Nobel Prize in 1932), and Walther Nernst (Nobel Prize in 1920). Other Nobel laureates studied or taught there as well.

In total, Göttingen boasts 44 Nobel Prize winners. So it will come as no surprise that the university offers museums of mathematics, physics, and chemistry. However, the best way to start a tour of Göttingen is with the dead. Göttingen's town cemetery (Stadtfriedhof) has more Nobel Prize–winning graves per square meter than anywhere else in the world.

The cemetery is located just west of the town center, and is a beautifully wooded area that seems made for contemplation (as does much of Göttingen). If you enter the cemetery from Kasseler Landstrasse in the northeast corner, one of the first graves you come to is that of Max Born, and to ensure there's no doubt he was a scientist, there's an equation on the grave: $pq - qp = h/2\pi i$.

Here, p is the momentum and q the position of a particle, and the formula shows that multiplication of these two values is not commutative (see also page 93); that is, pq is not the same as qp, otherwise pq – qp would be zero. The formula is an expression of Heisenberg's uncertainty principle, which states that it is not possible to observe both the location and the momentum of a particle at the same time.

Nearby is the grave of physicist Wilhelm Weber (he died in 1891), for whom the SI unit of magnetic flux (Wb, the weber) is named. Not far from the two physicists lies a chemist—Friedrich Wöhler, who synthesized urea [$(NH_2)_2CO$] in 1828 and became an almost accidental pioneer of organic chemistry.

Further into the cemetery and around the Kapelle (chapel) are three graves of scientific note: there's the chemist Otto Wallach, the mathematician Felix Klein (who disappointingly doesn't have a diagram of a Klein-4 Group on his headstone; see the upcoming sidebar for an introduction to group theory), and Karl Schwarzschild, whose work in astrophysics is honored by a globe atop the head stone.

Clustered on the other side of the cemetery is a celebrated group. There's a simple stone for mathematician David Hilbert, engraved with the words "Wir müssen wissen. Wir werden wissen." ("We must know. We will know.") Then it's a short walk to Max Planck's grave, where you'll find his name and constant engraved on another simple piece of stone.

Right next to Planck is Otto Hahn, who discovered nuclear fission—a small diagram on his tombstone shows a uranium atom absorbing a neutron (see pages 308 and 376). Next up is the physicist Walther Nernst, who lies next to chemist Adolf Windaus (who won the Nobel Prize for, among other things, showing that cholesterol turns into vitamin D3).

Then there's Max von Laue, the physicist best known for his work on X-ray diffraction by crystals (which would later become important in the understanding of the structure of DNA; see page 268). The group is completed by chemists Richard Adolf Zsigmondy and Gustav Tammann.

If you are not exhausted from the concentration of scientists in the Stadtfriedhof, then head over to the Albanifriedhof (also called the Cheltenhampark) to the east of the city center where one final great mind is buried—mathematician and scientist Carl Friedrich Gauss.

Practical Information

Göttingen's tourist office can provide information in English about reaching the two cemeteries: *http://www.eng.goettingen.de/*. To learn more about group theory and the Rubik's Cube, visit *http://www.usna.edu/Users/math/wdj/rubik_nts.htm*.

Group Theory

Group theory is a fine example of a piece of seemingly abstract mathematics that turns out to be useful in the real world. Even the celebrated 1980s puzzle the Rubik's Cube is based on group theory.

To mathematicians, a group is a set of elements combined with an operation that works with two elements from the set (see page 178). For example, the set of all integers (whole numbers, positive or negative) is written Z, and when combined with the operation + (normal addition), they form a group written (Z, +).

To qualify as a group, the set and operation must obey three rules: the operation must be associative; there must be an identity element; and there must be an inverse element for any element in the set.

Associativity means that when performing a number of operations on elements of the set, it doesn't matter which operations are done first. For the group (Z, +), that means that a sum like 2 + 3 + 4 can be worked out by doing 2 + 3 first and then adding 4, or by doing 3 + 4 first and then adding the result to 2.

The identity element leaves any element of the set unchanged when combined with it using the group's operation. The identity element in (Z, +) is 0, since when 0 is added to any number in the set, that number doesn't change (e.g., 2 + 0 is still 2).

Lastly, the inverse of an element is its "negative." In the group (Z, +), every number has a negative: for example, the negative of 2 is –2. In a group, every element must have an inverse element of this type—when an element and its inverse are combined using the operation, the result must be the identity (for example, 2 + –2 = 0).

One area in which group theory turns out to be useful is in understanding symmetry of objects. For example, it's possible to make a group from the rotations (Figure 22-1) and reflections (Figure 22-2) of a square. There are three possible rotations of a square: through 90° (r_{90}), 180° (r_{180}), and 270° (r_{270}). There are four possible reflections: horizontally (h), vertically (v), and through each of the diagonals (d_1 and d_2).

A group can be made by creating the set of symmetries of a square from all of these, plus an identity element (i) that does nothing to the square. All the symmetries can be placed in a set (S), which will contain i, r_{90}, r_{180}, r_{270}, h, v, d_1, and d_2.

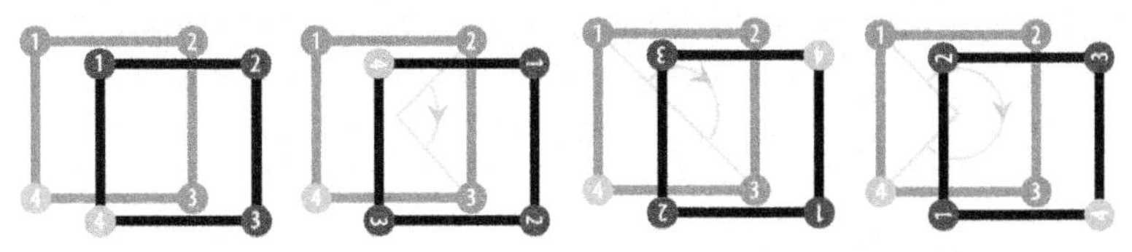

Figure 22-1. The rotational symmetries of a square: i, r_{90}, r_{180}, and r_{270}

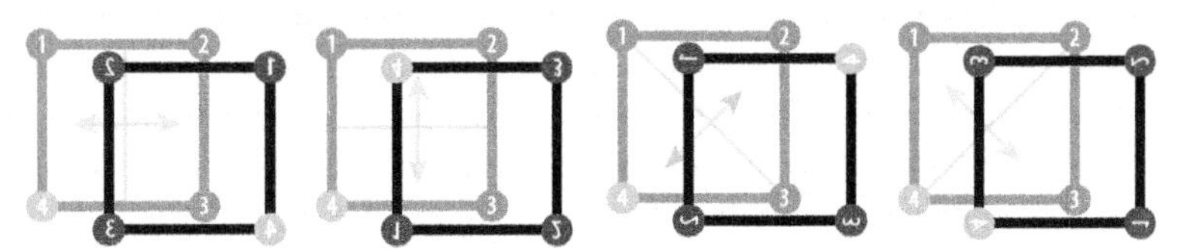

Figure 22-2. The reflective symmetries of a square: h, v, d_1, and d_2

An operation is needed to make a group—in the case of symmetries, this operation is simply "do this followed by that." For example, a rotation through 90° followed by a reflection around the horizontal can be written $h \leftarrow r_{90}$. The arrow is this group's operator ("do this followed by that") and the group can be written (S, ←).

(S, ←) does form a group—the operator is associative, there is an identity symmetry, and it's always possible to undo a rotation or reflection to get back to no change at all. All the possible combinations of rotations and reflections are shown in Table 22-1; this group table shows that every pair of symmetries is equivalent to one single reflection or rotation.

Table 22-1. Group table of (S, ←)

←	**i**	**r_{90}**	**r_{180}**	**r_{270}**	**h**	**v**	**d_1**	**d_2**
i	i	r_{90}	r_{180}	r_{270}	h	v	d_1	d_2
r_{90}	r_{90}	r_{180}	r_{270}	i	d_2	d_1	h	v
r_{180}	r_{180}	r_{270}	i	r_{90}	v	h	d_2	d_1
r_{270}	r_{270}	i	r_{90}	r_{180}	d_1	d_2	v	h
h	h	d_1	v	d_2	i	r_{180}	r_{90}	r_{270}
v	v	d_2	h	d_1	r_{180}	i	r_{270}	r_{90}
d_1	d_1	v	d_2	h	r_{90}	r_{270}	i	r_{180}
d_2	d_2	h	d_1	v	r_{270}	r_{90}	r_{180}	i

If you've managed to read this far, you're either a mathematician or you're thinking, "Great, so the symmetries of a square form a group. Now what?"

In real life (i.e., not in the heads of mathematicians), symmetry turns up inside crystals. Crystals form a variety of shapes that have symmetries, and understanding those shapes and the groups that they form allows you to understand the underlying structure of the crystal. From the structure, described mathematically by an appropriate group, the properties of crystals can be determined.

A special class of groups, called space groups, is used to determine the exact structure of crystals based on their possible symmetries. All possible space groups (different combinations of symmetries) have been determined—there are 230 of them—and they can be used to describe any crystal structure just by mentioning the group's name.

For example, simple table salt (NaCl) forms a crystalline structure that is described by the space group Fm-3m. The same space group also describes other crystals having the same structure as salt, such as calcium oxide (better known as quicklime).

So the next time you bump into a crystallographer at a party, you can impress him with that knowledge.

023

The Gutenberg Museum, Mainz, Germany

49° 59′ 59″ N, 8° 16′ 31″ E

Movable Type

This book exists because of Johannes Gutenberg's invention of movable type in about 1450. Gutenberg's life and printing revolution are commemorated in his home city of Mainz at the Gutenberg Museum.

Prior to 1450, printing was performed by carving each page to be printed on a wooden block. The block was inked and paper pressed against it. Gutenberg revolutionized the entire printing industry by separating the text into individual letters, spaces, and punctuation. Each letter was hand-carved onto a small metal block, which was then used to make a mold (called a matrix) by placing it against a piece of copper and hitting it with a hammer. The copper mold could then be used to make many identical letters.

A page was composed by choosing individual letters and laying them out to form words and lines. The laid-out page would then be inked and printed by using a screw press to press the paper to the page. The highlight of the Gutenberg Museum is the demonstration of this entire process using a reproduction of a 16th-century wooden screw press.

The museum also has two of the original Gutenberg Bibles. The Bible was the first major publication printed by Gutenberg using movable type and marked the start of his printing revolution in Europe. The museum is full of European books, as well as printed material from the Far East that predates Gutenberg, and illustrates the evolution of printing (and hand-copying) techniques.

There's an interesting collection of presses and typesetting machines from the earliest days of printing up to the 20th century. The museum's collection shows the progression from woodcut printing to movable type (the letterpress era), hot metal typesetting, phototypesetting, and finally, to digital methods. The museum also shows how book-binding has progressed through the centuries.

Letter Frequency

As letterpress typesetting developed, compositors had to quickly grab individual letters to make up a word. Before the introduction of machine typesetting, this was typically done by having a large number of letters available in a pair of boxes called cases (Figure 23-1).

Figure 23-1. Upper and lower cases; courtesy of Owen McKnight (addedentry)

Because small letters are used more frequently than large letters, they are kept in the lower case, closest to the compositor. The capital letters are kept in the upper case. As you can probably guess, this arrangement gave us the terms *upper* and *lower case*.

In the lower case, the letters are arranged by frequency, and the number of available pieces of type varies from letter to letter with the frequency of its appearance in English. The letter *e* is the most frequent letter and thus has the largest compartment in the lower case, containing the most pieces of type.

The frequency of letters in English comes into play again on the Linotype keyboard (Figure 23-2) used for hot metal typesetting. The Linotype operator would type the text to be typeset on this keyboard, which would put the appropriate metal type in place. The order of letters on the keyboard follows the letter frequency, starting with the letter *e* and continuing with *t*, *a*, *o*, *i*, *n*, *s*, *h*, *r*, *d*, *l*, *u*, and so on.

Figure 23-2. Linotype keyboard; courtesy of Marc Dufour

Knowing the frequencies of letters isn't just useful for the design of keyboards. Cryptographers use letter frequencies to break codes. In a simple substitution cipher, each letter is replaced by a different letter of the alphabet. By examining the frequencies of each new letter, it's possible to determine which is which. For example, if the letter *X* appears most frequently in an encrypted message, it's most likely to have been *E* in the original message.

Using trial and error, a cryptographer can decode the message by trying letters based on their frequency. Try that technique yourself on this coded message:

```
YMMDR AFYOD UDRAR UNEWU HEHDR AMYCD MEKRD DRACA YFYCZ ICDTA
KNAAC UVKNY ONUMM EHKYH TVNUD ROUHD RACIN VYLAV NUSDR ALULG
BEDUV YCSYM MFRED ACMUU BCRAF YCDRE NDOVE JAVAA DMUHK DRAFY
JACMU UGATM EGARE MMCLU SEHKI BVNUS PAREH TYHTS UCDUV DRALN
AFBNA VANNA THUDD UKMYH LAYDD RAS.
```

The Gutenberg Museum even has a shop worth visiting that sells unusual items such as a two-meter ruler with a scale consisting of the dates of various inventions. The shop also sells movable type—you can purchase entire metal letterpress fonts.

Practical Information

For information about the Gutenberg Museum, visit these two websites: *http://www.mainz.de/gutenberg/english/* and *http://www.gutenberg-museum.de/*.

024

Jantar Mantar, Jaipur, India

26° 55′ 29″ N, 75° 49′ 28″ E

An Enormous Observatory in Stone

In 1728 the Maharaja Sawai Jai Singh II commissioned the building of an observatory as part of the newly founded city of Jaipur. The Jantar Mantar observatory (Figure 24-1) was renovated in 1901 and is now a popular Jaipur tourist attraction and respite from the noise and heat of the city that surrounds it.

Figure 24-1. Part of Jantar Mantar

The Samrat Yantra is the largest sundial in the world at over 27 meters high, and is capable of telling the time, day or night, with an accuracy of about two seconds. Its design is slightly different from that of classical sundials, which consist of a stick (called the gnomon) that creates a shadow and a flat scale on which the time is read. The Samrat Yantra's gnomon is a huge triangle made of local stone. The gnomon's upper face is angled at 27° (the latitude of Jaipur), and the gnomon follows the local meridian, with the highest point pointing to geographical north. The shadow cast by the gnomon falls on a pair of marble-faced curving quadrants on the east and west sides of the Samrat Yantra. The quadrants are curved so that, unlike on a normal flat sundial, the hours are spaced equally apart.

Celestial Coordinate Systems

When talking about the position of celestial objects, there are two major coordinate systems: the equatorial and the horizontal. Both work on the concept of a celestial sphere. Since the Earth is roughly spherical, it's convenient to think of the stars as being positioned on the inside surface of a sphere sharing the same center as the Earth.

This celestial sphere is used to find the position of any celestial body, and the two coordinate systems work by using spherical coordinates. Spherical coordinates require three pieces of information: the distance to the celestial object, and a pair of angles relative to fixed axes. It's the determination of the two axes that defines the equatorial and horizontal coordinate systems.

The equatorial coordinate system is the most widely used and is based on projecting the Earth's poles and Equator onto the sphere. An imaginary line drawn through the Earth's axis of rotation (through the north and south poles) will intersect with the celestial sphere at the celestial north and south poles. The usual way of finding the northern celestial pole is to locate the star Polaris, which is very close to the imaginary pole; the southern celestial pole is found from the Southern Cross.

The plane that cuts through the Earth's Equator can be extended out to the celestial sphere, creating a celestial equator in line with the Earth's.

The equatorial coordinate system, then, is just a projection of the Earth's latitude and longitude measurements onto the celestial sphere. On the celestial sphere, latitude is called declination and is the angle from the celestial equator; longitude is called the right ascension.

On Earth, longitude is measured from the Greenwich Meridian, but in the equatorial coordinate system it is measured from where the Sun crosses the celestial equator on the March equinox (the Sun crosses the equator again during the December equinox, but going in the opposite direction).

Because the equatorial coordinate system is defined by the celestial equator and a point on it, it does not depend on the observer's location on the Earth's surface. The horizontal coordinate system, on the other hand, depends on the observer's latitude and longitude.

The position of any celestial object in the horizontal coordinate system is given by two angles: the altitude and the azimuth. The altitude is the angle of the object from the observer's horizon. An altitude of 0° indicates that the object is on the horizon, and an altitude of 90° indicates that the object is directly overhead (this is called the zenith).

The azimuth is the angle from a line parallel to the horizon that points due north. An azimuth of 0° indicates that the object is due north; 180° means that it is due south.

The horizontal coordinate system has the advantage that it is easy to observe—just find due north and the horizon, and you can determine the altitude and azimuth. Its great disadvantage is that it changes over time (as the Earth rotates) and from place to place.

The instruments at Jantar Mantar use both coordinate systems, and the Kapala Yantra can be used to convert between them.

The reason that the Samrat Yantra and the other instruments at Jantar Mantar are so enormous is that Jai Singh wanted to obtain the greatest accuracy possible. Because of the Samrat Yantra's massive size, its shadow can be seen moving at the rate of about 6 centimeters per minute. You can use the Samrat Yantra to tell the time at night by observing the position of a star from one of the quadrants and moving until the star just touches the top of the gnomon.

Another instrument, the Shasthansa Yantra, is essentially a darkened chamber with a pinhole through which the Sun's rays enter the chamber when the Sun is at its zenith. Inside the chamber is a scale that can be used to measure the declination and diameter of the Sun.

The most stunning instrument at the observatory is the Jai Prakash (also known as the Mirror of the Heavens). The Jai Prakash is a bowl-shaped instrument over 5 meters across whose interior is divided into marble-covered surfaces. You can enter the Jai Prakash in the spaces between the interior surfaces.

Suspended in the middle of the Jai Prakash is a metal plate with a small hole in the center for observing stars at night. During the day the plate casts a shadow on the interior of the bowl. At night an observer can find a star through the hole in the plate using a sighting device, and read the position of the star off the interior of the bowl. To make reading positions easy, there are actually two bowls with complementary surfaces and spaces for observers.

Another instrument, the Kapala Yantra, is an earlier, smaller version of the Jai Prakash that consists of a single bowl and lacks the easy access afforded by the spaces added to the later model.

The Ram Yantra consists of a pair of complementary cylinders that are used to measure the altitude and azimuth of celestial objects such as stars. In the center of each cylinder is a pole that has the same height as the cylinder's radius. During the day, the pole casts a shadow that can be used to determine the position of the Sun. At night, celestial bodies can be sighted over the top of the pole, and their positions can be read from the scales set into the floor and walls.

Other instruments are used to calculate the Hindu calendar (the Raj Yantra), locate the 12 signs of the zodiac (the Dhruva Yantra), find the altitude of celestial bodies (the Unnsyhsmsa Yantra), find the angle of any celestial body relative to the Equator (the Chakra Yantra), and observe heavenly bodies that are transiting the local meridian (the Dakshina Yantra). There is also a set of 12 additional sundials (the Rashivalayas Yantra).

Practical Information

Information about the Jantar Mantar observatory can be obtained at *http://www.tourismtravelindia.com/rajasthanportal/touristattractions/jantarmantar.html*.

025

Broom Bridge, Dublin, Ireland

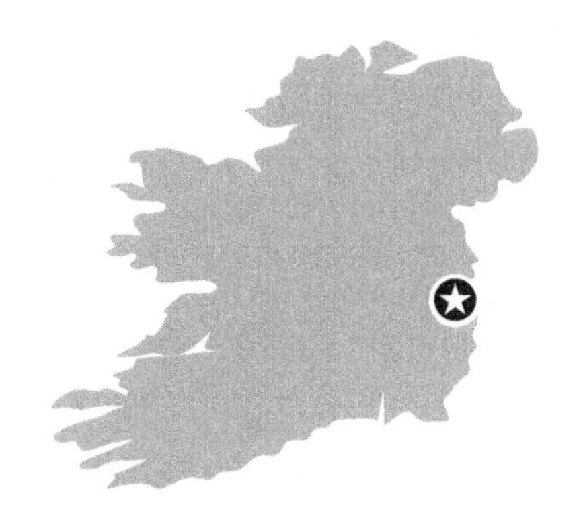

53° 22′ 22.8″ N, 6° 18′ 0″ W

Mathematical Equations As Urban Grafitti

Mathematicians have a habit of thinking of things in the oddest of places: the Irish mathematician Sir William Rowan Hamilton came up with the theory of quaternions while out for a walk with his wife in 1843. Crossing the Broom Bridge in Dublin, Hamilton scratched the quaternion multiplication equation Equation 25-1) into the bridge's stonework using a knife.

$$i^2 = j^2 = k^2 = ijk = -1$$

Equation 25-1. The equation of quaternion multiplication

Hamilton's mathematical vandalism is no longer visible on the bridge, but the bridge itself is still standing and a plaque (see Figure 25-1) was erected by the Irish premier Eamon de Valera in honor of Hamilton. The bridge crosses the Royal Canal in a dubious area of Dublin; the best way to visit it is probably not to try to recreate Hamilton's stroll, but instead to take a number 120 bus or a local Western Commuter line train to the Broombridge stop.

Figure 25-1. The plaque at Broom Bridge; courtesy of Robert Burke (robburke.net)

Complex Numbers

Mathematicians call the numbers we deal with every day (such as –2, 3.5, and 42) "real numbers." You can draw a number line that extends infinitely in each direction, where zero is in the middle, negative numbers are to the left, and positive numbers to the right. Any real number can be represented by a dot placed at the appropriate position along the line (Figure 25-2).

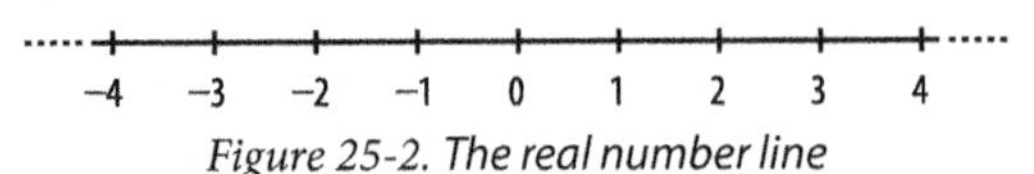

Figure 25-2. The real number line

Mathematicians also talk about complex numbers: these are not real numbers, but are based on working with the imaginary number, i. If you square a real number, say 2 to get 4, or 3 to get 9, then the square root is the number you first started with—the square root of 4 is 2, the square root of 9 is 3.

So, what's the square root of –1? There aren't any real numbers that when squared equal –1. To get around this problem, mathematicians just decided to make up a number (that's why it's called imaginary) and name it i. The square root of –1 is i, so you can write $i^2 = -1$. Obviously, i doesn't appear on the real number line; to get around that, mathematicians define an imaginary number line perpendicular to the real number line. These two lines make up the complex plane (Figure 25-3) and look just like the axes you might see on a graph.

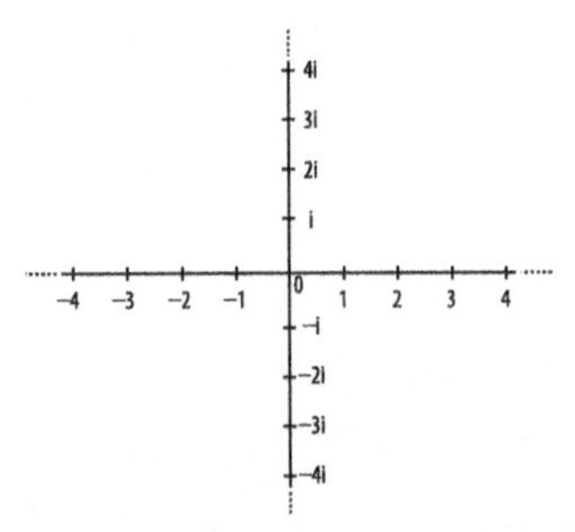

Figure 25-3. The complex plane

The complex numbers are numbers that appear on this plane. If you place a dot a distance, a, along the real number line (the x-axis) and a distance, b, along the complex number line (the y-axis), then you've found the complex number that's written a + bi (a and b are both real numbers). With the complex number you can take the square root of any number (for example, the square root of –13 is about 3.61i). Although i doesn't exist as a real number, it's a useful abstract concept that's typical of mathematical thinking—its creation eliminates the problem of not being able to take square roots of negative numbers.

Complex numbers obey the same rules as real numbers: addition and multiplication are commutative (i.e., it doesn't matter which direction you calculate—for example, $2 \times 3 = 3 \times 2$); addition and multiplication are associative (i.e., when you've got multiple numbers to add or multiply, it doesn't matter which pair you start with—for example, in calculating $2 \times 3 \times 4$, you can calculate 2×3 first or 3×4 first); and finally, multiplication is distributive over addition (i.e., you can "multiply out" a calculation—for example, $2 \times (3 + 4) = (2 \times 3) + (2 \times 4)$). Mathematicians call anything following the same rules as real numbers a *field*.

Because the complex numbers are a field, theroems used for any other field can be applied directly to them; once you know something's a field, a whole body of mathematics becomes applicable. Because of this, complex numbers are useful in all sorts of non-theoretical disciplines like electrical engineering and fluid dynamics. They are also the core of the beautiful mathematical fractal (a mathematical shape that you can zoom in on forever without ever coming to a smooth line) called the Mandelbrot Set (Figure 25-4).

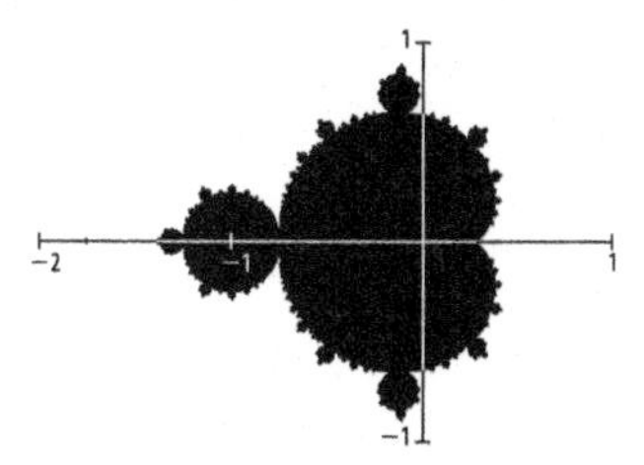

Figure 25-4. Mandelbrot Set with the real and imaginary number lines

Hamilton's quaternions are an extension of the complex numbers. A complex number has two parts to it: for example, in the number a + bi there's the real part, a, and the imaginary part, bi. A quaternion has four parts and would be written a + bi + cj + dk, where a, b, c and d are all real numbers and i, j, and k are imaginary numbers. To draw a quaternion would require four number lines, making a four-dimensional picture analagous to the complex plane.

What Hamilton worked out were the rules for adding, multiplying, subtracting, and dividing these quaternions. His inspiration on Broom Bridge was that the rule for multiplication of quaternions was not like simple multiplication of real (or even complex) numbers—that is, given two quaternions called q0 and q1, it's not the case that $q0 \times q1 = q1 \times q0$. That meant that multiplication of quaternions was not commutative.

Working with numbers (the quaternions) that were not commutative was a radical moment in mathematics, and although quaternions themselves have been overtaken by other mathematical ideas (such as vectors and matrices), they are still around in computer graphics. Quaternions provide a very fast way of calculating the rotation of an object on screen. Think of Hamilton the next time your video game spaceship does a fancy maneuver.

Once you have seen Broom Bridge, head back to the center of Dublin to visit Trinity College.

Practical Information

If you are using a GPS or Google Maps, the bridge can be found at 53° 22′ 22.8″ N, 6° 18′ 0″ W.

026

Tempio Voltiano, Como, Italy

45° 48′ 53.00″ N, 9° 4′ 31.00″ E

A Temple to Alessandro Volta

Alessandro Volta, the inventor of the modern battery and the man for whom the volt is named, was born in Como, Italy, and is honored in a temple and museum built on the shores of Lake Como.

The neoclassical Tempio Voltiano contains a collection of Volta's papers, letters, drawings, and instruments, including his Voltaic Pile, the first battery. The building itself has marble columns and mosaic floors, and was built in 1927 to commemorate the 100th anniversary of Volta's death. Unfortunately, much of Volta's original equipment was destroyed by an electrical fire in 1899 at an exhibition of his work. The temple at Como contains the surviving original equipment and reproductions made in the early 20th century. Of the 200 instruments at the temple, less than half are originals, but the museum is nevertheless a grand testament to his life and work.

In addition to inventing the battery, Volta was an active physicist. He performed experiments on igniting gases using electrical sparks, discovered methane gas in 1778, and was made chair of the physics department of the University of Pavia.

On display at the museum are Voltaic Piles (towers of alternating metal disks), batteries made from a set of liquid-filled beakers with electrodes joined together and dipped in the liquid, and a trough battery similar to a modern car battery. The museum also has paintings of Volta as well as his book collection (including a 1767 copy of Joseph Priestley's *History and Present State of Electricity*).

The museum has explanations in English.

Practical Information

Information about Como and the Tempio Voltiano is available, in English, from the Como Tourist Office at *http://www.turismo.como.it/en/*.

The Voltaic Pile

In 1800 Volta created the first battery, now called the Voltaic Pile (Figure 26-1). It consisted of a set of Voltaic Cells joined together in a series. Each Voltaic Cell was made from a disk of silver and a disk of zinc separated by a cloth soaked in brine (saturated salt water). Volta discovered that the more cells he piled together, the greater the electrical force—now called the voltage—generated.

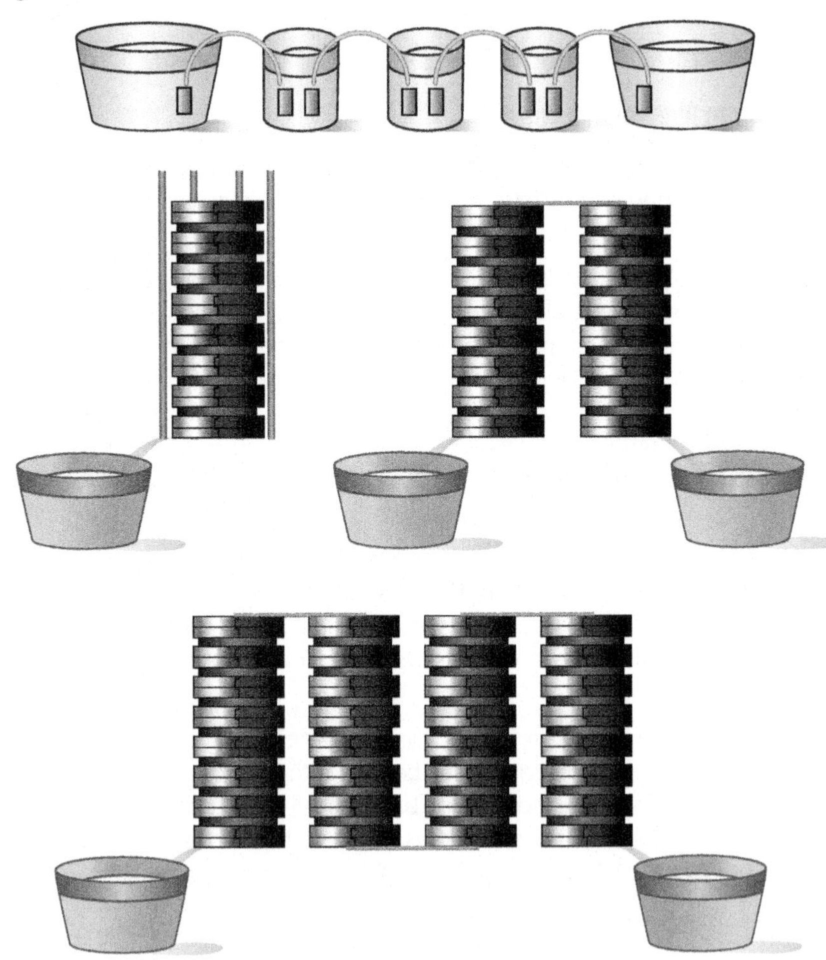

Figure 26-1. Recreation of Volta's 1800 drawing of the Voltaic Pile

He immediately wrote to the Royal Society in London and described the invention, which then appeared in the Royal Society's scientific journal *Philosophical Transactions*. This brought the Voltaic Pile to the attention of other scientists, who went on to discover the electrolysis of water (splitting it into hydrogen and oxygen by passing an electric current through it). Sir Humphrey Davy then used electrolysis to separate sodium, potassium, calcium, magnesium, and barium.

Volta's Pile was a major breakthrough: it provided a continuous current (Volta believed, incorrectly, that this was "perpetual"), and it didn't need to be recharged.

Volta didn't realize that the pile worked because a chemical reaction was taking place. He believed that the contact between different materials (the zinc and the silver) was creating electricity in the same way that static electricity is created when a piece of rubber and a piece of glass come into contact. This effect was well known at the time, but Michael Faraday (see Chapter 75) showed in the 1830s that a chemical reaction was happening inside the Voltaic Pile.

Critical to the operation of each Voltaic Cell was the brine, which we now call an electrolyte. The electrolyte contains ions (in Volta's case, from the salt) that are able to transfer the charge between the zinc and silver disks.

A modern Voltaic Cell can be made with a strip of zinc and a strip of copper dipped in diluted sulphuric acid. The sulphuric acid (H_2SO_4) breaks down in solution into a pair of positively charged hydrogen ions (H^+) and a negatively charged sulphate ion (SO_4^{2-}). Zinc molecules from the zinc strip lose two electrons and become Zn^{2+} ions, which then react with the sulphate ions to become zinc sulphate ($ZnSO_4$), which in turn dissolves in the solution. The freed electrons travel from the zinc strip through the electrical circuit connected to the cell. The electrons arrive at the copper strip and join up with the hydrogen ions to create hydrogen gas (H_2), which bubbles off the copper strip.

Eventually the zinc strip is entirely consumed, or the cell runs out of sulphuric acid.

027

Akashi-Kaikyō Bridge, Kobe, Japan

34° 36′ 59″ N, 135° 1′ 13″ E

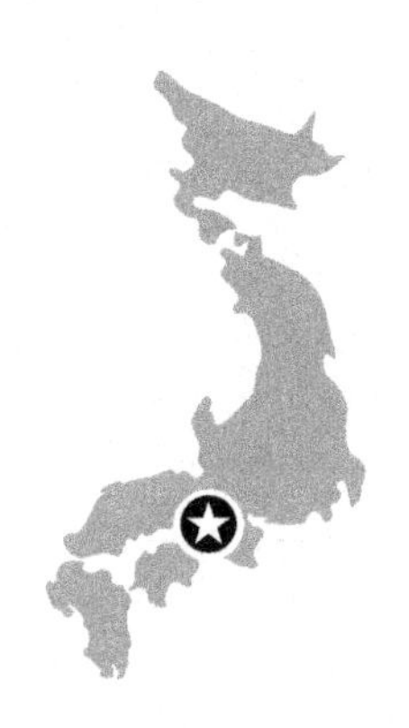

The World's Longest Suspension Bridge

There are suspension bridges all over the world, from the Clifton Suspension Bridge in Bristol, UK, to the Golden Gate Bridge in San Francisco, California. But only one can be the longest, and that honor goes to the Akashi-Kaikyō Bridge in Japan, which links the city of Kobe on the main Japanese island of Honshu with Awaji Island across the Inland Sea. The bridge was built to cross a dangerous waterway that had previously been plied by ferries and was the site of tragic sinkings in stormy weather.

The Akashi-Kaikyō Bridge has a main span of 1991 meters (by contrast, the main span of the Clifton Suspension Bridge is 214 meters, and that of the Golden Gate Bridge is 1,280 meters). The original plan was for a main span of 1,990 meters, but in 1995, after the main pillars were built but before the bridge was installed, a large earthquake hit the Kobe area and moved the pillars apart by an extra meter. Overall, the bridge is 3,911 meters long.

To help prevent swaying (especially in an earthquake), the bridge has pendulums inside the towers that act to dampen any movement (more on damping via pendulums in Chapter 34).

Avoiding corrosion is an important concern because of the salty air. The main cables consist of multiple cables bunched together, and dry air is pumped into the interior spaces to prevent any buildup of moisture. Painting the steel structure is also vital, and is achieved using a robot that can paint 500 square meters per day. Underwater portions of the bridge are protected from corrosion using sacrificial anodes (see pages 316–317).

As with other modern bridges (see, for example, Chapter 14), the Akashi-Kaikyō Bridge has extensive monitoring systems, with multiple anemometers measuring wind speed at the top of the towers and at the deck level. The exact position of the bridge is monitored using GPS, and accelerometers measure the movement of the towers and the anchors.

Visitors can, of course, drive over the bridge after paying the toll. But to get a real sense of its scale and construction, it's best to visit the Bridge Science Museum on the Kobe side of the bridge. There's a small park from which the bridge can be observed, and a tunnel that takes visitors underneath the bridge and out across the water to a glass-floored observation area.

It's also worth seeing the bridge at night, when it's illuminated in a pattern of ever-changing colors.

Practical Information

The Bridge Science Museum is situated on the Kobe side of the Akashi-Kaikyō Bridge, a short walk from the JR Maiko station. There is some information in English, but you might consider arranging for an interpreter to understand all the displays. The Japanese National Tourist Office has information about visiting the bridge at *http://www.jnto.go.jp/eng/location/sit/hyogo/2821.html*.

The Shape of Suspension Bridges

Before the deck of a suspension bridge is added, the main cables are draped across the pylons and hang down under their own weight to form a catenary (see page 408). But when the deck is added, the shape changes to a parabola. It only takes a few simplifying assumptions and a little bit of mathematics to understand why.

Start by considering the point where the curve of the suspension bridge cable is at its lowest, and use that point as the origin for a set of (x, y) coordinates (Figure 27-1). Here, (0, 0) is the lowest point; the x-axis runs parallel with the bridge's deck, and the y-axis runs vertically.

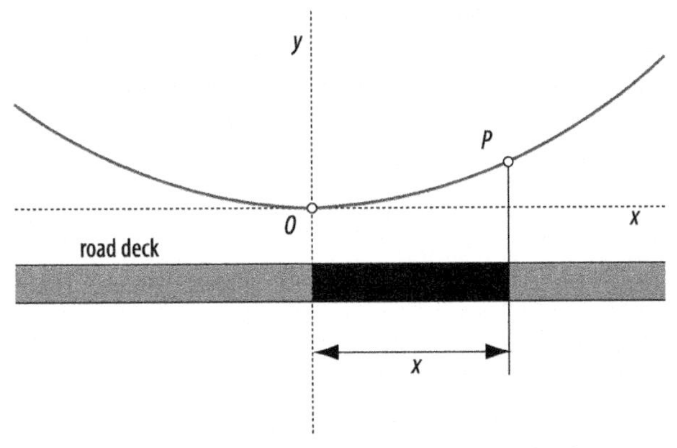

Figure 27-1. Mathematics of a suspension bridge

To simplify things, imagine that the cable has no weight (in fact, the cable is much lighter than the deck, so the deck's weight will be the factor that most influences the cable's shape). Also imagine that the deck is entirely uniform and that the weight of the deck acts on the cable uniformly (this isn't entirely true, because there isn't an infinite number of vertical rods linking the main cable to the deck, but it's close enough).

Now take a point P somewhere up the cable to the right of the origin at a distance x along the x-axis, and consider the portion of cable from the origin to P. To determine the shape of the cable, we need to determine the angle of this portion of cable (which we'll assume is short enough that we can pretend that it's straight).

There are three forces acting on the portion of cable (Figure 27-2). At the origin, there's a purely horizontal force coming from tension in the cable, which we'll call T. The weight of the section below the cable, W, pulls down on the cable uniformly. Finally, there's tension up the cable at P, which we'll call F.

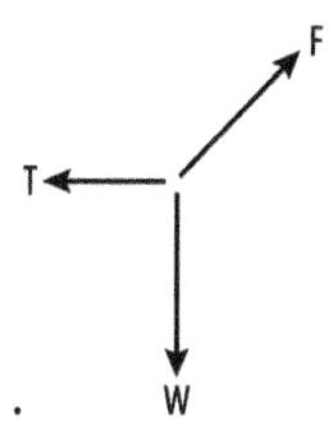

Figure 27-2. Forces on a small section of cable

Since the cable is not moving, these three forces must be in equilibrium, and therefore F must equal W/T. T is a constant force (i.e., it doesn't matter what point we choose for P) and W depends on the weight of the deck. Since we assumed that the deck has a uniform shape, we can specify a linear density D for the deck, and get a weight of Dx for the section between the origin and point P. So the force F is Dx/T.

Now imagine that the curve of the cable is defined by some unknown function f such that y = f(x) is the cable's equation. At a distance x, the cable is at a height y (refer back to Figure 27-1). The slope of the cable at x is the derivative dy/dx, which is df(x)/dy. Since we know the slope from the forces acting on the cable at distance x, we know the value of dy/dx: it's Dx/T (Equation 27-1).

$$\frac{Dx}{T} = \frac{df(x)}{dx}$$

Equation 27-1. Cable shape differential equation

It's just a matter of integrating both sides of this equation to recover f(x) (with an unknown constant c). Although the actual tension, T, in the cable is unknown, it's easy to see that the shape is a parabola (Equation 27-2).

$$f(x) = \int \frac{Dx}{T} dx = \frac{D}{2T} x^2 + c$$

Equation 27-2. Shape of a suspension bridge's cable

028

Akihabara, Tokyo, Japan

35° 41′ 54.38″ N, 139° 46′ 19.99″ E

Electric Town

In the aftermath of the Second World War, the area around the Sobu main railway in the Kanda district of Tokyo became the center of a thriving black market in radios and radio equipment such as vacuum tubes. Akihabara's position close to the docks where goods were flowing into Japan up the Kanda River, and to what is now Tokyo Denki University (where electrical manufacturing was being taught), made its back streets an ideal spot for trade in all types of electrical equipment.

Prior to the war, Akihabara was already a trading spot for many types of goods, but the explosion in demand for electronic gadgets made Akihabara what it is today: an enormous shopping area for everything from useless gizmos to the latest must-have electronics (see Figure 28-1). Akihabara's shops range from enormous department stores on the main street, Chuo Dori Avenue, to backstreet stalls with secondhand goods and spare parts.

Figure 28-1. Akihabara; courtesy of Michael D. Rubin

Goods available here include everything from the tiniest electronic gadgets and games to large electrical items like microwaves, air conditioners, and refrigerators. It's estimated that around 10% of Japanese electronics purchases take place in Akihabara. Prices are cheap by Japanese standards, but don't come looking for a bargain: instead, come to Akihabara to find something you can't get anywhere else. The best deals can be found in the side streets, away from the main avenue.

The largest stores include Laox, Akky, Sofmap, and Ishimaru. One of the objects you'll find in Akihabara that you can't get anywhere else in the world is a Japanese electric toilet seat. Most of the big stores have them, and you can choose from basic features such as a heated seat and built-in bidet to advanced gadgets such as music (perhaps even an iPod cradle), automatic opening and closing using a proximity sensor, and automatic flushing and washing. Another odd bathroom accessory is the Sound Princess, which emits the sound of a flushing toilet to disguise other embarrassing bathroom sounds.

Akihabara is also a paradise for mobile phone fanatics, with every imaginable phone and accessory available (including mobile phone decorations), and gamers come from miles around to see the latest games and controllers. Collectors of video games should head to Super Potato, where something like 50,000 used games are for sale. You can even find vendors still selling 1940s era radio equipment.

And of course, everything related to computing is available here. In fact, when it comes to electronics, if you can't find it in Akihabara, you probably can't find it anywhere—or at least, not in Japan.

A good starting point for a visit is the Radio Kaikan building, which is directly outside the Denki Gai exit from the station. Radio Kaikan is a microcosm of Akihabara, and a good base camp to get acclimated to the frenzied pace before you plunge into the fray outside.

Practical Information

Akihabara has a website with practical visitors information, including a map, in English at *http://www.akiba.or.jp/english/*. Akihabara is easily reached by taking the subway or Japan Rail to the Akihabara station. If you're staying overnight, the experience is not complete without a night in a capsule hotel, where you'll be slotted into a tiny room with a bed and a TV: *http://www.capsuleinn.com/*.

Akihabara News also has a website with information, reviews, news, and discussions in English at *http://www.akihabaranews.com/en/*.

Otaku and Cosplay

In addition to electronics, Akihabara has become a focal point for otaku culture. In Japanese, an *otaku* is an obsessive fan of any hobby or pastime, but particularly of anime (Japanese animation) and manga (Japanese comics). In Akihabara, otaku culture is expressed through cosplay and cosplay cafés.

Cosplay simply means costume play, and it involves people dressing up as characters from anime or manga. Akihabara stores sell costumes based on popular characters. These cosplay floors have a huge variety of well-made costumes for any anime- or manga-inspired fantasy.

But as you walk around Akihabara, the most striking aspect of cosplay may well be the colorfully dressed young women who invite people to visit cosplay restaurants. These *meido* attract customers by role-playing the part of a maid, complete with French maid–inspired outfits that are long on fantasy elements and typically short on skirt length.

When entering a cosplay restaurant, the patron is treated like the returning master or mistress of the house. The maids are deferential, and the customer is treated with great respect. One popular and high-quality cosplay spot is Café Mai:lish, where you'll be greeted with "Welcome home, master" and a curtsy. You can have your photograph taken with a maid dressed in a costume of your choosing.

Although almost all cosplay restaurants feature maids, there are some butler cafés (one, appropriately called Butlers Café, features non-Japanese men dressed as butlers) that cater to mostly female clients.

029

The Escher Museum, The Hague, Netherlands

52° 5′ 0″ N, 4° 18′ 52″ E

The Escher Museum

The Escher Museum in The Hague truly lets you enter the world of M. C. Escher through its large collection of his artwork and a clever optical illusion.

The museum contains almost all of Escher's prints, including his famous, never-ending waterfall, hands that appear to draw themselves on the page, and *Ascending and Descending*, where monks climb an infinite staircase. Escher's prints of transformation—where tiled fishes turn into birds and other animals metamorphosize into shapes, tilings, or even towns—are on display at the museum, as are rarely seen works such as his lithographs of the Amalfi coast.

For an additional fee, visitors can be photographed in a room that warps the viewer's perspective is such a way that one visitor appears tiny and the other huge. Because of the way the warped walls are painted, the mind is fooled into misjudging the two heights.

One of Escher's best-known prints depicts ants crawling around a Möbius strip (a surface with only one side). Much of Escher's work relies on unusual shapes (like the Möbius strip), impossible shapes (see sidebar), and optical illusions. The museum has a special exhibition on how optical illusions work.

Escher's work often has underlying mathematical meaning. His *Circle Limit IV* lithograph depicts devils and angels tessellated inside a circle. As the tiles approach the edge of the circle, they become smaller and smaller, depicting a mathematical limit and infinity in a finite amount of space.

The museum explains the meanings of Escher's work and the process used to create it. Information is available in English throughout the museum, and tours in English are available if booked in advance.

Practical Information

Full details in English are available on the Escher Museum's website at *http://www.escherinhetpaleis.nl/*.

Impossible Shapes

Escher's art is filled with impossible shapes—monks climbing a never-ending staircase, water flowing uphill to create a waterfall that powers the flow itself, and buildings with staircases in every direction.

Three of the basic impossible shapes used in Escher's work are the Necker Cube, the Penrose Triangle, and Penrose Stairs.

The Necker Cube (Figure 29-1) is a simple wire frame cube with nothing to reveal the depth of the image and which therefore can be seen from two different perspectives. In Escher's *Belvedere*, a drawing of a Necker Cube rests on the ground, and a boy is seen holding an Impossible Cube.

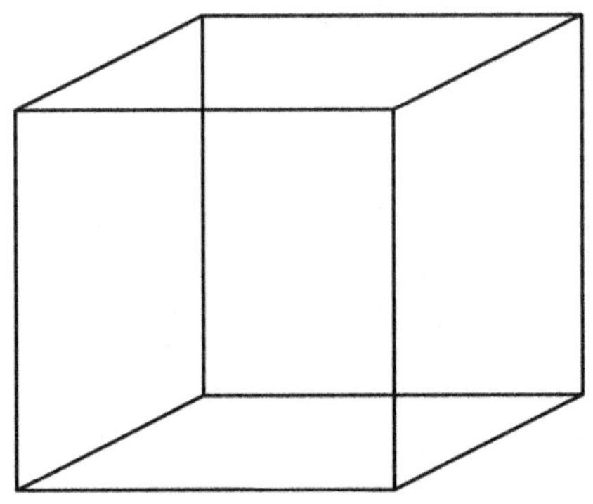

Figure 29-1. A Necker Cube

The Impossible Cube (Figure 29-2) appears to be solid, but closer examination shows that it cannot possibly be constructed from straight edges in our world.

Figure 29-2. An Impossible Cube

In the 1950s, the British mathematician Roger Penrose created the Penrose Triangle (Figure 29-3), an apparently rigid triangle with impossible edges and angles. The shape subsequently appeared in Escher's *Waterfall*, where a pair of Penrose Triangles form a gutter up which water flows.

Figure 29-3. Penrose Triangle

Penrose, together with his father, also created Penrose Stairs (Figure 29-4), which are featured in Escher's *Ascending and Descending*. Penrose Stairs continue in an infinite loop, with no top and no bottom.

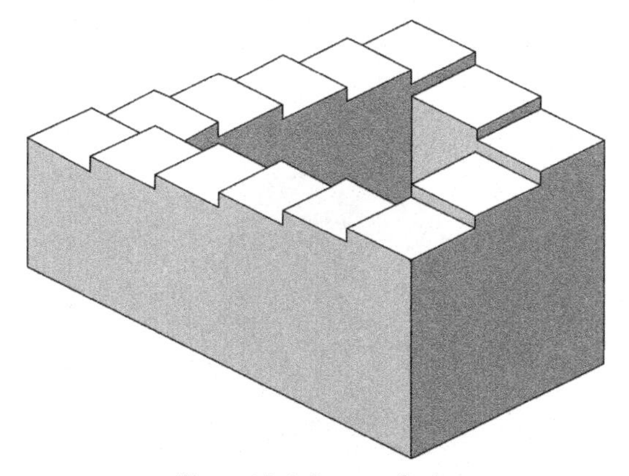

Figure 29-4. Penrose Stairs

030

Tesla Museum, Belgrade, Serbia

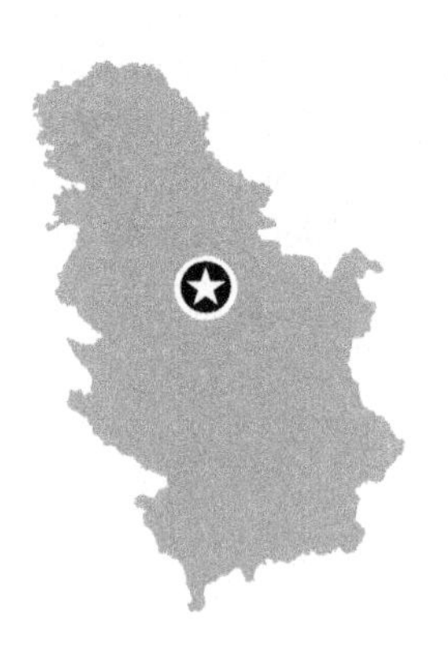

44° 48′ 19.38″ N, 20° 28′ 12.92″ E

The Man Who Lit Up the World

Thomas Edison is the name most people associate with electric light, but the man responsible for providing the electricity to lights and every other electrical device used today is Nikola Tesla (Figure 30-1). Tesla was born to Serbian parents in what is now Croatia, and lived in Hungary and France before moving to the U.S. in 1885. It was there that Tesla became the greatest electrical experimenter since Michael Faraday (see Chapter 75) and laid the foundations for modern electricity distribution.

Figure 30-1. Nikola Tesla

When Tesla moved to the U.S., he went to work in Edison's laboratory, where he redesigned Edison's DC generators and motors. He left after a disagreement about his pay, and after Edison had rebuffed Tesla's attempts to interest him in AC electricity generation.

Later, Tesla and Edison battled bitterly and publicly over AC and DC electrical power. Tesla had joined up with entrepreneur George Westinghouse to build AC power stations, while Edison was pushing DC power. Edison tried to show that AC was dangerous, and to prove it he carried out executions of dogs, cats, horses, cattle, and even an elephant using AC power.

Ultimately, Tesla was proved right: AC power is easier to generate (the generators are simpler, cheaper, and more reliable), it can be transmitted much further (DC power was limited to short distances and necessitated power stations close to consumers), and its voltage can be converted using a simple transformer.

But Tesla's inventions were not limited to AC power. Along with Marconi, he shares the honor of inventing radio (see Chapter 62), and he worked on wireless transmission of electricity, remote controls, vertical take-off and landing aircraft, directed-energy weaponry, robotics, spark plugs, and more. In all, he was awarded over 300 patents.

Nevertheless, Tesla died destitute, in room 3327 of the New Yorker Hotel in New York City. Two thousand people attended his funeral.

After his death, Tesla's nephew and heir (who was also the Yugoslav ambassador), Sava Kosanović, arranged for Tesla's personal effects to be removed from the U.S. and returned to Yugoslavia. Today the Tesla Museum in Belgrade houses his complete collection of books, writing, and objects, as well as his cremated ashes on display in a golden sphere.

The museum explains many of Tesla's inventions, including AC power, and is the definitive place to understand Tesla's life and work.

Practical Information

Information about the museum and details of Tesla's life are available at *http://www.tesla-museum.org/*. If Serbia is too far away to visit, there's also a memorial to Tesla on the Canadian side of Niagara Falls, where Tesla's hydroelectric power plant was situated. See *http://www.teslasociety.com/victoria.htm*.

AC Versus DC

Direct current, or DC, is simple: it's the type of electricity that batteries supply. In a DC circuit, electricity flows in one direction only—for example, from the positive terminal of a battery through a circuit to the negative terminal. Alternating current, or AC, changes direction cyclically, typically in the form of a sine wave (Figure 30-2).

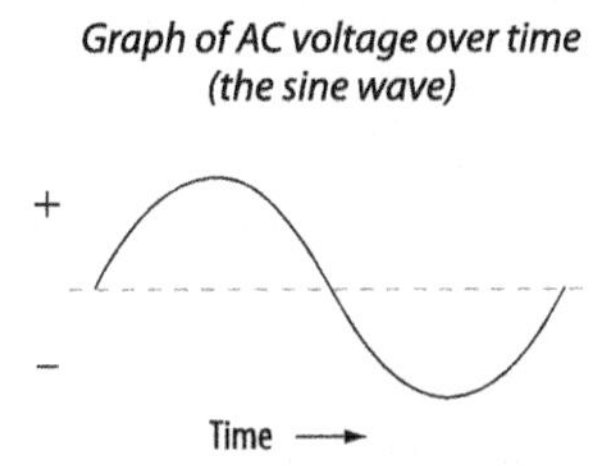

Figure 30-2. AC voltage varying with time

AC varies in voltage from a positive maximum to a negative minimum over time. To generate AC power, a current can be induced in a pair of coils using a rotating magnet. The current varies as the magnet rotates. Since the magnet does not touch the coils, AC generators are reliable and simple. DC generators, on the other hand, require a more complex mechanism, with rotating brushes touching metal connectors that are used to change the direction of the current to keep it positive.

It's simple to change the voltage of AC using a transformer (Figure 30-3). A basic transformer consists of a pair of coils, separated either by air or, more commonly, by some ferromagnetic material such as a bar of iron. Because the AC voltage varies over time, it creates a changing magnetic field around the coil it is connected to. This magnetic field induces an AC voltage in the other coil. The ratio of the number of windings of cable in the two coils determines the change in voltage (and current) across the transformer.

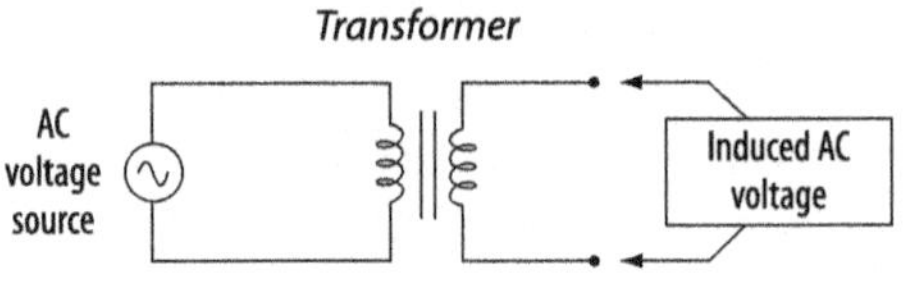

Figure 30-3. An AC transformer

But AC's biggest advantage is in power transmission. Because AC's voltage can be increased or decreased using transformers, it's possible to choose the most appropriate voltage for a given situation (see Figure 30-4); that is, power transmission can use a very high voltage that is then reduced by a transformer before entering a home. Generators in power plants can produce power at a lower voltage than the transmission line, with the voltage being increased before transmission by another transformer.

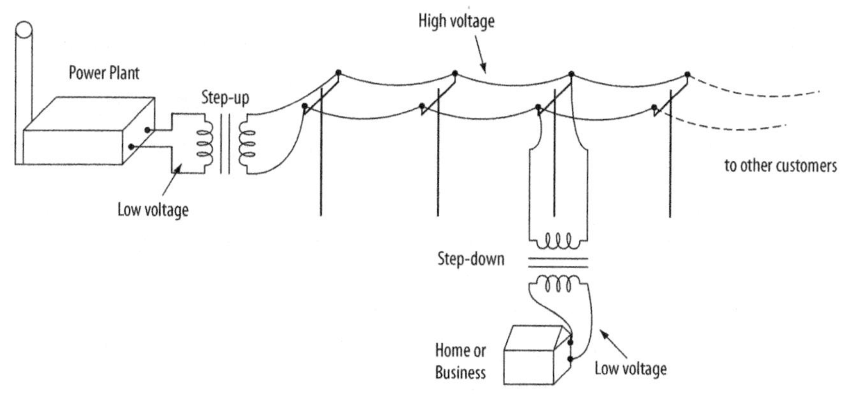

Figure 30-4. Power transmission using AC

Because the voltage can be changed so easily, AC is able to take advantage of the fact that power loss in a cable is proportional to the square of the current. By increasing the voltage (perhaps to hundreds of thousands of volts), with a corresponding decrease in current, power can be transmitted over great distances. It can then be transformed to a lower voltage (and higher current) for delivery.

AC can also be converted to DC, using a simple device called a rectifier. Small electrical appliances (like cell phones) usually operate on DC, and power adapters both convert the AC supply to a low voltage and turn it into DC.

031

Solúcar PS10 Power Station, Sanlúcar la Mayor, Spain

37° 26′ 34.8″ N, 6° 15′ 0″ W

The Tower of Power

The city of Seville in southwestern Spain gets a lot of sunshine—over 320 days of nine or more hours of sunshine every year. In the peak of summer, the temperature can soar to 50°C, and the Sun shines for over 15 hours a day. The perfect weather makes Seville an ideal spot for solar power.

Drive 20 kilometers to the west of Seville and you'll come to the small town of Sanlúcar la Mayor, where the Spanish company Abengoa has constructed two solar electricity–generating stations and is in the process of constructing more. The Sevilla PV station uses photovoltaic panels that track the Sun and provide power for about 500 homes, but if you drive out to Sanlúcar la Mayor, the first thing you'll be struck by is a 115-meter tower rising from the landscape and apparently being sprayed with beams of white-hot gas.

The tower is at the center of a field of heliostats (mirrors that track the movement of the Sun) that focus the bright Spanish sunlight onto a receiver near the tower's top. The reflected sunlight is so intense that water vapor and dust in the air glow white. All that's needed to complete the scene is a maniacal James Bond villain atop the tower.

This tower is at the center of the Solúcar PS10 power station (see Figure 31-1). At the top of the tower is a solar receiver that is heated by sunlight to create saturated steam at 257°C. The steam is then used to drive a turbine that generates electricity. Make sure you're wearing sunglasses when you look up to the top; the tower's brilliant white glow is very intense.

The surrounding field contains 624 Sanlúcar 120 heliostats, each with a 120-square-meter surface for reflecting sunlight. Overall, the system converts about 17% of the sunlight energy into 10 megawatts of electricity. In comparison, the nearby Sevilla PV field generates about 1.2 megawatts with an efficiency of 12%.

Figure 31-1. Solúcar PS 10; courtesy of Alejandro Flores (afloresm)

Right next door to the PS10, an even larger project is under construction—the Solúcar PS20, which will generate 20 megawatts of electricity. The entire Sanlúcar la Mayor project will eventually generate 300 megawatts of electricity from towers, photovoltaics, solar troughs, and Stirling engines.

The Solnova 1 station, also currently under construction, uses parabolic troughs to focus the Sun's rays on a pipe containing a synthetic oil that is heated to almost 400°C. The hot oil is then used to create steam, which drives turbines to generate power. When complete, Solnova 1 will generate 50 megawatts of electricity with an efficiency of 19%.

Finally, Abengoa is planning to build a solar plant based on Stirling engines. Large parabolic mirrors will focus the Sun's rays onto a Stirling engine at the center of each mirror. A Stirling engine is a simple closed engine that works by the repeated heating and cooling of a gas enclosed in a pair of cylinders with a pair of pistons—one piston moves when the gas is heated, the other when the gas is cooled. Between them, the pistons turn a generator to make electricity.

Practical Information

Sanlúcar la Mayor is easily reached by road from Seville, following the A-477 north to Los Ranchos de Guadimar. The Solúcar PS10 power station is 3 kilometers northwest of Los Ranchos de Guadimar, and is easily visible from the road at 37° 26′ 34.8″ N, 6° 15′ 0″ W.

Photovoltaics

The Solúcar PS10 station works by first converting sunlight into steam, and then converting that steam into electricity. A more direct method, as used by the small Sevilla PV station, is to go straight from sunlight to electricity using photovoltaics. All photovoltaic panels, more commonly referred to simply as solar panels, share the same technology whether they are used to generate power for homes or to supply electricity for a pocket calculator—they are semiconductors that exploit the photoelectric effect (see page 131) to make electricity.

A typical solar panel consists of a sandwich made from a pair of conductors (see Figure 31-2) whose middle is filled with a semiconductor (such as silicon) made to produce the core component of all semiconductors—a PN junction (see page 325), where N-doped and P-doped silicon meet.

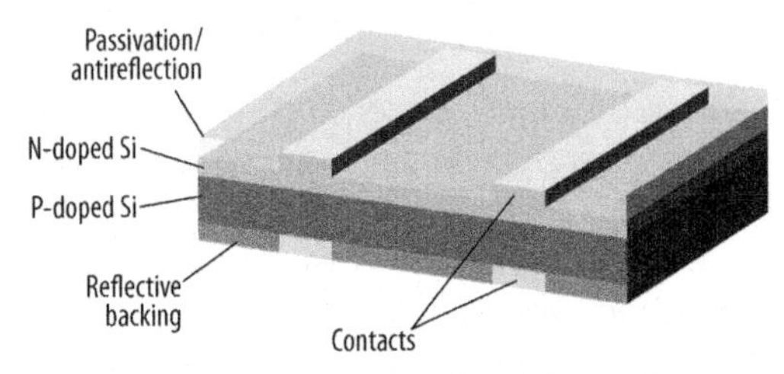

Figure 31-2. Layers of a solar panel

The N-doped silicon (which has an excess of electrons) is pointed toward the sunlight, and the conductors are connected to a device to be powered. The conductor facing the Sun must be transparent (either made of a transparent material, or filled with holes) to allow sunlight through. The sunlight strikes the N-doped silicon, and the photoelectric effect causes some of the electrons to receive additional energy from the incident photons.

When an electron receives enough energy, it is knocked loose from the atom that it was attached to and is able to move within the semiconductor. Because a load is applied across the PN junction, these freed electrons will tend to move from the N-doped region across the P-doped region (which has too few electrons) and around the circuit.

The electrons can only move in one direction because the PN junction creates a diode—a component through which current can only flow in a single direction. This happens because an area called the depletion layer is created where the N-doped and P-doped silicon meet.

When not connected to any load, there's a natural balancing of electrons across the PN junction (Figure 31-3). Since the P-doped silicon is missing electrons, electrons from the N-doped silicon near the junction between the two are attracted across (with a corresponding movement of a "hole"—a space for an electron—from the P-doped side to the N-doped side). This process is completed when all the electrons and holes near the junction have migrated across. This leaves a negatively charged region on the P-doped side, and a positively charged region on the N-doped side. That difference in charge creates an electric field that's responsible for the ability of the semiconductor to conduct in only one direction.

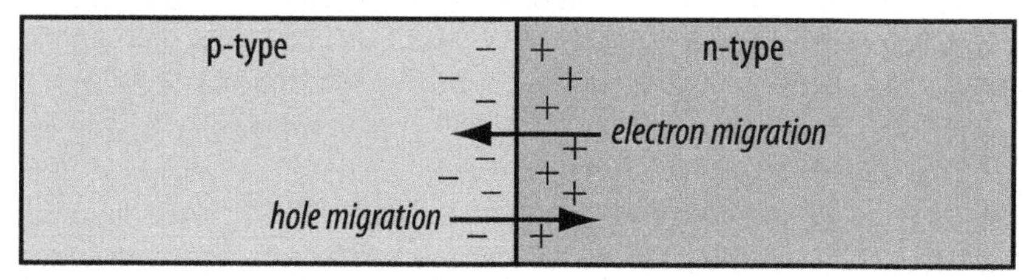

Figure 31-3. The depletion layer

To break free from the atoms, the electrons need to receive a specific amount of energy from the photons hitting the solar panel. If the photon's energy is too low, it will simply pass through the solar panel and not generate any electricity. Some photons will reflect off the surface of the panel and be lost. The rest will be used to make electricity. Currently, the best solar panels have an efficiency of up to 41% in laboratories.

032

CERN, Geneva, Switzerland

46° 14′ 3″ N, 6° 3′ 10″ E

High-Energy Physics

CERN, the European Organization for Nuclear Research, is the largest particle physics laboratory in the world and operates six particle accelerators plus the new Large Hadron Collider. Visiting CERN takes half a day—it is simply enormous, and most of it is underground. All visits are guided tours by people working at CERN, and the exact itinerary is determined by which parts of CERN are off-limits because of ongoing work.

Typical tours include the LINAC and LEIR accelerators; the COMPASS particle detector; the AD (anti-proton decelerator), which is used to make anti-matter; the CAST and AMS experiments used to find particles coming from the cosmos; parts of the Large Hadron Collider; and/or the computing resources of CERN.

Since the creation of CERN in 1954, the laboratory has been home to the Nobel Prize winners of 1976, 1984, 1988, and 1992. CERN has also been at the forefront of high-energy physics, with the discovery of neutral currents (1973) and the W and Z bosons (1983), and the creation of anti-hydrogen (the anti-matter form of hydrogen).

CERN is also the birthplace of the World Wide Web. Tim Berners-Lee and Robert Cailliau built the prototype Web with a simple server and browser designed to give anyone easy access to documents. CERN is also a center for the development of grid computing—the detectors used to find particles produce an enormous amount of data (the Large Hadron Collider's ATLAS detector will produce around 1 petabyte of data per year) that requires very large grids of computers for processing.

Before or after you've toured CERN, check out an exhibit that's not to be missed—Microcosm, CERN's interactive science center. Microcosm explains the physics studied at CERN, the equipment CERN uses, the Large Hadron Collider, and the importance of computing to CERN's work. The computing exhibit even includes the first web server: the NeXT workstation of Tim Berners-Lee.

There's also the Globe of Science and Innovation, which houses temporary exhibitions and is one of the few parts of CERN that is both fascinating to see and above ground.

Practical Information

Visiting CERN requires you to book months in advance, but visits are free. Be sure to carefully read the information about what to wear and CERN's safety rules before arriving, to avoid being refused entry. Full details of how to visit CERN are available at *http://outreach.web.cern.ch/*.

The Higgs Boson, the Large Hadron Collider, and the Origin of Mass

The Standard Model of particle physics is used by physicists to describe most, but not all, of the interactions between particles from which matter is made. The Standard Model is able to explain the electromagnetic force between particles, the weak force, and the strong force. It fails to explain gravity.

The Standard Model includes three types of particles: fermions, gauge (or vector) bosons, and the Higgs boson. The fermions consist of the quarks and the leptons. There are six quarks (called up, down, charm, strange, top, and bottom) and six leptons (the electron, the muon, and the tau, plus three neutrinos: the electron neutrino, muon neutrino, and tau neutrino).

The quarks group together in threes to form other particles. The proton consists of an up, up, down quark triplet (Figure 32-1), and the neutron an up, down, down triplet. The proton, neutron, and electron are the building blocks of matter.

Figure 32-1. A proton is made of two up quarks and one down quark

There are four gauge bosons: the gluon, the photon, and the W and Z bosons. These are responsible for the forces between particles. The photon is responsible for the electromagnetic force, the W and Z bosons for the weak force (which governs radioactivity), and the gluon for the strong force (which binds particles together).

All the particles in the Standard Model have been observed or detected by experiments except one: the Higgs boson. The missing piece in the Standard Model is an explanation of the mass of matter. When physicists composed the various theories that make up the Standard Model, there was one problem: the gauge bosons should have been massless. The photon has no mass (and moves at the speed of light), as does the gluon, but the W and Z bosons are heavy (relatively speaking).

The Higgs boson explains the mass of the W and Z bosons (and all matter) by postulating a field that permeates the cosmos, known as the Higgs field. The Higgs boson is thought to interact with the Higgs field and give other particles a mass, with more interaction with the field leading to a greater mass. The Higgs field is akin to a viscous fluid—if an object tries to move through the fluid, there's resistance to its motion. The Higgs boson moves through the Higgs field and this resistance leads to mass.

No one has observed the Higgs boson, but the Large Hadron Collider at CERN is hoping to find it.

The Large Hadron Collider is the biggest particle accelerator in the world—it's housed inside a 27-kilometer circular tunnel. The collider sends a pair of beams of protons in opposite directions around in a loop, until the beams reach an energy of 7 TeV. They then collide head on.

When the protons collide, they break apart, releasing the particles from which they are built. Detectors around the collider are designed to pick up traces of the particles in the hope of finding the Higgs boson.

033

Historisches Museum Bern, Bern, Switzerland

46° 56′ 35″ N, 7° 26′ 57″ E

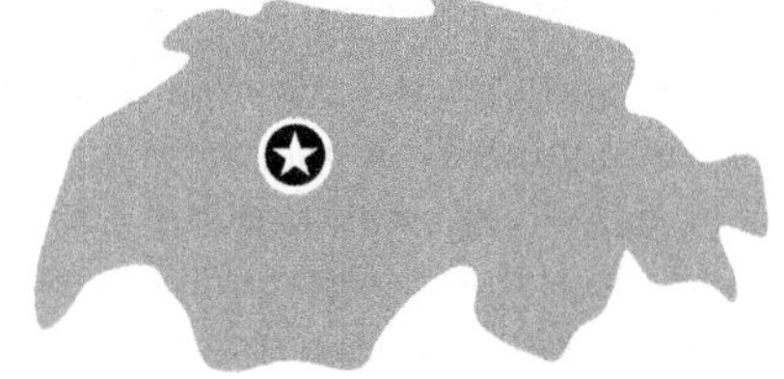

Albert Einstein's Annus Mirabilis (Miracle Year)

In 1905, Albert Einstein was living in Bern, Switzerland. He was working at the patent office as a clerk examining patent applications. That same year he submitted his doctoral thesis and published four papers that changed physics forever. These Annus Mirabilis papers covered the photoelectric effect, Brownian motion, special relativity, and the relationship between matter and energy.

Einstein won the Nobel Prize in 1921 for his explanation of the photoelectric effect. It had been observed that when matter absorbs light, it emits electrons. Einstein explained that light was not absorbed continuously, but rather in discrete packets of energy (termed quanta). This insight was one of the underpinnings of the wave-particle duality of matter, and was later shown experimentally to be correct.

The paper on Brownian motion helped put an end to the idea that atoms were a mere theoretical device. The paper showed that it should be possible to observe the motion of atoms and molecules under a microscope. The French physicist Jean Perrin used this paper as the motivation to study Brownian motion and was able to determine experimentally that Einstein was correct, publishing his results in the book *Les Atomes* in 1913.

The paper discussing special relativity introduced the notion that the speed of light was fixed and independent of the motion of the light source; the paper also expanded on Newton's laws of motion. While Newton was correct for speeds that he was able to observe, Einstein showed that different laws apply at speeds close to the speed of light.

Einstein's final paper introduced the famous equation $E = mc^2$ and showed the equivalence between mass and energy. The equation helped predict the energy possible in a nuclear explosion, and had the surprising side effect of showing that when a system gains energy, it gains mass and vice versa (for example, a compressed spring has greater mass than an uncompressed one).

Einstein's incredible year is celebrated in Bern at the Historisches Museum Bern with a special, permanent exhibition called the Einstein Museum. It is also commemorated at the Einstein Haus.

The Historisches Museum Bern is a general history museum with exhibits of Swiss archeology, and the work of artists from the Bern area. The only scientific portion is the special Einstein exhibition within the main museum. It's a large space dedicated to explaining Einstein's life and work, from his childhood through his study at ETH Zurich, his work in Bern, the theories of special and general relativity, and his years at Princeton.

The entire exhibition is presented in English, and there's an English-language audio guide. As well as displaying objects relating to Einstein's life, the exhibition contains well-reasoned discussions of relativity, the Michelson-Morley Experiment (see sidebar), and the muon experiment (which examined muons created by cosmic radiation hitting the Earth's atmosphere and verified that Einstein's prediction of time dilation—the idea that time is not constant—was a reality). All in all, the exhibition is not to be missed.

Also in Bern is the Einstein Haus. During his Annus Mirabilis, Einstein was living in this apartment at Kramgasse No. 49 with his wife and son. The apartment has been restored to its 1905 state and has a small exhibition concerning Einstein's scientific achievements.

Practical Information

The Historisches Museum Bern website is at *http://www.bhm.ch/*, and the Einstein Haus website is at *http://www.einstein-bern.ch/*. Both have English-language sections.

The Michelson-Morley Experiment

During the 19th century, physicists believed that light needed to travel through a medium; they believed that even a vacuum was actually filled with a mysterious substance they termed the *luminiferous ether*. They postulated that since sound waves need air to travel through, and waves travel on the surface of water, there must be an equivalent medium transporting light.

This ether was predicted to be present everywhere, but was completely undetectable. In 1887 two American scientists, Albert Michelson and Edward Morley, set out to detect the ether by exploiting the motion of the Earth through the imagined ether wind. The ether itself was assumed to be static, but as the Earth orbited the Sun, it would move through the ether at different angles, creating an apparent ether wind.

Michelson and Morley's experiment used interference to try to detect a slight difference in speed between two beams of light that had traveled at right angles to each other for exactly the same distance. They thought that they would see a difference in speed between these two beams because the light had traveled at different angles to the ether wind. By rotating the experiment, they expected to maximize this effect, with one beam moving with the ether wind and the other "cross wind."

For their experiment, Michelson and Morley placed a granite slab into a large bowl of mercury. The slab floated in the mercury, which damped vibrations while allowing the apparatus to be rotated in different directions across the ether wind. On the granite slab they placed a light source directed at a half-silvered mirror set at 45° (Figure 33-1).

Half the light passed straight through the mirror, hit another mirror, and was reflected straight back to the half-silvered mirror, where half of it bounced off into the eyepiece of a telescope. The other half of the light bounced off at 90° and hit another mirror, which bounced it straight back to the half-silvered mirror. From there, half the light passed through the mirror and into the telescope.

The two scientists expected to see an interference pattern, because although the two parts of the reflected light had traveled exactly the same distance, they thought that the speed of the light would be different and hence waves would be arriving slightly out of phase. Moreover, as they rotated the apparatus, they predicted that the interference pattern would change as the relative speeds changed.

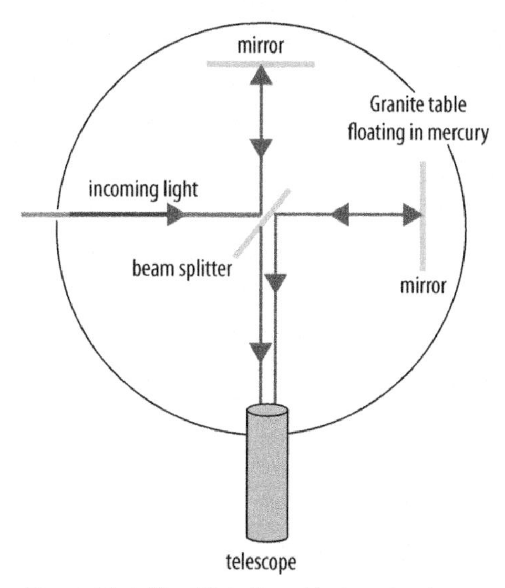

Figure 33-1. The Michelson-Morley experiment

The entire apparatus was enclosed in a wooden box to prevent air currents and rapid changes in temperature from affecting the outcome. Michelson and Morley rotated the slab slowly and measured the fringes of interference seen, and they performed the experiment at different times of the day to try to understand the effect of the Earth's rotation on the strength of the wind.

They found that the ether wind was much, much weaker than expected. It was so weak, in fact, that what they observed was likely due to inaccuracies inherent in their equipment. Since 1887, the experiment has been repeated multiple times with more and more accurate equipment pushing down the upper bound on the speed of the ether wind each time. Today, the ether wind is assumed to not exist at all.

Some call the Michelson-Morley experiment the most famous failed experiment of all time: it did fail to find what the scientists were looking for, but in true scientific fashion, they honestly reported their results and attempted to interpret them.

Little did they know that eight years later, in 1905, Einstein would predict that no matter what the circumstances, the speed of light would be fixed.

034

Taipei 101, Taipei, Taiwan

25° 2′ 1″ N, 121° 33′ 52″ E

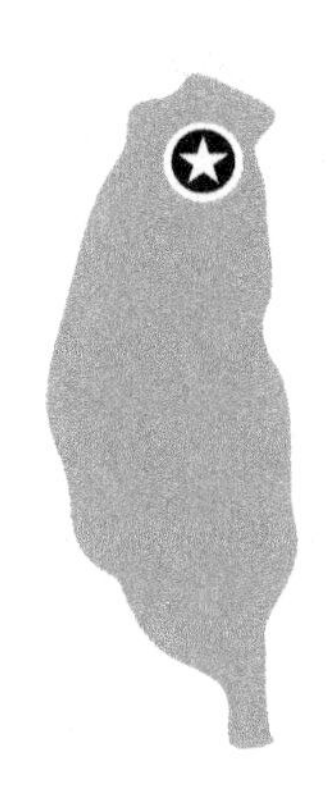

The 660-Tonne Golden Ball

The Taipei 101 is the tallest occupied building in the world, with 101 floors overlooking Taipei's business district. But Taipei is prone to both typhoons and earthquakes, so the skyscraper contains a 660-tonne, gold-colored pendulum near the top to prevent the building from swaying and vibrating. It is the largest and heaviest such pendulum in the world.

Many skyscrapers contain such devices, called tuned mass dampers, for the same purpose, but the Taipei 101 pendulum is unusual because it is on public view (Figure 34-1). It hangs between the 87th and 91st floors, and there are public viewing areas on the 88th and 89th floors. It's even visible from the restaurant and bar. Two other tuned mass dampers, located in the building's pinnacle, are not on display and are tiny by comparison: they weigh only 6 tonnes each.

Figure 34-1. Taipei 101's main tuned mass damper; courtesy of Mingszu Liang

The ball is made of forty-one 12.5-centimeter steel plates welded together for a total size of 5.5 meters. It is attached to the building by eight steel cables, each capable of supporting the ball's entire weight. In normal use the ball can move up to 35 centimeters in any direction and cuts building vibration by 40%. In a major typhoon, the ball is designed to move up to 1.5 meters; hydraulic bumpers below the ball absorb its energy and prevent it from moving too far.

When the building sways in one direction, the ball opposes the movement by swinging the opposite way. The movement of the ball pushes (and pulls) on the hydraulic bumpers and causes them to heat up, absorbing the energy from the motion of the building. The pendulum is tuned by adjusting the length of the cables holding it. By changing the period of the pendulum (the time it takes to swing back and forth), it can be tuned to match the motion of the building.

The building sways back and forth with a specific period: when it is blown in one direction by the wind, it will try to return to the upright position because of the stiffness of its construction. Taipei 101's period is about seven seconds, and the golden ball is tuned to match the seven-second period, thus resisting the building's natural motion. Controlling the movement of the skyscraper makes it more comfortable for the occupants of high floors (who would otherwise be working on floors that move up to 12 centimeters laterally), and reduces wear and fatigue on the building.

In normal wind conditions, you'll have to watch carefully for the ball's small movements. If it starts moving violently, then it's probably time to end the tour!

Practical Information

Visiting information for the Taipei 101 tower is available, in English, on the tower's website: *http://www.taipei-101.com.tw/*. The Observatory (where you can see the ball) is open from 10 a.m. to 10 p.m. daily, and tickets cost NT$400 per person (about 12 USD). When you are done looking at the pendulum, visit the 91st-floor outdoor observatory for a spectacular view from an altitude of 390 meters.

The Pendulum

One surprising thing about pendulums is that the period (the time it takes to swing back and forth, starting and ending at the same position) is independent of the weight of the bob (the weight at the end of the pendulum). The great Dutch scientist (and inventor of the pendulum clock) Christiaan Huygens discovered that the period of a pendulum depends only on its length: the longer the pendulum, the longer the period.

As long as the starting angle Θ is small (less than 60°), the pendulum swings back and forth with the same period, independent of the weight and the starting angle (Figure 34-2). This was first discovered by Galileo Galilei and refined by Huygens. Neither man was thinking of stabilizing buildings; they were thinking of making clocks. Since the motion is independent of weight and angle, it is possible to keep time by simply keeping a pendulum swinging. The pendulum can be moved by a mechanism (such as a spring) without affecting its period.

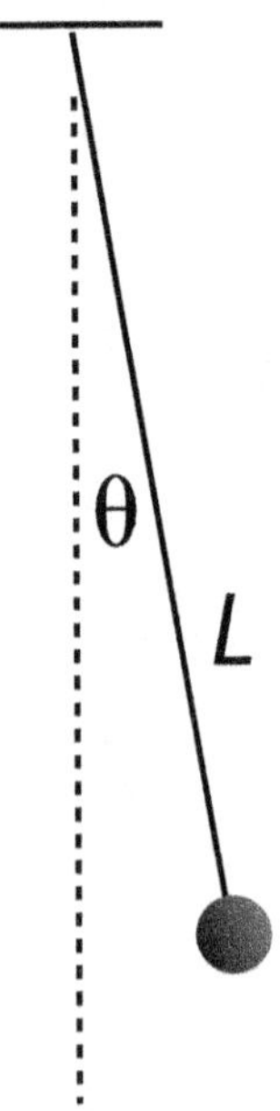

Figure 34-2. A simple pendulum of length L started from angle Θ

Huygens discovered the formula relating the period, T, of a pendulum to its length, L, and gravity, g (Equation 34-1).

$$T = 2\pi\sqrt{\frac{l}{g}}$$

Equation 34-1. Huygens's Law

Flipping that equation around makes it possible to determine the length of a pendulum needed for a specific period (Equation 34-2).

$$l = \frac{g}{\pi^2} \times \frac{T^2}{4}$$

Equation 34-2. Determining the length of a pendulum

Knowing that Earth's gravity, g, is 9.81 ms^{-2}, a pendulum that has a period of one second would need to be 25 centimeters long, about right for a mantel-piece clock.

The pendulum in Taipei 101 needs to match the building's period of seven seconds. Using Huygens's Law shows that the pendulum needs to hang about 12 meters, thus the need for a four-story-high hole in which to house it.

035

14 India Street, Edinburgh, Scotland

55° 57′ 19.17″ N, 3° 12′ 19.78″ W

James Clerk Maxwell

Scotland has produced a great number of famous scientists and inventors, including Alexander Graham Bell (Chapter 4), Lord Kelvin (Chapter 73), John Napier (Chapter 57), John Logie Baird (page 452), and James Watt (page 299). But in the world of mathematics and physics, one name stands above them all: James Clerk Maxwell.

Einstein described Maxwell's contributions as "the most profound and the most fruitful that physics has experienced since the time of Newton" because in 1864 Maxwell showed, in the paper *A Dynamical Theory of the Electromagnetic Field*, that light is actually formed from electromagnetic waves. He also suggested that there might be other types of radiation obeying the same laws, and it wasn't long before other types of radiation were discovered: radio waves were found by Hertz in 1886, X-rays by Röntgen in 1895, and gamma rays by Villard in 1900.

And, above all, Maxwell's important theoretical step underpins Einstein's 1905 work on relativity. But completely changing physics wasn't enough for Maxwell's prodigious talent—he also made a major contribution to thermodynamics and the kinetic theory of gases.

Since Maxwell's advance was entirely theoretical, there aren't any inventions or apparatus to visit. There is, however, the James Clerk Maxwell Foundation, which is headquartered in Maxwell's family home on India Street in Edinburgh.

The building is home to the Foundation and shared with the International Centre for Mathematical Sciences, but its interior is filled with objects related to James Clerk Maxwell, his family, and his work. And there are examples of one of Maxwell's little-known talents—writing poetry.

The house itself and its location attest to the fact that Maxwell's parents were well off. His father was an advocate (similar to a lawyer), but luckily allowed his son to forgo following in his footsteps and let him study at Edinburgh and Cambridge universities.

Maxwell also took what is considered to be the first color photograph. Working with a photographer, he took a picture of a piece of tartan ribbon tied into a bow using a simple camera and three color filters (one each of red, green, and blue). He then projected the three images on top of one another to reproduce the original scene. The photographic plates he used are on display at 14 India Street. To this day, computer and television screens reproduce images using mixtures of red, green, and blue light.

Practical Information

You will find the website of the James Clerk Maxwell Foundation at *http://www.clerkmaxwellfoundation.org/*. Note that you must make an appointment to visit the house, as it is not open to the public for unscheduled visits. Visits are free, but it's good karma to make a donation to keep the foundation alive.

Waves and Particles

It would be nice to have a thorough discussion of Maxwell's equations in this book, but to really do them justice we'd need another 50 pages (Maxwell's 1864 paper runs to 54 pages of dense theory and mathematics) and the willingness to navigate a minefield of partial differential equations over vectors. If you are interested in the details, the HyperPhysics website is a good starting point.

But even without the mathematical detail, the result of Maxwell's equations—that light and other radiation are electromagnetic waves—is the foundation for much of modern life. And it was only half right: light behaves as both a wave and a particle.

Maxwell's calculations showed that light was formed from a pair of waves, at right angles to each other (Figure 35-1). One wave is formed from an electric field that varies in amplitude (i.e., its strength varies), and the other wave is a varying magnetic field. In fact, the electric wave creates the magnetic wave and vice versa. The two waves are in phase with each other and have the same wavelength (the distance between two peaks of the wave).

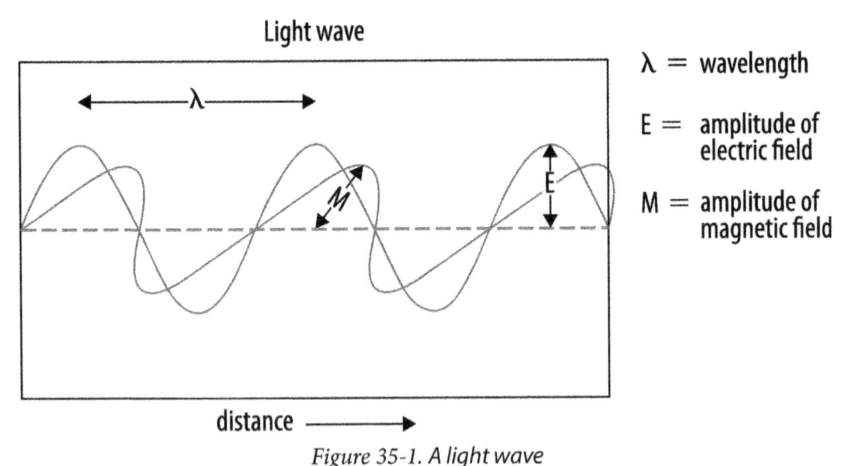

Figure 35-1. A light wave

Different wavelengths cause electromagnetic waves to have different properties (see Figure 35-2). Visible light has a wavelength of between 400 and 700 nanometers; blue light has the shortest wavelength, and red the longest. To one side of the visible light spectrum are ultraviolet waves (roughly between 200 and 400 nanometers), which are responsible for sunburn, and on the other side is infrared (between about 700 nanometers and 1 millimeter), which is used for night vision, remote controls, and thermal imaging (where you can see the heat being given off by an object).

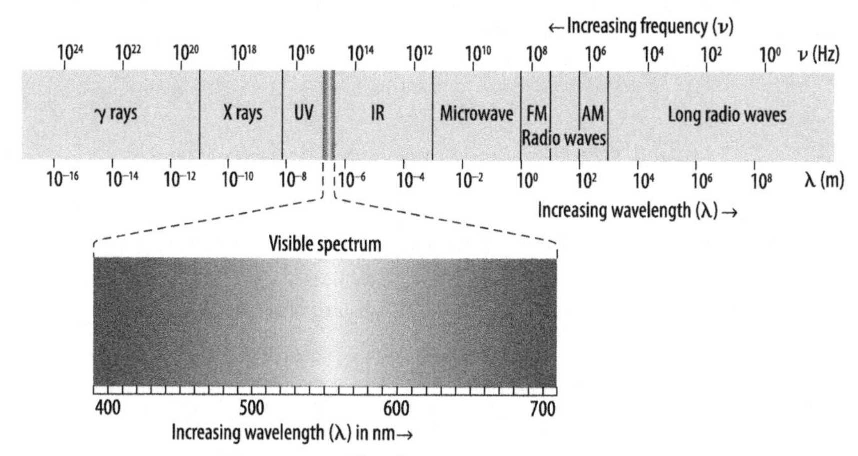

Figure 35-2. The electromagnetic spectrum

X-rays and gamma rays have wavelengths shorter than ultraviolet; microwaves and radio waves sit to the other side of visible light with wavelengths longer than infrared. Normally when talking about radio waves, the frequency of the waves is given (e.g., a radio station might be announced as KQED 88.5 FM, where 88.5 is the frequency in MHz). The frequency, f, is just the number of peaks of the wave per second. For electromagnetic waves (which move at the speed of light, c), frequency is related to the wavelength, λ, by the formula in Equation 35-1.

$$f = \frac{c}{\lambda}$$

Equation 35-1. Wavelength and frequency

But Maxwell's equations did not answer all the questions about radiation. In particular, the photoelectric effect—where electromagnetic radiation is absorbed by matter, causing the emission of electrons—could not be explained from the wave theory. Maxwell's equations predicted that this release of energy (in the form of electrons) should be proportional to the intensity of the electromagnetic radiation, but it was actually proportional to the frequency. The effect was well known—in 1901 Nikola Tesla (Chapter 30) applied for a patent for a device that charged a capacitor from an electric plate exposed to sunlight.

The missing link was discovered by Einstein in one of his Annus Mirabilis papers in 1905 (see also Chapter 33). He showed that light also behaves as discrete packets of energy (now called photons) and that the energy, E, of these "quanta" of light was proportional to the light's frequency, f (see Equation 35-2). This went a long way toward explaining the photoelectric effect—light hitting a piece of matter was being absorbed in discrete units (photons) with specific energy related to the light's frequency, and this energy was sufficient to knock an electron out of its orbit around an atomic nucleus.

$$E = hf$$

Equation 35-2. Energy and frequency (h is Planck's constant)

More electrons can be forced out of a piece of matter by increasing the light's intensity, but they will all have the same energy determined by light's frequency. The energy of each photon is partially used to detach the electron, and the remainder turns into the kinetic energy of the photon itself.

Maxwell introduced the electromagnetic wave explanation for light and other radiation, and Einstein showed that in fact the same radiation also behaves as if it were discrete units of energy, like particles. These two apparently opposing views are not contradictory—they are the basis for the wave-particle duality of matter, which says that all matter behaves as both waves and particles.

This duality has been experimentally determined for particles other than photons. Because the wave theory has the simple result of explaining diffraction and diffraction patterns, physicists have looked for diffraction patterns caused by other particles.

The 1937 Nobel Prize in Physics was awarded for the successful detection of diffraction of electrons (thus verifying that electrons have wave-like properties). In the electron diffraction experiment, a beam of electrons was directed at a nickel crystal. A movable detector (a galvanometer) was used to measure the resulting spread of electrons around the crystal. By looking at the peaks and troughs in the diffraction pattern caused by the crystal, the wavelength of the electrons was determined.

036

Air Defence Radar Museum, RAF Neatishead, England

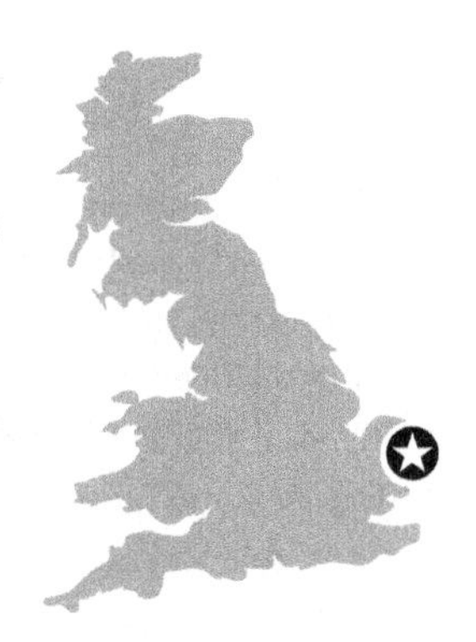

52° 42′ 52.2″ N, 1° 28′ 13.6″ E

It's Bigger Than You Think

RAF Neatishead is the site of the first secret air defense radar in Britain, and today houses a museum that covers the history of radar from its invention through the Second World War and up to the end of the Cold War. Everything in the museum is authentic, including the Cold War operations room where you are invited to sit at one of the terminals and try to deal with a four-minute warning of a nuclear attack. The early-warning equipment was still in use in 1993.

The history starts with the Second World War. You'll see the operation rooms that were in use during the Battle of Britain and the Blitz, where information from early radar systems was translated as fast as possible into positions marked on maps on large tables.

Britain's first early-warning system, the Chain Home coastal radar stations, consisted of 110-meter-high towers that broadcast radio signals with a wavelength of 10 to 15 meters. The radio signals bounced off approaching aircraft, and by timing the delay between transmitting a pulse of radio waves and receiving the bounced signals, it was possible to calculate the distance. The Chain Home system was also capable of determining the direction of approaching aircraft by comparing the bounced signals received at two different antennas. The differing signal strength could be used to determine the direction, and a second pair of antennas could find the elevation.

Once you knew the distance, elevation, and direction, it was possible to accurately detect aircraft up to 200 kilometers away, and Chain Home was used to great effect during the Battle of Britain and the Blitz. (For more on radar, see Chapter 100.)The museum also contains a timeline describing the events that led to the development of radar, and a collection of original radar equipment and vehicles.

The Cavity Magnetron

The original Chain Home system used a long wavelength, and therefore couldn't distinguish small objects. In addition, the transmitters were large and had to be ground based. In early 1940, two scientists working at Birmingham University invented a device that could produce a powerful radio signal on a very short wavelength—the cavity magnetron.

With the cavity magnetron, which was small enough to fit into a briefcase, it was possible to install radar in aircraft and to detect very small objects such as a submarine periscope sticking out of the water. The cavity magnetron produced radio waves with a wavelength in centimeters and was also used for mapping terrain to provide accurate information for bombing, tracking Nazi German V-1 flying bombs, and directing the artillery fire.

The cavity magnetron (Figure 36-1) works by creating a rotating electrical field using a combination of a hot cathode to emit electrons, and permanent magnets to make the electrons rotate inside a cavity.

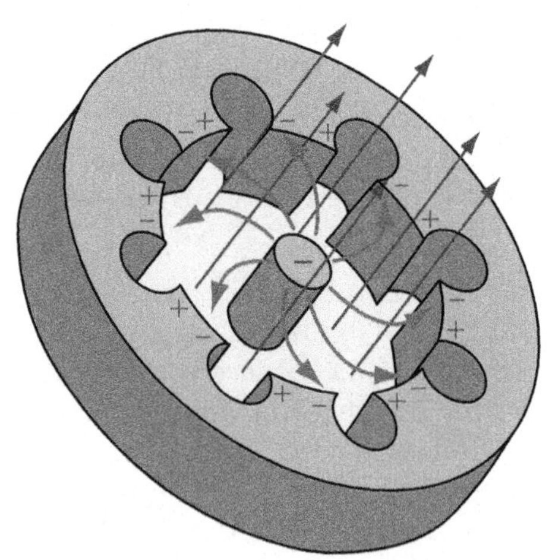

Figure 36-1. Cavity magnetron

The hot cathode is in the center of the cavity magnetron and is connected to a DC supply. The cathode heats up and emits electrons that are then attracted to the anode, which is ring-shaped and surrounds the cathode. Between the anode and the cathode is an empty space.

A magnetic field, produced by permanent magnets, passes through the cavity and acts on the electrons—instead of flying straight off the cathode to the anode, the magnets force the electrons to swirl around in the cavity. As the electrons swirl around, they create a pattern similar to the spokes of a bicycle wheel.

The anode has small circular holes cut into it, which open onto the main cavity (Figure 36-2). As the electron spokes swirl around in the main cavity, they create a charge at the openings of the holes, and current flows around each hole.

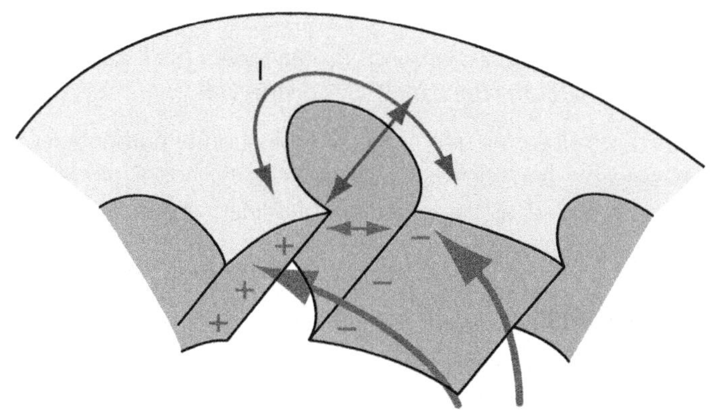

Figure 36-2. A single hole

Each hole acts like a resonant LC circuit (for more on resonance, see Chapter 62): the circular portion is an inductor with current flowing around it, much like a coil of wire, and the gap where the hole opens is a capacitor. Together the inductor and capacitor resonate to produce an electromagnetic field with a very high frequency. This field can be used in a radar system by extracting it—a simple extraction method is a small loop antenna placed inside one of the holes, which picks up the field and can be connected to a large radar antenna.

Most people have a cavity magnetron in their homes without even realizing it—it's the device that makes a microwave work.

Outside the museum there's a Bloodhound missile system. The Bloodhound was an early surface-to-air missile system guided by radar. It used semi-active radar to find its target—instead of carrying a complete radar system itself, the Bloodhound relied on a ground-based radar to illuminate the target. The missile would pick up the reflected radar signal and follow it to the target without having to actually create the radar signal in the first place.

The Cold War exhibit is the eeriest part: it's unchanged since 1993, and with the power on and the background sound effects provided, it appears that the operators might have just stepped out for a cup of tea. There's also one of 800 bunkers, where red telephones stand by to report on a nuclear strike.

Much of the museum is hands-on, and you can try out the fighter control computers that were used until the early 1980s by the RAF.

What makes this museum a gem is the combination of authentic equipment and knowledgeable and friendly staff. The free, two-hour guided tours are strongly recommended, as they explain in fascinating depth the equipment on display.

Practical Information

Visiting information is available from the Air Defence Radar Museum's website at *http://radarmuseum.co.uk/*.

037

Albury Church, Albury, England

51° 13′ 11.83″ N 0° 28′ 44.26″ W

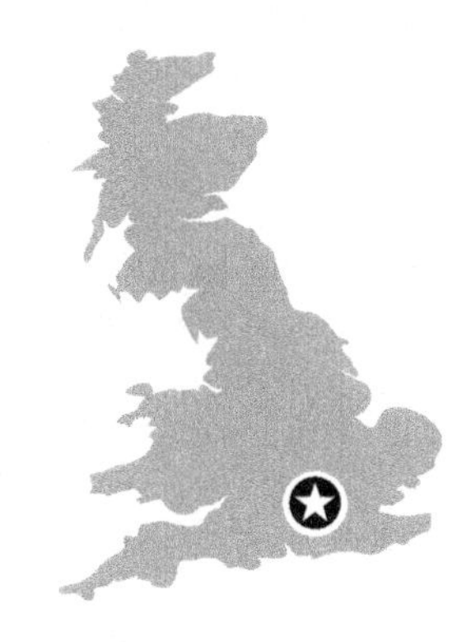

William Oughtred

For almost 50 years (until his death in 1660), the British mathematician and clergyman William Oughtred was rector of Albury Church near Guildford, England. During that half-century, Oughtred volunteered his time to teaching mathematics to interested students. These students included mathematician John Wallis (who later became a cryptographer and was involved in the invention of calculus) and Sir Christopher Wren (celebrated architect, astronomer, and founder of the Royal Society).

Prior to becoming a clergyman, Oughtred was a fellow of King's College, Cambridge, which he had attended since he was 15 years old. He became a fellow at the age of 21. Oughtred's dedication to mathematical education led him to provide not only free tuition, but free lodging as well. Accounts of his life indicate that he liked to study late into the night, and was often still in bed the next day at noon. His church salary of £100 per year seems to have adequately provided for Oughtred and his wife, leaving him free to study, publish, and teach (when he was not busy tending to his congregation).

Oughtred published books designed to help students of mathematics, and that brought together his knowledge of the subject in a compact form. The books, written in Latin, also included many mathematical symbols of Oughtred's own devising; today, the only symbol that remains in common use is the × for multiplication.

Oughtred's most important contribution to mathematics, however, was the invention of the slide rule. Prior to Oughtred, various types of rule had been used (often in conjunction with a pair of calipers) to perform complex calculations. Galileo Galilei is usually credited with the invention of the sector during the 16th century, which was used for simple calculations (multiplication and division) as well as complex ones (trigonometry and square roots).

The Slide Rule

The simplest slide rule of all can be made by taking a pair of rulers and using them to do addition and subtraction. For example, to add the numbers 3 and 4, align the 3 on one ruler with the 0 on the other. Then find the 4 on that ruler, and the number it is aligned with on the other ruler is your answer: it will be the number 7 (Figure 37-1).

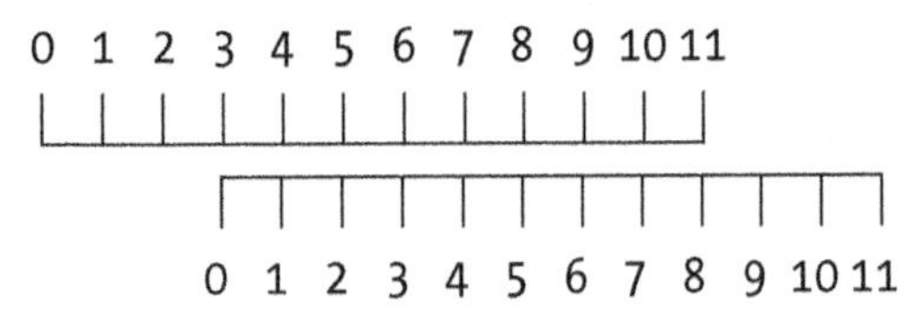

Figure 37-1. Adding 3 and 4 with a pair of rulers

Subtraction is performed by aligning the two numbers to be subtracted (with the larger on top) and then reading the result at the 0 position (Figure 37-2).

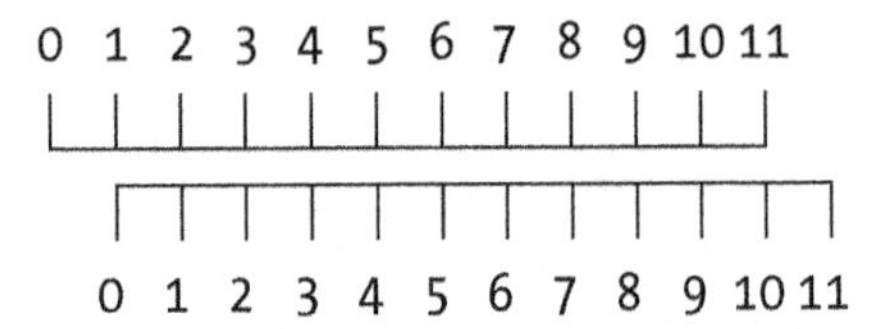

Figure 37-2. Subtracting 4 from 5 with a pair of rulers

Slide rules turn multiplication into addition and division into subtraction by using two laws of logarithms. The first law says that the logarithm of the product of two numbers is the sum of the logarithms of the numbers (Equation 37-1).

$$\log(a \times b) = \log(a) + \log(b)$$

Equation 37-1. Turning multiplication into addition

The second law says that the logarithm of the division of one number by another is the difference between the logarithms of the two numbers (Equation 37-2).

$$\log(a \div b) = \log(a) - \log(b)$$

Equation 37-2. Turning division into subtraction

The same system of addition and subtraction is used on a slide rule, but the scale on the rulers is changed from linear (numbers are spaced equally, so the distance of any number from 0 is proportional to the number) to logarithmic (the distance from the 1 is proportional to the logarithm of the number). So, for example, on a slide rule the number 3 is a distance of log(3), the number 4 is a distance of log(4), and so on. Logarithmic scales get closer together the larger they are.

Using a slide rule, multiplication of 2 and 3 works the same as addition on a pair of simple rulers: the 2 is aligned with the 1 on one ruler, 3 is found on the other ruler, and the corresponding number, 6, is read off. The distance is the sum of the two logs, log(2) and log(3), which can be found at a distance of log(6) on the scale because of the first law of logarithms described previously (see Figure 37-3).

Figure 37-3. Multiplying 2 by 3 on a slide rule

Division is performed by aligning the two numbers to be divided and reading the result from the 1 position on the slide rule. This is similar to doing subtraction using the simple rulers shown earlier.

Commercial slide rules (which are still available to collectors via eBay) feature standard logarithmic scales for multiplication and division, and additional scales for trigonometry, roots, and raising a number to a power.

But the invention of the logarithm by John Napier in the early 1600s (see Chapter 57) laid the theoretical foundation that made the slide rule possible. The logarithm turned a problem of multiplication or division into one of addition or subtraction. Since addition and subtraction are easy to calculate, this made multiplication and division simple. By placing logarithmic scales on a pair of rods or a pair of concentric circles, accurate multiplication and division could be done without calipers by sliding the rods to align the numbers to be multiplied or divided, and then reading the result directly from the rule.

The slide rule was further refined in 1675 by Sir Isaac Newton, who added a movable cursor: a line used to mark a position on the slide rule. Others (including steam pioneer James Watt) improved the slide rule by adding scales that enabled fast calculation of square and cube roots, and in 1851 Amédée Mannhein, a French military officer, created a standardized slide rule with four scales (labeled A, B, C, and D). This would remain in use until the slide rule's demise, which was brought about by the widespread availability of electronic computers and calculators in the 1960s.

For 300 years, the slide rule was the method of performing arithmetic for any engineering or scientific pursuit. Albert Einstein calculated using one, Wernher von Braun used them to do his rocketry calculations, and the Apollo astronauts took slide rules with them as a back-up calculating mechanism. Even the F-16 Fighting Falcon aircraft was designed using slide rules.

The parish of Albury has changed radically since Oughtred's time. The original Saxon church where Oughtred preached was closed in 1840 and was replaced by three other churches in the parish: St Michael's (a converted 19th-century barn), the Catholic Apostolic Church, and St. Peter and St. Paul (located in the heart of Albury).

However, the church building itself is preserved by the Churches Conservation Trust and is open to the public, as well as being occasionally used for services. Sitting in the peaceful Albury Park (which includes a large, privately owned home used for conferences and other functions), the church is nestled among trees and is a stone's throw from the River Tilling. The church door dates back to the 13th century, and the interior has parts dating from the 15th through the 19th centuries. A small plaque in the church indicates that Oughtred is buried in the chancel, but there are no other markings of his grave.

Visit in January or February, and the church is surrounded by snowdrops. Any time you choose to visit, it's a beautiful spot to contemplate the life and work of a self-taught mathematician, and to ponder the logarithm and the slide rule. It was an invention made by candlelight that, in the hands of engineers, helped to usher in the Steam Age, the Jet Age, and the Space Age.

Practical Information

For information about visiting the Albury Church, go to the Albury Parish website at *http://www.alburychurches.org/*. For more on William Oughtred and the slide rule, visit the Oughtred Society at *http://www.oughtred.org/*.

038

Alexander Fleming Laboratory Museum, London, England

51° 31′ 2″ N, 0° 10′ 23″ W

Fleming's Cramped Laboratory

Antibiotics and penicillin are likely the first things that come to mind when the name Alexander Fleming is mentioned. And, certainly, Fleming did accidentally discover in 1928 that the fungus Penicillium notatum prevented the bacteria staphylococcus from spreading in a culture dish. But the story of antibiotics is much larger than Fleming's discovery, and this museum is the place to get the complete picture.

The museum, spread over four floors, has an in situ reconstruction of Fleming's 1928 laboratory (Figure 38-1), complete with his microscope, samples of Penicillium notatum prepared by Fleming himself, a culture plate showing the effect of penicillin mold on Staphylococcus aureus (the most common cause of staph infections) and on B. Coli (a bacteria that resides happily in pigs and can jump over to humans, resulting in severe bowel problems), and a culture dish showing penicillinase (the enzyme responsible for penicillin resistance).

While Fleming is the star of antibiotic research, he wasn't the first to produce an antibiotic. His 1929 paper on his discovery, which shows that he tested penicillin against a range of pathogens, was largely ignored. And Fleming never produced penicillin in a quantity suitable for treating an infected person. That honor goes to Florey and Chain (page 209), who shared the Nobel Prize with Fleming in 1945.

Prior to penicillin's arrival on the market, the antibiotic salvarsan (first marketed in 1910) created by the German scientist Paul Ehrlich was successfully used in treating syphilis. Another antibiotic, Prontosil, was largely created by another German, Gerhard Domagk, and was marketed in 1935 as a treatment for streptococcus. It was rapidly discovered (at the Institut Pasteur; see Chapter 11) that Prontosil was metabolized into sulfanilamide, and that sulfanilamide was the actual antibiotic; this led to a whole range of "sulfa" antibiotics that were active against a wide range of diseases in the early 1940s. Penicillin itself became widely available in 1942. Subsequently, streptomycin (which is used to treat plague) came on the market in 1944.

Figure 38-1. Fleming's laboratory today; courtesy of the Alexander Fleming Laboratory Museum (Imperial College Healthcare NHS Trust)

Penicillin is the most famous antibiotic, in part because of its role in the Second World War and especially because of its ability to deal with sexually transmitted diseases. The museum puts Fleming's work in the context of other antibiotic research, and explains the difficulty that was encountered in mass-producing penicillin in time for D-Day.

In addition, the museum is especially interesting because the enthusiastic curator, Kevin Brown, has written the definitive guide to Alexander Fleming in his book *Penicillin Man* (The History Press).

Practical Information

The somewhat tricky-to-find website for the Alexander Fleming Laboratory Museum is at *http://www.imperial.nhs.uk/aboutus/museumsandarchives/*. The museum itself is not tricky to find: it's at St. Mary's Hospital in London, right next to Paddington Station. All visits to the laboratory are escorted, and your guide will provide commentary along the way.

Penicillin

Penicillin works by preventing the correct growth of the cell walls of the bacteria it is attacking. Penicillin is one of a group of β-lactam antibiotics that all operate in the same way. As the bacteria multiply, each individual bacterium first grows in length and finally splits into two new cells. With penicillin present, the cell wall of the growing bacterium does not form correctly, and the bacterium bursts open. It's the β-lactam component of the antibiotic that causes the weakness in the cell wall.

β-lactam is a ring of three carbon atoms and a nitrogen atom found in the middle of the β-lactam antibiotics (see Figure 38-2).

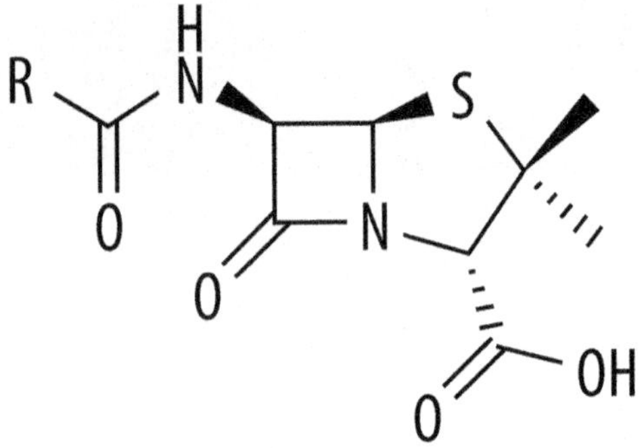

Figure 38-2. Penicillin; β-lactam is the square in the middle

As a bacterium is dividing, it creates a new cell wall. The part of that wall that gives it strength is called the peptidoglycan layer, which is a mesh built from two different substances. Long chains of two alternating sugars are linked to form the mesh. To increase the strength of the mesh, cross-links are made by an enzyme called transpeptidase.

Penicillin prevents the transpeptidase from making the cross-links. It is able to do this because the β-lactam ring binds to the transpeptidase right where the enzyme would have been. The mesh cell wall is still created, but without the cross-links, it is weak and eventually ruptures.

039

Anderton Boat Lift, Northwich, England

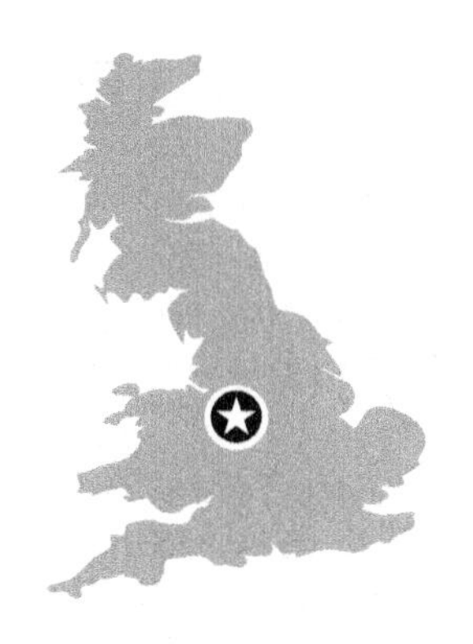

53° 16′ 22.08″ N, 2° 31′ 49.8″ W

A Hydraulic Boat Lift

The Anderton Boat Lift is a link between two British waterways that raises and lowers boats 15 meters using a pair of hydraulic rams. It is the oldest working boat lift in the world and was constructed in 1875. The lift was closed in 1983 because of corrosion, but it has has been fully restored to working order and since 2002 has been open to water traffic and visitors.

The lift was constructed by an engineer named Edwin Clark, who went on to build other boat lifts in Europe. His Belgian boat lifts have also recently been restored to operation. A modern boat lift in Scotland opened in 2002 and is also open to visitors (see Chapter 72). But the Anderton Boat Lift is the oldest.

The lift links the River Weaver and the Trent and Mersey Canal. Initially the plan was to build a series of locks to connect the two waterways, but a suitable site could not be found. Also, an enormous amount of water would have been lost operating the locks, because the canal is 15 meters higher than the river. Loss of canal water was also a problem on another canal in England, the Birmingham Canal, where James Watt built a steam engine to pump water back up past a series of locks (see Chapter 78).

The boat lift originally consisted of a pair of cassions (water-tight boxes) that were mounted on hydraulic rams filled with canal water. The rams were linked together by a small tube, which allowed water to pass between them. The higher of the two cassions was filled with water until it weighed more than the other. That forced the cassion down, the water flowed to the other ram, and the other cassion was raised.

The Hydraulic Press

The Anderton Boat Lift is actually a form of hydraulic press, albeit one where there's no pressure difference between the two cylinders. In a hydraulic press, a pair of cylinders are linked together in a closed system. The system is filled with a fluid such as water or oil, and a pair of pistons move (or are moved by) the liquid.

When one piston is pressed down, the other piston rises. Pascal's Principle says that an increase in the pressure of the liquid is transmitted to every part of the liquid (since the fluid is completely enclosed in the press). Thus a force applied to one piston is transferred to the other.

In the Anderton Boat Lift, the two pistons are exactly the same size, and the force of the descending cassion is transferred to the ascending cassion over the course of about three minutes. In a hydraulic press the piston sizes can be varied, changing the transmission of force so that a small force applied to one piston becomes a much larger force on the other (Figure 39-1).

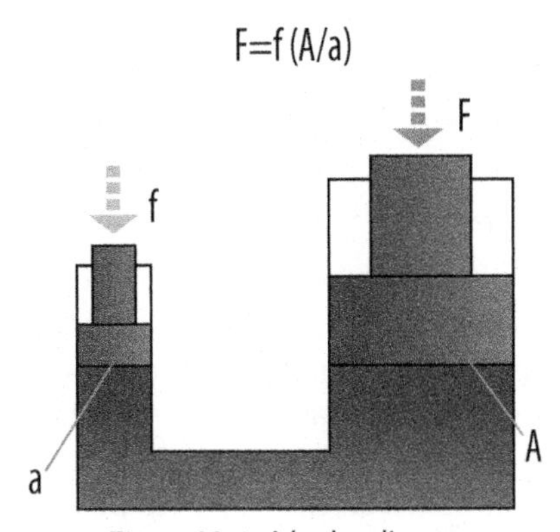

Figure 39-1. A hydraulic ram

For example, if a force f is applied to a piston, then the force F resulting at the other piston is proportional to the ratio of the areas of the two pistons ($F = f \times A / a$).

A force applied to a small piston is magnified by the ratio of piston areas and is applied to the large piston. The distance moved by the small piston is greater (by the same area-to-area ratio) than the distance moved by the large piston. A force over a distance at the small piston becomes a larger force over a smaller distance at the large piston.

All hydraulic presses work on this principle. In the Anderton Boat Lift the pistons are the same size, the ratio of the piston areas is 1, and the force of the descending cassion is equal to the force on the ascending side.

In 1908, the lift was converted to use electric motors and counterweights, because the original hydraulic rams had become unreliable and needed fairly frequent repair. When the lift was restored, the system of electric motors was left in place, but the original hydraulic system was replaced with a new system. That system uses modern techniques, oil instead of water as the hydraulic fluid, and electric pumps to move the oil between the cylinders. The structure of the lift is made of cast iron, and during restoration it was completely dismantled and put back together.

Today, it is possible to visit the lift (and its associated exhibition center) and take a 40-minute boat trip through the lift itself. The boat trip can be extended to include a short sail down the River Weaver once the lift has been navigated.

Practical Information

Details of the Anderton Boat Lift, directions to the site, and information on boat trips are available at *http://www.andertonboatlift.co.uk/*.

040

Bletchley Park, Bletchley, England

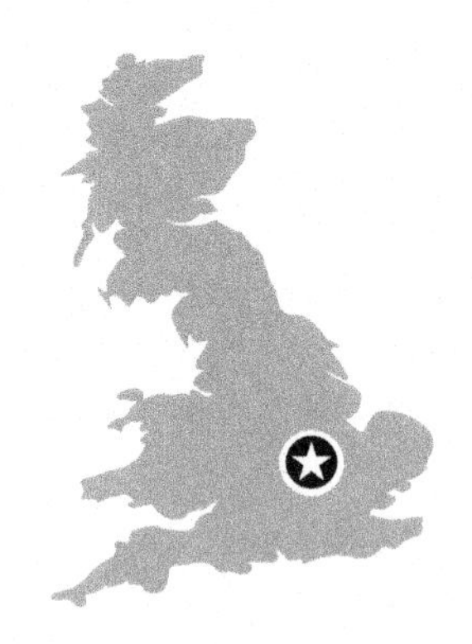

51° 59′ 47.44″ N, 0° 44′ 33.94″ W

ULTRA

Draw a straight line between the British universities of Oxford and Cambridge, and halfway between the two you'll find the grand house at Bletchley Park. Back in 1939, Bletchley Park was close to the intersection of the Oxford-to-Cambridge railway line and the main west coast railway leading out of London. It was the perfect spot to bring together the brightest code-breaking minds in the country.

The brains that worked at Bletchley were recruited for their skill at mathematics, crossword puzzles, bridge, and languages. They worked throughout the Second World War in total secrecy, and at the end of the war their work was hidden and destroyed to prevent discovery. It was not until the 1970s that the secret work of Bletchley Park was revealed, and its role in the Allied triumph in the Second World War was understood.

The code breakers at Bletchley Park had succeeded in breaking the code used for almost all Nazi German communications: the Enigma (see sidebar). The information gathered from this successful code-breaking was so sensitive that it required a new designation and become known as ULTRA (for Ultra Secret). The most famous cryptographer at Bletchley Park was the mathematician and computer scientist Alan Turing (Chapter 66), who played a key role in breaking the Enigma code and went on to create the theory that underpins today's computers. He was joined by many other men and women who read everything from messages sent by Nazi German infantry units to the most secret messages between Adolf Hitler and his high command.

But cracking Enigma required more than just theory and raw brainpower—it required machines. Massive Enigma-cracking machines called Bombes were constructed at Bletchley. These machines were electromechanical, with spinning rotors connected to electrical circuits that could indicate when a code was broken just by completion of the circuit.

After years of neglect, Bletchley Park was rehabilitated in the early 1990s, and is today a museum that commemorates the work of the thousands of men and women who cracked Enigma and other codes. There are displays of historical encryption machines (including the Enigma), a complete explanation of the work of Bletchley Park, a tribute to Winston Churchill (who described the people at Bletchley as "The geese that laid the golden eggs—but never cackled"), and an amusing exhibition dedicated to the role of carrier pigeons during the war.

There's even a new Bombe that has begun breaking Enigma messages again, and was switched on in 2007 by the WRNS (Women's Royal Naval Service) operators who had been in charge of the machines during the Second World War. The Bombe was reconstructed using the few parts and plans that survived destruction, and the memories of those who worked on it. It's on display at Bletchley, sitting just where it would have in the 1940s.

Also at Bletchley is the newly created National Museum of Computing (Chapter 58). It has on display one of the first electronic computers, the Colossus, which was used at Bletchley to break Nazi Germany's highest-level code, the Lorenz (see Chapter 58).

Practical Information

Bletchley Park's website is at *http://www.bletchleypark.org/*.

The Enigma Crib

Throughout the Second World War, Nazi Germany used the Enigma machine (Figure 40-1) to send encrypted messages. The machine consisted of a keyboard and a set of lights, one light for each letter. When a key was pressed, a light would illuminate, indicating the encrypted letter corresponding to the key.

Figure 40-1. The Enigma machine; courtesy of Tim Gage

The core of the machine was a collection of rotors. Each rotor had 26 electrical contacts on each side; each contact on the righthand side of the rotor was linked to another contact on the lefthand side. Multiple rotors (three, four, or five) were placed on a spindle, and when a key was pressed it would allow current to connect to a contact on the rightmost rotor. The current would pass through the rotor to a corresponding contact on its lefthand side, which would connect to the next rotor, and so on. In this way, the signal would flow through the rotors until it reached the lefthand end coming out of some apparently random contact.

At the lefthand end was a reflector that consisted of 26 contacts, wired in pairs, that sent the signal back through the three rotors by a different route. Finally, the signal would emerge from the righthand side to light up one of the light bulbs indicating the encrypted letter.

When a key was pressed, the lefthand rotor would move one position (out of 26); once it had rotated through a complete turn, the next rotor would move one position and so on, like a car's odometer. This rotation meant that the same letter would be encrypted as different letters each time it was pressed.

The reflector in the Enigma machine meant that letters were encrypted and decrypted together in pairs—if A was encrypted to F, then F was encrypted to A. That meant that there was no difference between the encryption and decryption processes: it was enough to set the rotors correctly, and the machine could be used to encrypt or decrypt.

But the reflector also introduced a flaw—no letter could be encrypted to itself—which was exploited at Bletchley to help speed up the cracking process. If code breakers could guess some of the original message being sent, they could use the rule that no letter encrypts to itself to find parts of the encrypted message that matched the guessed portion.

This turned out to be the key to breaking Enigma. The Enigma machine operators around Europe would send messages in the same format (especially weather reports), use the same phrase repeatedly (e.g., "Nothing to report"), or use common words that could be guessed (such as numbers spelled out). These likely phrases were referred to as "cribs."

Suppose that the message HELPPLEASERESPONDQUICKLY was encrypted using an Enigma machine to SVRIORABNHDTKARYANVGMXAT, and the code breaker guesses that the word PLEASERESPOND appears in the original message. The code breaker can eliminate a large number of possible positions for PLEASERESPOND in the encrypted string of letters by sliding PLEASERESPOND along SVRIORABNHDTKARYANVGMXAT, letter by letter, eliminating any positions where a letter in the encrypted portion lines up with the same letter in the original message. Figure 40-2 shows the positions that can be eliminated (they are marked with stars).

```
SVRIORABNHDTKARYANVGMXAT
PLEASERESPOND
 PLEASERESPOND
  PLEASERESPOND
   PLEASERESPOND*
    PLEASERESPOND
     PLEASERESPOND
      PLEASERESPOND*
       PLEASERESPOND
        PLEASERESPOND*
         PLEASERESPOND
          PLEASERESPOND*
           PLEASERESPOND
```

Figure 40-2. Eliminating positions using a crib

There were 12 possible positions originally, but 4 can be eliminated because of the reflector. So with 8 possible positions, a code breaker could try out all the possible initial settings of the rotors to see which one correctly decrypted the matching segment. This could be repeated for each of the non-eliminated positions.

But to complicate the scrambling further, each Enigma machine had a Stecker board, which had literally billions of possible settings and made such manual decryption impossible. The board consisted of 26 holes with an associated set of 13 cables. Any pair of letters could be swapped by connecting the two holes in the Stecker board with a cable. Unlike the rotors, this scrambling was fixed, but the number of possible combinations was enormous.

To defeat the Stecker board, the code breakers at Bletchley were forced to use mathematical tricks and build a high-speed Enigma cracking machine called the Bombe. The mathematical tricks allowed the Bombe to quickly test possible Stecker-board settings for any particular crib, eliminating those that were impossible to quickly find the wheel settings used.

041

British Airways Flight Training, Hounslow, England

51° 28′ 19″ N, 0° 24′ 34″ W

As Close As You're Likely to Get

Most airlines don't let customers take the controls of their jets. But British Airways will—or at least, they'll let you fly in the same flight simulators they use to train their own pilots. At the British Airways Flight Training center, visitors can fly a Boeing 737, 747, or 777.

The flight training center, located close to Heathrow Airport, is used by British Airways to train airline pilots and crew. There are flight simulators for all types of British Airways Boeing aircraft and the Airbus A320, and cabin simulators that are used to train cabin crew on the handling of emergency situations. Like the cockpit simulators, some of the cabin simulators are mounted on hydraulics, and they can simulate different emergency situations such as fire (with smoke filling the cabin), decompression, and exiting via the slides.

For budding pilots, the cockpit simulator experience starts with a briefing by a British Airways pilot. The briefing introduces the flight controls and the computer systems and instruments. Then it's into the cockpit to start the engines. The pilot guides you through the takeoff, cruise, and landing (or perhaps many landings, depending on how successful your first attempt is). After landing, the procedure for shutting down the engines is explained and performed.

British Airways offers two different training times—one hour and three hours. To really get the most out of the experience, the three-hour flight is preferable: one hour flies by (no pun intended) when it's your first time in the cockpit of a complex aircraft. Of the three Boeings available, the easiest to fly is undoubtedly the 777, because of its fly-by-wire and automatic systems.

The flight simulator experience can only be enjoyed by one or two people at the same time. However, British Airways does offer corporate events where larger groups can fly aircraft in the simulators and take part in a simulated emergency landing and smoke-impaired evacuation in a Boeing 737 cabin simulator, followed by jumping down an Airbus A320's emergency slides.

Instrument Landing System

Modern commercial airliners all have the ability to land when visibility is very poor (something that's quite common at airports like London Heathrow). To land almost blind, pilots use the Instrument Landing System (ILS), which can guide a plane down to the runway, and can be used by an autoland system to complete the landing all the way to coming to a halt.

An ILS has three major components: the localizer, the glideslope, and the markers. The localizer and the glideslope transmit radio signals from the runway, which the pilot uses to line up with the runway's center line and fly at the correct angle of descent. There are typically three markers—outer, middle, and inner—and as the pilot flies toward the runway, the markers give an indication of critical points in the descent.

The localizer gives horizontal positioning information, which the pilot can use to determine if the aircraft is to the left or right of the runway and adjust accordingly. The glideslope gives vertical information, which tells the pilot whether the aircraft is flying above or below the correct sloping line necessary for a correct landing.

Both the localizer and the glideslope work in the same way—they transmit a pair of signals that overlap along the line the pilot should be flying. The localizer transmits an AM (amplitude modulation) signal with a carrier wave of around 110 Mhz (the exact frequency is determined by the airport localizer transmitter). This signal is transmitted from the left and right of the runway with an overlap down the center line. The lefthand signal is modulated at 90 Hz, the righthand signal at 150 Hz, and the depth of modulation diminishes the further the pilot drifts from the center line.

An AM signal starts with a carrier wave with a set frequency (the frequency that you tune to in a car radio, for example). The amplitude of this carrier wave is modulated (varied) to match another waveform (typically a sound wave). The varying amplitude carries the information to be transmitted (see Figure 41-1).

The depth of modulation (the percentage by which the carrier wave is varied) specifies how much the amplitude of the carrier wave can be altered when it is modulated.

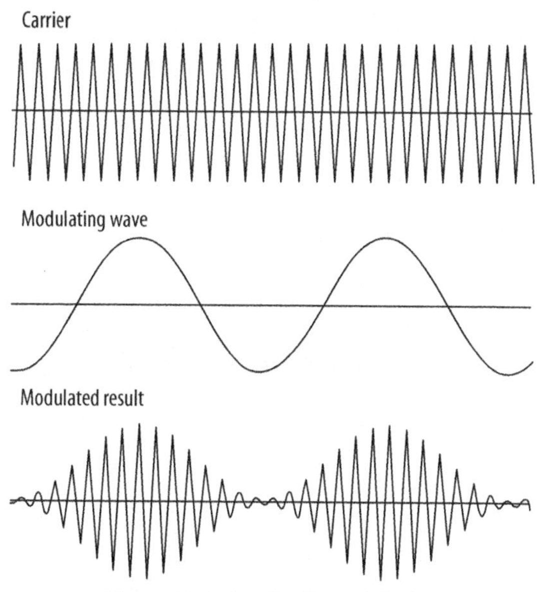

Figure 41-1. Amplitude modulation

To follow the localizer signals, an instrument such as the Horizontal Situation Indicator (HSI) measures the amount of modulation in the two signals and compares them (see Figure 41-2). When the signals are identical, the aircraft is flying along the center line; when one dominates, the aircraft is off course and the HSI shows the deviation. On the HSI there's a symbolic aircraft and a Course Deviation Indicator (the bar in the center of Figure 41-2), which shows whether or not the aircraft is lined up with the runway.

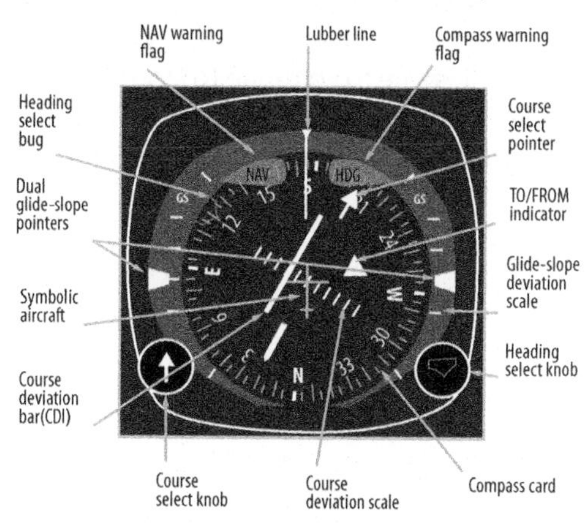

Figure 41-2. Horizontal Situation Indicator

The glideslope (also shown on the HSI, and illustrated with the localizer in Figure 41-3) works in the same way. A pair of signals on the same carrier frequency (around 330 Mhz) modulated in the same manner are used to guide the plane down at the right angle. On the HSI, the glideslope is indicated by a pair of pointers that align with the symbolic aircraft's wings when the aircraft is descending correctly.

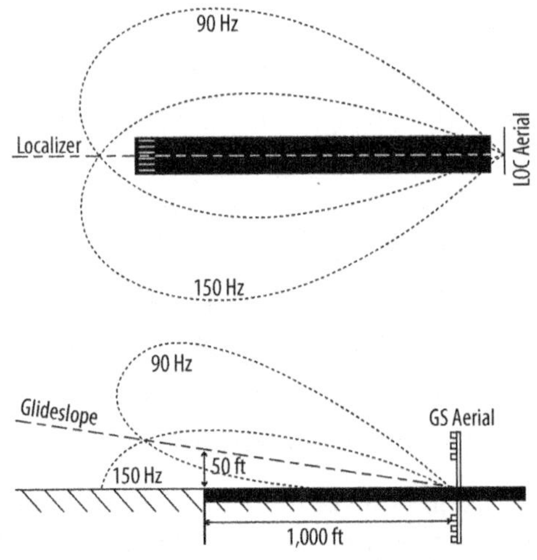

Figure 41-3. Localizer and glideslope

To help the pilot determine the distance to the runway, the markers transmit distinctive Morse Code signals almost directly upward. The outer marker (the furthest from the runway) indicates the start of the final approach to landing (at this point, everyone is sitting down with their seat belts fastened and their tray tables stowed). The middle marker is used by pilots to make a decision about whether it is safe to continue to land (for example, whether the pilot has enough visibility for the final moments of landing). The inner marker indicates that the aircraft is about 30 meters from the end of the runway.

Practical Information

Information about British Airways Flight Training is available at *http://ebaft.com*. Because these simulators are also used for real pilot training, you'll need to book your session months in advance.

042

Bunhill Fields Cemetery, London, England

51° 31′ 25.59″ N, 0° 5′ 21.06″ W

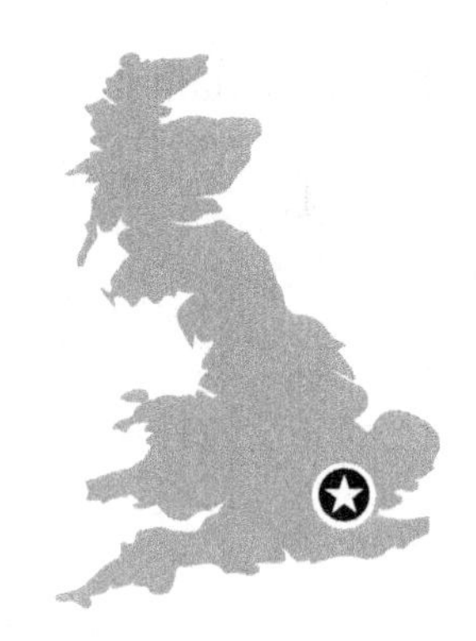

The Non-conformists

Bunhill Fields Cemetery in London has been the burial place of non-conformists (Christians who were not members of the Anglican church) from the 17th century to the 19th century. Many well-known writers of the time are buried there, including John Bunyan (who wrote *Pilgrim's Progress*) and Daniel Defoe (author of *Robinson Crusoe*). But scientists visiting Bunhill Fields Cemetery should head straight for the grave of the mathematician and Presbyterian minister Thomas Bayes.

Thomas Bayes lived in the 18th century, and his greatest mathematical work, *An Essay Towards Solving a Problem in the Doctrine of Chances*, appeared after his death thanks to the efforts of Bayes's friend, the philosopher Richard Price. The essay is extremely difficult to read—it's written in the style and language of the time, and covers subtle points of statistics. It was published in the scientific journal *Philosophical Transactions of the Royal Society of London* in 1763, and remained mostly unnoticed until Laplace rediscovered the essay's most important theorem in 1774.

Bayes's Theorem gives mathematicians a way of updating a probability when new information comes along. For example, say that 70% of the pupils in a school are boys, and 30% are girls. The girls have a choice of uniform (trousers or skirts), and the boys just wear trousers. If a mathematician encounters a pupil at random, then he knows that the chance of that pupil being a girl is 30%.

Now suppose the mathematician, who is deep in thought and staring at the ground, only notices what's covering the pupil's legs. If he sees a pair of trousers, he can calculate the probability that the pupil is a girl using Bayes's Theorem. He's updated his original estimate (which was 30%) based on new information, and comes up with the answer of 18%.

The cemetery is in the City of London, making it easily accessible and a welcome break from the bustling financial district. It's close to the Old Street and Moorgate London Underground stations. Around 120,000 people are buried at Bunhill, yet it is a popular lunchtime spot for office workers and the occasional statistical tourist.

Practical Information

Information about the cemetery and its opening times can be found at the City of London's website at *http://www.cityoflondon.gov.uk/*. The website has a downloadable map you can use to find Bayes's tomb, or you could just go for a random walk through the four hectares of greenery and graves.

Bayes's Theorem

Bayes's Theorem turns up in many different areas of life, from spam filters to interpreting blood test results. The theorem shows how to calculate the conditional probability of two events—for the two events A and B, the conditional probability of A given B is the probability that A will happen knowing that B has occurred. This conditional probability is written P(A|B).

Bayes's Theorem enables the calculation of P(A|B) from the probabilities of the individual events occurring (the probability of A occurring is written P(A), and the probability of B is P(B)). The reverse probability—the probability that B will happen knowing that A has occurred—can also be calculated, and is written P(B|A).

Bayes's Theorem is shown in Equation 42-1.

$$P(A|B) = \frac{P(B|A)\,P(A)}{P(B)}$$

Equation 42-1. Bayes's Theorem

The best way to understand Bayes's Theorem is through an example. Let's say that a new and deadly disease has been discovered that affects one out of one million people. If you have the disease, the blood test for it is 100% accurate (if you have the disease, the blood test will be positive); if you do not have the disease, the blood test will clear you 99.99% of the time. That is, if you do not have the disease, there's a 0.01% chance, or 1 in 10,000, that the blood test will misdiagnose you.

So, should you take the test? What will it say? How useful are the results of the test?

First, this situation can be split into two events. Call A the event that you are healthy, and B the event of a positive test result. What you want to know is, "what is the probability that I am healthy, given that I have a positive test result?" This is written P(A|B).

Bayes's Theorem shows how to work that out, by finding P(A), P(B), and P(B|A). P(A) is the probability that you are healthy. This is known because the disease strikes one in a million people, so P(A) is 99.9999%. P(B|A) is the probability that you are healthy if the test gives you a positive test result: that's 0.01%.

All that remains is to calculate P(B), the probability of getting a positive result. That's the probability that you have the disease and get a positive test result (100% × 0.0001%), *plus* the probability that you don't have the disease and still get a positive test result (99.9999% × 0.01%). So P(B) is 0.00100999% (that is, positive results don't happen very often).

Combining these with Bayes's Theorem gives a P(A|B) of 99.0098%. So if you get this "100% accurate" test and it says you are positive, then you are over 99% likely *not* to have the disease.

This apparently counterintuitive result comes about because people tend to discount the likelihood of being healthy, and focus on the test's accuracy when they are ill. But the small rate of false diagnoses becomes a big number when dealing with a disease that affects a very small number of people.

While you are visiting the cemetery, you can pass the time by trying to work out the right thing to do in the following situation using Bayes's Theorem.

You are standing in front of three unmarked tombs, one of which contains the body of Thomas Bayes. The cemetery groundskeeper, who knows who is in which tomb, asks you to pick a tomb at random. You tell him your choice, and he then opens one of the other two tombs to show that it does not contain Thomas Bayes. (If you had picked Bayes's tomb in the first place, the groundskeeper would make a random choice between the two possible tombs he could have opened.)

The groundskeeper then says, "Do you want to open the tomb you originally picked, or switch to the remaining tomb?" What's the best strategy for coming face-to-face with Thomas Bayes?*

*The best strategy is to switch to the other tomb.

043

Down House, Downe, England

51° 19′ 51.6″ N, 0° 3′ 9.5″ E

Charles Darwin's Home and Garden Laboratory

Down House, Charles Darwin's home from 1842 until his death in 1882, sits on seven hectares of greenery and greenhouses. After his years of travel on HMS *Beagle*, Darwin married and settled with his family at Downe, where he continued his studies surrounded by a garden filled with plants for observation and experimentation. It was here that he wrote his masterpiece *On the Origin of Species by Means of Natural Selection*.

Today, Down House is owned by English Heritage and has been restored to its original state. The gardens contain Darwin's greenhouses with their collections of orchids and carnivorous plants. Outside the drawing room is a well-stocked flower garden with the original sundial. The lawn is filled with rare fungi, and the walls support different varieties of lichen. There's also a wild "weed garden," which was used by Darwin to confirm the idea of natural selection.

Darwin built a circular sandy path in 1846, and used it daily for walking and thinking (Figure 43-1). He called it his "thinking path," and walked around it counting circuits by using a pile of stones. Strolling in the shade of the path's trees today, it's not hard to imagine Darwin taking a daily constitutional and pondering natural selection surrounded by his own nature collection.

The exterior of the house is covered in trellises, with ivies and creepers that match those present when Darwin lived there. The interior of the house is equally interesting. Two rooms have been taken over by a shop and a tea room, but the rest of the house is preserved and furnished as it was in Darwin's day.

Darwin's old study contains many of his books, instruments, and notes. In the corner by the fireplace is the comfortable armchair where Darwin sat to write. For a writing surface, he used a cloth-covered board resting on the chair's arms. The chair has wheels attached, so Darwin could scuttle around the room without getting up.

Figure 43-1. Darwin's thinking path; courtesy of Michael Spry

Also in the room are Darwin's microscopes. There's a model of the *Beagle* and a small exhibition explaining Darwin's theory of evolution. Once you've exhausted all that the house and garden have to offer, you can take tea in the tea room, or head into Downe and find the George and Dragon pub, where Darwin attended meetings of the Downe Friendly Society (he was the society's treasurer).

Practical Information

Visiting information is available from the English Heritage website at *http://www.english-heritage.org.uk/server/show/nav.14922/*.

Natural Selection

Darwin knew nothing of genetics—he was 100 years too early for that (see Chapter 71)—but he observed that the characteristics of plants and animals changed from one generation to the next, and that selective breeding by humans had led to a great variety of plants and animals with varied characteristics.

Today the characteristics of an organism are known as the phenotype—anything observable about an organism including its morphology, biochemical properties, and behavior. Darwin reasoned that some changes to the phenotype would be favorable: they would increase the likelihood of the organism surviving.

In *On the Origin of Species by Means of Natural Selection*, Darwin writes:

> Owing to this struggle for life, any variation, however slight and from whatever cause proceeding, if it be in any degree profitable to an individual of any species, in its infinitely complex relations to other organic beings and to external nature, will tend to the preservation of that individual, and will generally be inherited by its offspring.... I have called this principle, by which each slight variation, if useful, is preserved, by the term of Natural Selection, in order to mark its relation to man's power of selection.

His theory had three essential parts: that offspring have variations that may or may not be favorable; that traits may be inherited from parents by offspring; and that competition for limited resources (such as food) creates competition between individuals.

And he argued that this process of natural selection would lead to distinct species. In 1837, Darwin sketched in a notebook his first thoughts on this tree of life (Figure 43-2). The sketch depicts successive generations inheriting useful traits from their parents, and slight variations under competitive pressure creating new species.

Darwin starts this seminal drawing with the simple words "I think." Probably the biggest understatement of all time.

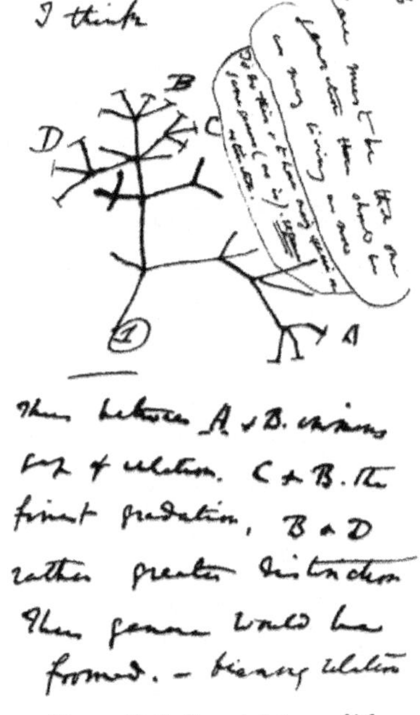

Figure 43-2. Darwin's tree of life

044

Edward Jenner Museum, Berkeley, England

51° 41′ 24.96″ N, 2° 27′ 26.68″ W

From Variolation to Vaccination

Prior to Edward Jenner's invention of vaccination, smallpox was killing about 10% of the population of Europe every year. Those that survived were often left blind or seriously scarred. The only practical way of preventing smallpox infection was through variolation—deliberately infecting a person with a (hopefully mild) strain of smallpox.

Variolation had been in use since at least 1000 AD in the Far East, and later across the Middle East. It was introduced to Britain in 1721 by Lady Mary Wortley Montague, the wife of the British Ambassador to Turkey, where she had seen variolation practiced.

To variolate a patient, dried smallpox pustules from a mild case of smallpox were either blown up the patient's nose or placed in a small incision in his hand. Smallpox is caused by a virus that has two forms: Variola major and Variola minor. If a patient was variolated with either one, he would become immune to the other, and by choosing a mild case of smallpox the physician hoped to be choosing Variola minor, which had a much lower death rate than Variola major. Unfortunately, at the time there was no way of distinguishing the two.

Variolation killed about 2% of patients, but protected about 80% from smallpox once they recovered. Given the high fatality rate of smallpox (about 30%), many people took the gamble of variolation.

Edward Jenner was a doctor and scientist who lived in the small town of Berkeley. He observed that milkmaids would frequently catch cowpox from the cows they were milking, and that they did not catch smallpox. Cowpox causes skin blisters in humans, and stimulates an immune system reaction that leads to protection against the similar smallpox virus.

Jenner conjectured that there was a connection between cowpox and smallpox immunity, and in 1796, he deliberately infected the eight-year-old son of his gardener with cowpox taken from a rash on a milkmaid's hand. He scratched the boy's arm and rubbed in the infectious material; the boy quickly developed cowpox and recovered.

Two months later Jenner variolated the boy to try to infect him with smallpox. The disease did not develop; Jenner tried again a number of times, and finally confirmed that infection with cowpox protected against smallpox. He then attempted to publish his findings, but the Royal Society rejected his paper, saying that he should "be concerned for his reputation and esteem among his colleagues."

Jenner quickly published his results at his own expense in the book *An Inquiry Into the Causes and Effects of the Variolae Vaccinae, a Disease Discovered in Some of the Western Counties of England, Particularly Gloucestershire, and Known by the Name of the Cow Pox*. Within two years, vaccination was common in Europe and by 1800, was being practiced in North America.

Edward Jenner's former home in Berkeley has been turned into a museum of his life and of modern immunology. There's a reproduction of Jenner's study based on an inventory taken on his death in 1823. But the real interest lies in the modern exhibit, created by the British Society for Immunology, that covers the science of immunology and vaccination.

In the garden surrounding the house is his Temple of Vaccinia, a small structure where he vaccinated the local poor free of charge.

Jenner is known for his vaccinations, but he only became a fellow of the Royal Society after publishing his observations on the life of the cuckoo. The museum helps visitors understand the importance of vaccination, but also the life of a man who was at once a local doctor and an avid scientist, and who met his future wife when a balloon experiment ended in her father's garden.

Practical Information

You can find visiting, historical, and immunology information on the museum's website at *http://www.jennermuseum.com/*.

Viruses

Viruses, unlike bacteria (see page 372), are not able to survive without infecting someone (or something) living; viruses are parasites that need to infect a cell to be able to reproduce.

They consist of a piece of genetic material (DNA or RNA) that is surrounded and protected by a layer of protein called the capsid. The shape of the capsid determines the overall shape of the virus (Figure 44-1). Common viruses are helical (the capsid forms a helix around the genetic material) and icosahedral (typically having 20 faces, each an equilateral triangle), and appear spherical under a microscope. Smallpox is an exception—it has a brick-like shape that is shared by all the pox viruses.

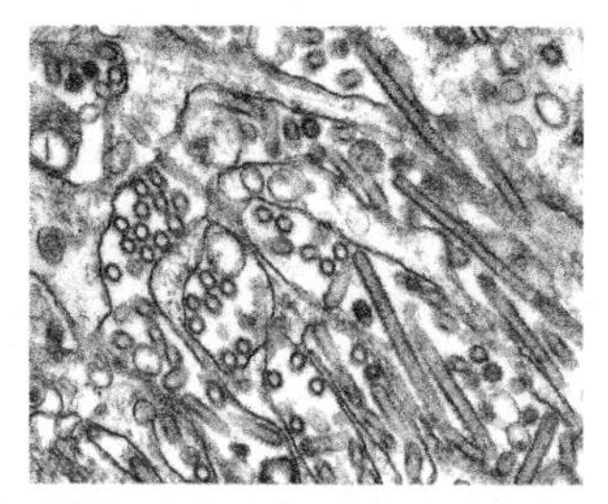

Figure 44-1. The H5N1 Avian flu virus is rod-shaped; courtesy of CDC/Cynthia Goldsmith, Jacqueline Katz, and Sharif R. Zaki

Some viruses coat themselves in material obtained from the host they are attacking to form an additional barrier. For example, both HIV and influenza have a viral envelope surrounding the capsid.

As well as protecting the genetic material, the capsid is the starting point for the life cycle of a virus. It provides the proteins necessary to attach to a host cell. The composition of the capsid determines the type of cells the virus can infect (and more broadly, the species).

Once attached to a cell, the virus penetrates the cell membrane either by merging with it or by endocytosis (essentially tricking the cell into thinking the virus is a useful nutrient that would normally by allowed through the wall). Some viruses are able to inject their genetic material into the cell interior directly. If the capsid enters the cell, it then has to be removed (called uncoating) so that the raw genetic material can be released.

Once in the cell, the genetic material uses the cell's own replication mechanism to make new copies of the virus. Once all the cell's raw material has been used up, the virus is done with the host. The virus can leave the cell by killing it off, by passing through the cell wall and taking some of the wall with it (the route taken by smallpox), or by the endocytosis process in reverse.

045

Elsecar Heritage Centre, Elsecar, England

53° 29′ 41.72″ N, 1° 25′ 5.20″ W

129 Years of Steam

From 1795 to 1923, one of the world's first steam engines ran at the Elsecar New Colliery in Yorkshire, pumping water from the mine. Steam pumping allowed deeper mining of coal than had previously been possible. In 1923 the steam engine was replaced by electric pumps, although it briefly ran again in 1928 when the electric pumps became overwhelmed by flooding in the mine. Happily, the steam engine was not destroyed, and today is the only surviving example of a Newcomen Beam Engine still in its original location.

The steam-powered pump at Elsecar was invented by Thomas Newcomen in 1712, and until James Watt invented his steam engine in 1763, the Newcomen Beam Engine was the way to get mechanical power from steam. Newcomen's engine was also the first steam engine to enter widespread use: over 100 were installed throughout the UK and across Europe.

Newcomen's engine was inefficient—only about 1% of the energy in the steam was actually converted to mechanical work—and required a large amount of coal to operate. Since the engine was primarily used for extracting water from coal mines, its inefficiency was relatively unimportant—a large supply of coal was always on hand.

The Newcomen Beam Engine at Elsecar is open to the public by appointment, but it is not in working order. The main cylinder is cracked, and restoring the engine to working order would require replacing the cylinder. To see a working Newcomen engine, you'll have to travel either to the Science Museum in London (see Chapter 77) or the Henry Ford Museum in Dearborn, Michigan (see Chapter 102).

Still, the Newcomen engine at Elsecar is well worth the visit: here, the engine's importance to the Industrial Revolution is clear. The engine made coal mining safer and collieries more productive; the coal that was mined could power steam engines for all sorts of mechanical work, and coke made from coal was used to smelt iron ore (see Chapter 74).

The Heritage Centre itself is located in historic buildings in the preserved village of Elsecar. Elsecar was an industrial village centered on coal mining and a foundry. The buildings of the Heritage Centre were part of the ironworks and colliery workshop; they now contain shops selling antiques and crafts, and a tea room.

The Newcomen Beam Engine is outside the Heritage Centre, and just around the back of the buildings is a preserved steam railway. The Elsecar Railway operates steam and diesel trains along a small track on most Sundays and some Saturdays.

Practical Information

The Elsecar Heritage Centre's website provides visiting information: go to *http://www.elsecar-heritage-centre.co.uk/*. To visit the Newcomen Beam Engine, call ahead to make an appointment.

Newcomen Beam Engine

The Newcomen Beam Engine (Figure 45-1) is relatively simple—it relies on creating a vacuum using condensed steam, and then letting atmospheric pressure move the piston that controls the beam connected to the pump.

The engine has a steam boiler at the bottom that creates low-pressure steam for use in the piston. Directly above the boiler sits the piston.

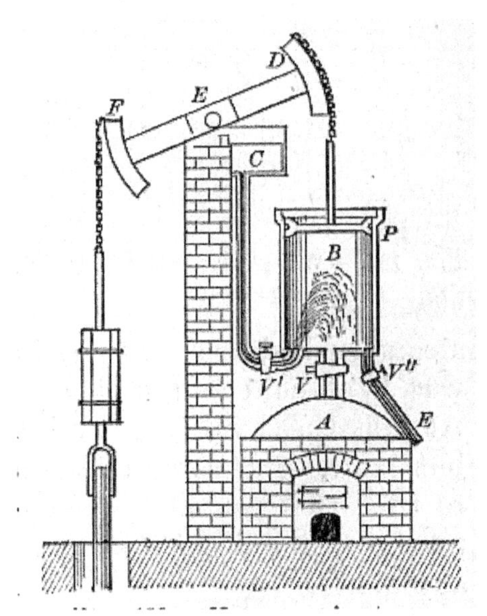

Figure 45-1. Diagram of a Newcomen Beam Engine

In Figure 45-1, the beam is weighted so that it naturally falls to the left, pulling the piston up. As the piston rises, steam enters the cylinder. When the piston has reached the top, the valve V' is opened, letting in a spray of water. The water spray cools the steam in the cylinder and it condenses, creating a partial vacuum.

Atmospheric pressure on the top of the piston then forces the piston down, raising the beam's lefthand side and performing work such as pumping water. When the piston reaches the bottom, the water inside is let out through valve V'' and the cycle starts again, with the beam dropping under gravity and pulling up the piston, letting in new steam.

Newcomen engines didn't move quickly—they produced 10 to 20 strokes per minute—but they were capable of producing a great deal of energy (for the time). Newcomen engines could produce between 5 and 20 horsepower, depending on the size of the cylinder and the length of the piston's movement.

046

Farnborough Air Sciences Museum, Farnborough, England

51° 16′ 56.28″ N, 0° 45′ 10.8″ W

Making Aircraft Move

The Farnborough Air Sciences Museum has a collection of early aviation equipment focusing on the science of flying. It's a small and concentrated dose of aircraft and aviation equipment, based on an almost discarded collection of historical artifacts from the Royal Aircraft Establishment at Farnborough.

The most important exhibit is an original jet engine built from Sir Frank Whittle's design. Whittle patented the jet engine in 1930, and in 1941 the first British jet-powered aircraft, the Gloster E.28/39, was flying. The W2/700 jet engine (see Figure 46-1) on display is the ancestor of all modern jet engines and was the first production jet engine in Britain.

Figure 46-1. Whittle W2/700 jet engine

The Jet Engine and Newton's Laws of Motion

A jet engine, also often called a turbojet, is a form of reaction engine, which is an engine that makes a vehicle move by exploiting Newton's Third Law: every action has an equal and opposite reaction (for more on Newton, see Chapter 69). At the back of a jet engine is a nozzle through which hot, pressurized air is expelled at very high speed. This creates an equal and opposite force acting on the engine and pushing the vehicle forward.

In a turbojet, the high-speed air is created by sucking in air from the atmosphere and using it to burn fuel. The core of a turbojet has three essential parts: a compressor, a combustion chamber, and a turbine (see Figure 46-2). These parts are all found in the gas turbine engines commonly used as power for tanks, hovercraft (see Chapter 50), and helicopters, but in a turbojet, it's the hot gases leaving the fourth part, the jet nozzle, and not the rotation of the turbine itself, that provides power.

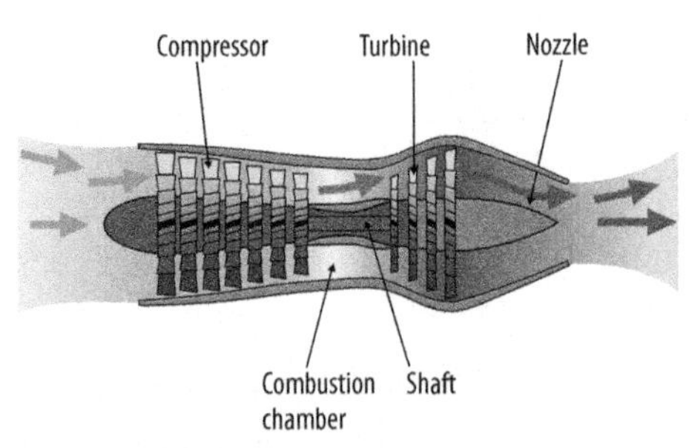

Figure 46-2. Turbojet engine parts

The compressor and turbine are linked by a shaft: when the turbine rotates, it turns the compressor. As air enters the jet engine, it first encounters the compressor, which increases the air pressure up to 40 times. The compressed air then enters the combustion chamber where jet fuel is injected and ignited.

The burning fuel increases the temperature of the compressed air, and a high-temperature, high-pressure flow of air hits the turbine, causing it to turn (which turns the compressor). The air and other gases then leave the engine through the nozzle, and as the air expands (and its pressure drops to normal atmospheric pressure), it accelerates. This high-speed air produces the thrust by Newton's Third Law.

Newton's Second Law (that the force, F, required to accelerate a mass, m, is proportional to the mass and the acceleration, a: $F = ma$) can be used to calculate the thrust generated by a jet engine. The mass being accelerated in a jet engine is the air entering it.

If a mass m of air enters the engine with velocity V and exits with (greater) velocity v and takes time t to pass through, the resulting thrust F is proportional to the mass of air sucked in and the difference in the air's speed between the intake and nozzle. This equation is shown in Equation 46-1.

$$F = \frac{mv}{t} - \frac{mV}{t}$$

Equation 46-1. Calculating the thrust of a jet engine

The large jet engines used in commercial aircraft actually employ two different thrust-generating techniques: they consist of a standard turbojet engine as just described, but also have a very large fan in front of the compressor. These engines are known as turbofans.

The turbine drives the compressor as before, and compressed air and fuel is burnt and ejected to create thrust as in a turbojet. But the turbine also drives a large fan, which accelerates the air entering the engine; the air passes around the engine and exits without passing through the turbojet (see Figure 46-3). Nevertheless, Equation 46-1 applies, as this bypass air (so called because it bypasses the turbojet) is moving faster than when it entered and creates thrust.

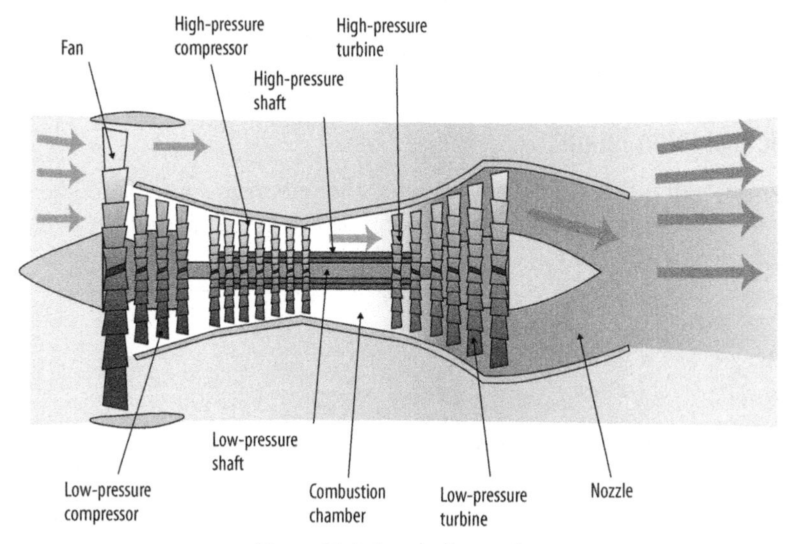

Figure 46-3. A turbofan engine

The fan accelerates a large amount of air by a small amount, producing additional thrust. This acceleration is provided almost for free—it requires very little extra fuel to power the fan—and increases engine efficiency.

When you look into a jet engine on a commercial flight, you'll see the huge blades of the fan; the turbojet is hidden from view.

There's an early afterburner (called at the time an "augmenter") that was designed to fit behind the Whittle engine and generate extra thrust by injecting fuel into the hot gases escaping from the jet engine.

There's also a display of a helicopter rotor, which explains how the blades rotate and pitch to provide lift and motion.

The museum has a significant collection of wind tunnel models, including models that illustrate the performance of the Concorde at low speeds. There's a collection of shock cones used for wind tunnel research of air intake shapes for supersonic aircraft. There's also a working wind tunnel demonstration.

And of course, the museum has a collection of aircraft. There's the cockpit of a 1971 Hawker Siddeley Trident passenger jet—the Trident was a pioneering aircraft because of its speed (Mach 0.88) and its ability to land "blind" using automatic landing equipment. The cockpit has been restored, with all the instruments working, and is set up for visitors.

There's also a Sea Harrier "jump jet" capable of vertical takeoff and landing, a British Lightning T5 jet fighter from the 1960s that's capable of Mach 2, and a Puma Helicopter.

The museum is entirely run by expert volunteers, and many of them are on hand to explain the exhibits.

Practical Information

The museum is open on the weekends, and entrance is free of charge. You can find full details from the Farnborough Air Sciences Trust website at *http://www.airsciences.org.uk/*.

047

Gonville and Caius College, Cambridge, England

52° 12′ 21.4″ N, 0° 7′ 3.34″ E

Stained-Glass Scientists

The first thing you need to know about Gonville and Caius College is how to pronounce it. Gonville is easy, no surprises there. But Caius is another matter. John Keys was the second founder of the college; in 1529 he entered what was then Gonville Hall at just 18 years old. He later became a Fellow of the college, and finally Master (after having spent a chunk of his own fortune restoring it).

Along the way he Latinized his last name, turning Keys into Caius without changing the pronunciation. Thus, Gonville and Caius is pronounced "Gonville and Keys"; within Cambridge the college is usually referred to as simply Caius.

Caius has an illustrious list of alumni. They include William Harvey (who discovered blood circulation), George Green (mathematician), John Venn (who popularized the Venn diagram), Charles Sherrington (Nobel Prize–winning neurophysiologist), R. A. Fisher (probably the greatest statistician ever), Sir James Chadwick (who discovered the neutron), Francis Crick (DNA; see Chapter 71), and Stephen Hawking.

Six of these alumni are honored with stained-glass windows representing their greatest contribution in the college Hall where students and Fellows eat.

John Venn's famous Venn diagram (Figure 47-1) is used to show intersecting sets of items and is represented by three colored circles. Each circle overlaps the other two, and all three overlap in the middle with appropriate changes in color.

R. A. Fisher was not only a great statistician, but also a celebrated geneticist and evolutionary biologist. His window shows a 7×7 Latin Square that uses colors instead of the typical numbers (Figure 47-2). A Latin Square is a square grid in which numbers are placed so that each number only appears once per row and column. (Many people will recognize this as a sort of Sudoku.)

Figure 47-1. Venn's window diagram; courtesy of Derek Ingram, Gonville and Caius College

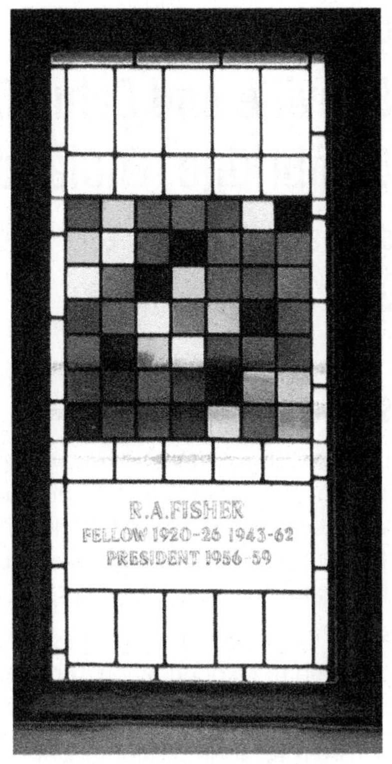

Figure 47-2. Fisher's Latin Square; courtesy of Derek Ingram, Gonville and Caius College

Fisher used Latin Squares in statistical analysis and in particular in agriculture. By dividing a field into a square grid, it's possible, for example, to try out seven different fertilizers using a 7×7 Latin Square. By associating one fertilizer with each of the seven different numbered squares, it is possible to use the square to eliminate differences in soil, drainage, sunlight, etc. because each fertilizer will be applied once per row and column.

Francis Crick's window shows the structure of DNA. When the window was installed, Crick insisted that it not be visible from the outside at night so that the DNA would only be seen coiling in the right direction.

George Green's window shows something that at first glance looks like it might be bacteria, but it's in fact a representation of Green's Theorem (which you'll be familiar with only if you studied calculus to a high level). Green's Theorem shows how an integration over an enclosed area (the bacteria-like shape in the window) can be turned into an integration over just the line that forms the shape's boundary. Although this may seem totally abstract, it is used directly in the planimeter (see page 67), which can calculate the area of a shape on a graph or map just by tracing around its edge.

Sir James Chadwick's window shows an alpha particle (a helium nucleus) hitting a beryllium atom and causing the emission of a neutron (and creation of a carbon atom).

Finally, there's Charles Sherrington's window, which shows "two excitatory afferents with their fields of supraliminal effect in a motor neuron pool of a muscle." It looks like some sort of plant, but it actually shows the motor neurons that control muscle movements. Sherrington won the Nobel Prize in 1932 (with Edgar Douglas Adrian) for discovering the function of neurons.

Gonville and Caius is open to the public (and free of charge), but unfortunately the public does not have access to the Hall where the windows are installed. To see them, you'll have to make friends with a Cambridge University student who's willing to accompany you, or make do with a virtual tour.

The college is right next to Trinity College, where Newton lived (see Chapter 69), and a short walk from the Eagle Pub, where Crick and Watson announced the "secret of life" (see Chapter 71).

Practical Information

The website of Gonville and Caius College is at *http://www.cai.cam.ac.uk/*, and you can take an interactive tour at *http://www.cai.cam.ac.uk/map1.php*.

Set Theory and Transfinite Numbers

To mathematicians, a set is a collection of objects with no duplication. For example, a mathematician might state that U is the set of colors on the American flag (red, white, and blue), I is the set of colors on the Italian flag (red, white, and green), and J is the set of colors on the Japanese flag (red and white). The objects inside a set (in this case, the colors) are called its elements.

A Venn diagram shows the relationship between two sets. Each set is represented by a circle containing the set's elements, and where the sets have common elements, the circles overlap (this is called the intersection of the sets). A Venn diagram of the sets U and I shows that the common colors are red and white (Figure 47-3).

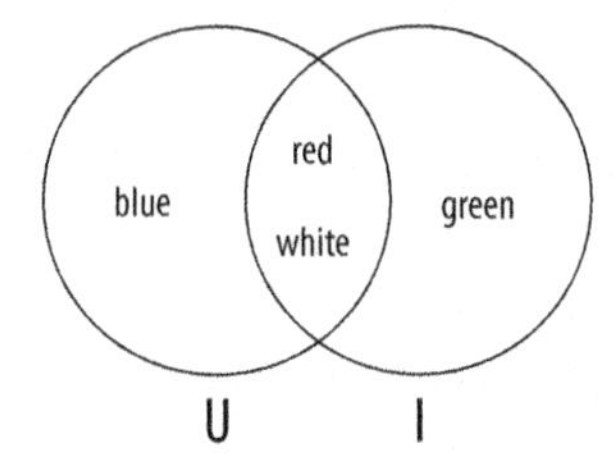

Figure 47-3. Venn diagram of U and I

All the colors of the American and Italian flags taken together form another set (with elements red, white, blue, and green). When two sets are merged together, the resulting set is called their union.

It's also possible for one set to be entirely contained within another. For example, J is entirely contained inside U, so we can say that J is a subset of U. In the Venn diagram, the circle for J is completely inside the circle for U (Figure 47-4).

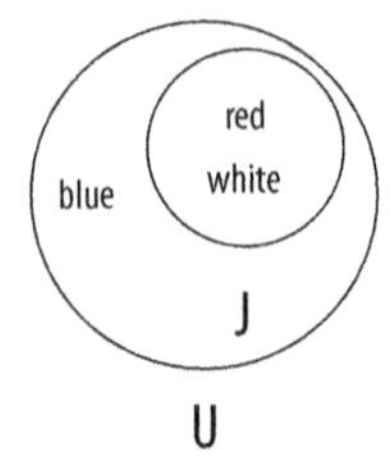

Figure 47-4. Venn diagram of U and J

The main creator of set theory, which formally introduced the ideas of sets, subsets, intersections, and unions, was the German mathematician Georg Cantor. Cantor used set theory to answer questions about infinity, and came up with the idea that there's more than one infinity and that some are bigger than others. This was quite a controversial idea at the time.

The sets U, I, and J are all finite; that is, they have a finite number of elements (U and I have three elements, and J has two). But it's possible to create a set that has an infinite number of elements. For example, the set N is the set of all natural numbers (all whole numbers—0, 1, 2, 3, ...). Obviously there is an infinite number of them.

The set Z is the set of all integers—it's the set N plus all the negative whole numbers as well (0, 1, –1, 2, –2, ...). One of Cantor's questions was, "Are there more integers than natural numbers?" or equivalently, "Are there more elements in Z than in N?" The answer is no—they both have the same number of elements.

To establish this, mathematicians use a one-to-one correspondence. If you watch a child count a bag of apples, she'll point to each apple in turn, counting out loud (1, 2, 3, etc.) until she reaches the last apple. The child has set up a one-to-one correspondence between the apples and the numbers: each apple is assigned a single number in the sequence 1, 2, 3, etc.

The same thing can be done for infinite sets. A set is countable if there's a one-to-one correspondence with the natural numbers. If there's a way to assign a natural number to each element of a set, then the set has the same size as the natural numbers (the same number of elements).

So, Cantor's question about Z comes down to, "Can a one-to-one correspondence be made between Z and N?" Doing so is fairly easy: first connect 0 in N to 0 in Z. Then assign each even number to the corresponding number divided by –2: so 2 in N is assigned to –1 in Z, 4 is assigned to –2, 6 to –3, and so on. That gives a one-one correspondence between the even numbers and the negative numbers. Finally, the odd numbers in N are each assigned a positive number in Z by adding 1 and halving the result so: 1 in N is assigned to 1 in Z, 3 is assigned to 2, 5 is assigned to 3, and so on.

So Z and N have the same number of elements (albeit an infinite number). Cantor's next question was, "Are all sets countable?" The answer to that is no. Cantor's brilliant argument is called "diagonalization," and shows that the set of real numbers (the set containing every number, including all the fractions and decimal numbers) is not countable.

Suppose there was a one-to-one correspondence between N and R (the set of real numbers). It would therefore be possible to assign each natural number to a real number. Cantor's diagonalization is best understood with a picture of this imaginary one-to-one correspondence. In Figure 47-5, the natural numbers are on the left and each corresponding real number is on the right.

```
0 0.12342878234792839 55...
1 0.39849235720349348 90...
2 0.23247882734823940 33...
3 0.23489273582389230 93...
4 0.23948293489203234 94...
5 0.90949328974823849 23...
6 0.58948939857823748 92...
                     .
                     .
                     .
  0.2039944...............
```

Figure 47-5. Cantor's diagonalization

There's an easy way to get a contradiction from this by making a new number different from all the others. In the figure, the first digit of the first number is 1, the second digit of the second number is 9, the third digit of the third number is 2, and so on. In this way, it's easy to build up a new number (such as the number at the bottom of Figure 47-5) that differs digit by digit from every number in the table, just by picking digits that differ from the digits on the diagonal.

But that means there's a real number that's not in the table, and that means that the original assumption (that N and R have a one-to-one correspondence) is false. And that means that R is not countable and has more elements than N. So Cantor had established two different infinite numbers—the number of elements in N and the number of elements in R. He called these new numbers "transfinite," because the idea of having more than one infinity was, at the time, controversial (especially to theists who believed in an infinite God).

048

Goonhilly Satellite Earth Station, Goonhilly, England

50° 2′ 53″ N, 5° 10′ 55″ W

The "Arthur" Parabolic Dish

In 1962, NASA launched the world's first telecommunications satellite into orbit around the Earth. On its first day in space, the satellite relayed television pictures from the U.S. to northern France, and a few weeks later the world's first parabolic dish was switched on at Goonhilly in southwest England. The French received signals using a horn antenna, but the British were the first to use the parabolic shape that has since become familiar for satellite communication and radio telescopes.

The dish, dubbed Arthur, measured 26 meters in diameter and weighed over 1,100 tonnes. In comparison, the satellite, Telstar, was tiny, weighing under 78 kilograms and with a diameter of 86 centimeters. But between Arthur and Telstar, history was made when U.S. president John F. Kennedy gave a live transatlantic press conference, and the first telephone calls and faxes were routed through space.

Telstar only worked for about a year; it was badly affected by the U.S. Starfish Prime high-altitude nuclear test explosion (see page 197) that occurred the day before its launch, and it eventually stopped transmitting. Nevertheless, it remains in orbit to this day, and provided the proof that international space-based communication was possible.

Arthur had a happier fate—it is still in use, and the Goonhilly site became a major telecommunications point with multiple satellite dishes and transatlantic cables. And Arthur is the ancestor of the small satellite dishes that are attached to homes around the world and share the same parabolic shape.

The Parabolic Antenna

In Marconi's original experiments (Chapter 62), his transmitting antenna sent signals in all directions, and his kite-borne receiving antenna picked up signals from any direction: neither antenna was directional. In contrast, the parabolic dish antenna installed at Goonhilly to communicate with the first Telstar satellite is highly directional.

The surface of Arthur is in the form of a parabola. Parabolas show up in many other walks of life—the curving shape of a suspension bridge (page 100), the path of a ballistic missile, and the flight path of the Zero-G aircraft (Chapter 111) are all parabolas. The most familiar parabola, at least from school, is the shape formed by the equation $y=x^2$.

But in general, a parabola is defined as all the points that are equidistant from a specific line and a specific point. The line is called the directrix, L, and the point the focus, F. In Figure 48-1, the lines FP_1 and P_1Q_1 are the same length, as are FP_2 and P_2Q_2, and FP_3 and P_3Q_3.

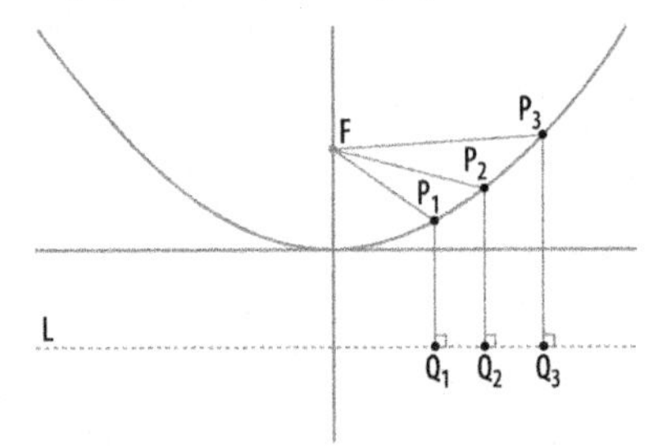

Figure 48-1. A simple parabola

It's the focus that matters in the design of a parabolic antenna, because the actual receiver or transmitter is placed at that point. Parallel radio waves hitting the surface of the parabolic dish will bounce off and hit the focus, thus concentrating the incoming signal (which may be very weak, having traveled from an orbiting satellite; see Figure 48-2).

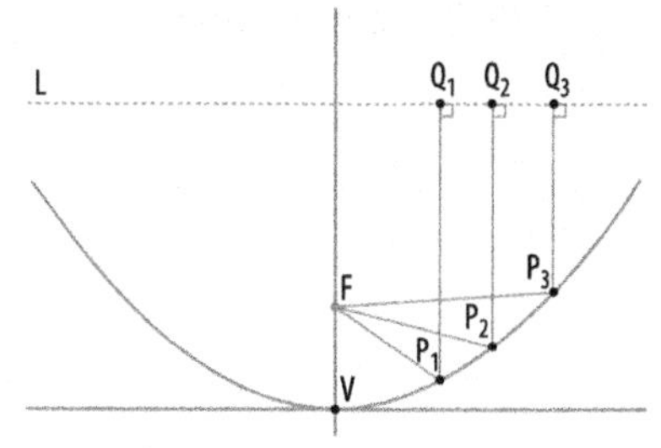

Figure 48-2. Parallel radio waves bounce toward the focus

Equally, if a transmitter is placed at the focus and sends radio waves toward the parabolic surface, those waves will bounce off the dish and be reflected as parallel waves.

Since the waves entering or leaving the dish are all parallel, communicating with a distant satellite involves pointing the dish directly at it. Because the Telstar satellite was not in a fixed position in the sky (it was orbiting the Earth approximately every 160 minutes), Arthur had to move to follow it. And Telstar's movement meant that transatlantic transmission was only possible for about 20 minutes per orbit (when the satellite was visible in both the UK and North America).

Even the much smaller satellite dishes used to receive TV transmissions at home are parabolic. But the receiver is not mounted in the center of the dish: it is typically mounted below the dish so that it does not obscure any radio waves hitting the dish. The receiver is still at the focal point of the parabola, but the parabola has been sliced asymmetrically. Since the satellite is in a geostationary orbit (meaning it does not move relative to the Earth), the dish does not have to move.

Today, Arthur is a protected national monument and the historic center of a tourist attraction called Future World @ Goonhilly. Watch out, though—the rest of the site is a tourist trap and has little scientific interest, so if you are planning to visit Goonhilly, make sure that the tours of Arthur are available. That tour is definitely worth it, however. You'll be awed by Arthur's great size, and you can climb up underneath the dish and see the equipment that enables it to rotate and lock onto a satellite signal.

There is admittedly one fun, geeky thing to do at Goonhilly once you've seen Arthur: you can try out the Segway "personal transporter" on an adventure trail. The Segway may not have been a commercial success, but it's well worth taking the opportunity to experience the sensation of a machine that balances itself (and you).

Once you are done with Goonhilly, head off to the nearby site of Marconi's first transatlantic radio transmission at Poldhu (see Chapter 62).

Practical Information

Details of the Arthur dish and Future World @ Goonhilly are available at *http://www.goonhilly.bt.com/*.

049

Greenwich, London, England

51° 28′ 44.76″ N, 0° 0′ 0″ E

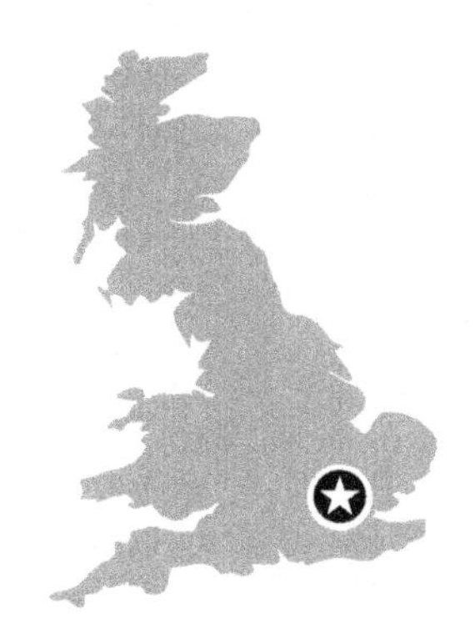

Royal Observatory and National Maritime Museum

It's no exaggeration to say that Greenwich is the center of the world: after all, the Prime Meridian of the Earth runs right through the Royal Observatory at Greenwich. A meridian is an imaginary north/south line that passes through both of the Earth's poles; the Prime Meridian defines 0° of longitude and the time basis for all the world's clocks. Greenwich Mean Time is now officially called Universal Time.

The Royal Observatory has been at Greenwich since 1675 and was of great importance in maintaining Britain's naval power: observations of planets, moons, and stars were needed for the nautical almanacs used for navigation. Close to the Royal Observatory is the National Maritime Museum, which contains the clock that revolutionized navigation in 1759—John Harrison's H4 marine chronometer.

Also on display at the National Maritime Museum are three clocks that Harrison built prior to the H4: logically enough, these are called the H1, H2, and H3. These three clocks are still running today; the H4 still works, but is only wound on rare occasions to avoid wearing out its components.

Highly accurate clocks were (and still are) vital to marine navigation, because longitude is found by comparing the time difference between noon on the Prime Meridian and noon on a ship at sea. Since the Earth rotates once every 24 hours through 360°, each hour of difference accounts for 15° of longitude. If a ship is sailing at the Equator, 1° of longitude (which corresponds to a time difference of 4 minutes) is 111 kilometers. Knowing the time at Greenwich accurately was—and remains—vital to knowing a ship's location.

But keeping time accurately was an enormous challenge, so much so that in 1714 the British government offered a £20,000 prize for a method of determining longitude to within 56 kilometers. Harrison ultimately won most of the money by building a clock capable of keeping accurate time in the hostile world of a ship.

In a sea trial of the H4 sailing from England to Jamaica, the clock was found to be off by only 5 seconds (which corresponds to a navigation error of less than 1 kilometer). The 1.5-kilogram, 12-centimeter H4 proved itself able to keep time despite the movement of the ship, the salty air, battering by waves and wind, and changes in temperature.

The museum traces the history of timekeeping with the H4 and over 400 chronometers and watches. The Prime Meridian, longitude 0°, is visible outside the observatory and marked by a stainless steel bar; visitors can straddle it with one foot in the western hemisphere and the other in the east. At night, a green laser traces the meridian through the air.

Just before 1 p.m. each day, a large ball atop the Observatory is raised into the air. At precisely 1 p.m. the ball drops, and you can set your watch by it while standing on the meridian. Originally, this was done so that ships anchored in the port of London could set their clocks by watching for the visual signal from Greenwich.

Practical Information

The observatory and museum are inside Greenwich Park, which affords a wonderful view over London and a good place for a picnic lunch. Entry to the Royal Observatory, the National Maritime Museum, and Greenwich Park is free. Information about visiting the museums and special events is available from *http://www.nmm.ac.uk/*.

Finding Your Longitude

Today you'd likely use a GPS to find your longitude, but it's also possible to do so with an accurate watch set to Greenwich time and some simple tools. If you are on dry land, you can use a sundial to find out precisely when noon occurs at your longitude, and compare the time difference between local noon and Greenwich noon (see Chapter 8).

But if you're on a ship at sea, a sundial obviously isn't very practical because the ship moves. A better method is to observe the Sun using a sextant. A sextant determines the angle between the horizon and some celestial body, such as the Sun. The horizon is viewed through the sextant's eyepiece, and the arm of the sextant is swung until the Sun and horizon line up in the eyepiece. This is achieved by a pair of mirrors that reflect the Sun's image (see Figure 49-1).

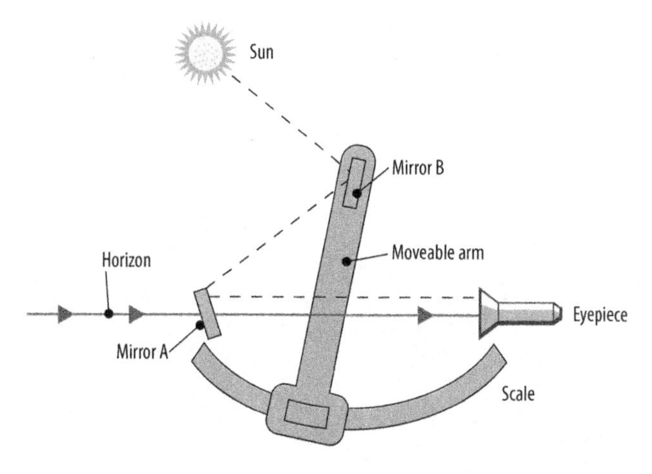

Figure 49-1. A sextant

With the Sun and horizon aligned, the angle of the Sun above the horizon is read directly from the position of the arm on the sextant's scale.

To find the longitude, start a few minutes before local noon and measure the angle of the Sun. Note the angle and the time. Then watch as the Sun reaches its zenith and begins to fall. Note the time when the Sun again reaches the angle originally recorded.

Local noon will have occurred halfway between the first and second time readings. The difference between local noon and Greenwich noon gives your longitude. If local noon occurs before Greenwich noon, you are to the east of Greenwich; if it occurs after Greenwich noon, then you are to the west.

050

Hovercraft Museum, Lee-on-the-Solent, England

50° 48′ 28.08″ N, 1° 12′ 32.01″ W

The World's Biggest Collection of Air-Cushion Vehicles

The hovercraft seems like the kind of invention that would have had the inventor exclaiming, "It's so crazy, it just might work" (at least in the movie adaptation). And anyone who has had the pleasure of flying in a hovercraft across the English Channel from Britain to France will recall the rapid, bumpy ride.

Unfortunately, cross-Channel hovercraft travel ended in 2000 due to competition from the Channel Tunnel. But the hovercraft live on at the Hovercraft Museum (Figure 50-1).

Figure 50-1. The Hovercraft Museum; courtesy of The Hovercraft Museum Trust

The museum is housed in a disused Royal Navy Air Station and has the best collection of hovercraft in the world. There's everything from a tiny, one-man Winfield to massive passenger and military machines.

The museum also has many of the parts of the large SR N4 Mk II hovercraft that used to cross the Channel, carrying up to 278 passengers and 36 cars. On display at the site are two complete SR N4 Mk IIs that are not currently owned by the museum; access to these craft is restricted, but just standing next to one is inspiring. The SR N4 Mk II was powered by four Rolls Royce gas turbine engines, and could make the trip from Dover to Calais in as little as 22 minutes.

The museum also has preserved a Royal Navy BH7, which was capable of traveling at 65 knots. It's the only large British military hovercraft still in existence.

The last surviving model used for development of the SR N1, the first practical hovercraft, is on display and in working order.

In addition, there's an extensive collection of model hovercraft that were used for research purposes, and a reproduction of hovercraft inventor Sir Christopher Cockerell's original experiment. It was this experiment, using a pair of coffee cans and the motor from a vacuum cleaner, that convinced him that this crazy idea just might work.

The museum also houses a collection of models by John Thornycroft. Thornycroft was a boat designer who worked on air lubrication of ships using a concave hull, which would contain a bubble of air under the ship. These air-lubricated designs are predecessors of the hovercraft, and led Thornycroft to work on hydrofoils.

Practical Information

The Hovercraft Museum is only open to the public by appointment (or on one of its rare open days). Visitors are welcome, but you'll need to call at least a week in advance and organize a tour. Tours last a couple of hours and take in 50 to 60 hovercraft. Full details are available from *http://www.hovercraft-museum.org/*.

If the museum awakens in you the desire to ride in a hovercraft, there's a commercial service crossing from nearby Southsea to the Isle of Wight. The trip takes 10 minutes.

How the Hovercraft Works

The hovercraft works by flying on top of high-pressure air blown underneath the craft by its engine. Sir Christopher Cockerell's experiment with a pair of inverted coffee cans and a vacuum cleaner showed that the most efficient way to operate a hovercraft was by blowing air around the edge of the craft.

In that experiment, Cockerell used cans of different sizes to create a small space at the edge between them. Air was blown into the top of the outer can and out around the gap. To measure the effectiveness of his hovercraft, Cockerell placed it on a kitchen scale and put weights on the top. By directing the air around the edge, he was able to lift three times the weight possible with a single can.

In a real hovercraft, the air cushion is contained inside curtains typically made of rubber. These curtains are able to compensate for the changing terrain, or sea, across which the hovercraft flies while keeping the high-pressure air inside. When the curtain is completely full of air, some air begins to escape from under the curtain and the hovercraft will float (see Figure 50-2).

Figure 50-2. Hovercraft air flow

The curtain also means that more air can be trapped under the hovercraft. This allows it to rise higher than in Cockerell's original experiment, and to cross over rougher terrain or sea without losing its air cushion.

Since hovercraft are essentially frictionless, making them move is relatively easy. Typically, hovercraft design includes large fans directed away from the direction of travel to blow the vessel along.

051

Jodrell Bank Observatory, Cheshire, England

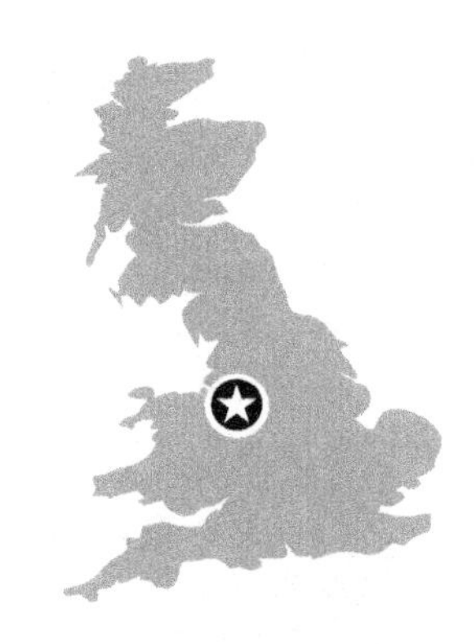

53° 14′ 10.5″ N, 2° 18′ 25.7″ W

The Big Ear

The Jodrell Bank Observatory's enormous Lovell Telescope may have been relegated to third place in the category of biggest movable radio telescope, but it has an illustrious history, is open to the public, and is set in a bucolic location on the Cheshire plain.

The observatory started life as a botanical garden run by the University of Manchester. In 1945, physicist Sir Bernard Lovell brought surplus Second World War radar equipment to the site, in order to escape interference from the electric trams running through Manchester, 30 kilometers to the north.

Lovell had worked on radar during the war and planned to reuse radar equipment to study cosmic rays entering the Earth's atmosphere from space. By the end of 1945, Lovell had shown that odd echoes appearing on military radar were, in fact, caused by meteor trails in the upper atmosphere. He quickly realized that much larger and more sensitive telescopes were needed, and a series of dishes was built in 1947, 1957, and 1964.

The 1957 telescope, then called the Mark I and now called the Lovell Telescope, was switched on just in time for the Soviet Sputnik 1 mission on October 4. Sputnik 1 and its booster rocket went into orbit, and while the Lovell Telescope couldn't track the satellite itself, it was able to track the booster rocket using radar, since Lovell had included a powerful radar system as part of the telescope's design.

The Lovell Telescope later communicated with the U.S. Pioneer 5 mission, and tracked many U.S. and Soviet space missions including Luna 9 (see sidebar). But the telescope's main use was in radio astronomy, and included pioneering work on the discovery of quasars (a QUASi-stellAR radio source) using interferometry (see Chapter 107). Ever since, the telescope has taken part in interferometry observations as part of the MERLIN and VLBI experiments, and remains in use 50 years (and a few upgrades) after its inauguration.

Luna 9 and the Fax Machine

The Lovell Telescope's greatest public moment came in 1966 when it intercepted images transmitted by the first probe to land on the Moon (without crashing). The British press scooped the Soviet Union by publishing the images before any had been officially released (Figure 51-1).

Figure 51-1. Luna 9 image intercepted by the Lovell Telescope; courtesy of Jodrell Bank Centre for Astrophysics, University of Manchester

On February 3, 1966, the Soviet Luna 9 probe landed, using an inflated air bag to soften its touchdown. Almost immediately, it started to transmit television images of the lunar landscape. The Soviet Union had announced when the probe would transmit images, and the times coincided with periods when the Moon was visible to the Lovell Telescope (which, at the time, was the largest dish in the world).

This led Sir Bernard Lovell to remark, "I am sure the Russians intended us to do the recordings." And the transmissions were in the format used for newspaper wire photos—a form of facsimile.

Back in the 1930s, newspapers had started to carry photographs sent by wire. Photographs were taken using ordinary cameras and the film developed in the normal way. The photograph was then wrapped around a cylinder. The cylinder rotated and slid sideways, so that the photograph could be scanned line by line by shining a light on it and measuring the reflected light using a photocell. The current from the photocell was amplified, turned into sound, and sent down a normal phone line.

At the receiving end, the sounds were turned back into a varying beam of light, which scanned across a rotating cylinder of the same diameter to produce a new negative. This negative was then developed, and the resulting picture could then be added to the newspaper.

Essentially the same technique was used by the Luna 9 lander, using a television camera to produce the scanned lines to be transmitted by radio. When Jodrell Bank heard the transmissions, a member of the staff recognized them as sounding just like the facsimile signals used for wire photos. The *Daily Express* newspaper lent Jodrell Bank a suitable facsimile machine, and images from the Moon appeared in the newspaper before the Soviet Union's TASS news agency had released its own.

Within hours the story was out, with accompanying pictures of the Moon.

The observatory and telescope have a small visitor center with an exhibition explaining the history of the site. Adjacent to the telescope is a 14-hectare arboretum. There are a number of trails throughout the arboretum, including the Solar System Trail, which has the Sun and planets laid out to scale among the trees. And a special path around the Lovell Telescope, known affectionately as the Big Ear, brings you face-to-face with its 76-meter dish.

Since the telescope is close to Manchester, it's possible to pack in three visits in one day by adding trips to the Sackville Street Gardens (Chapter 66) for the Alan Turing statue, and taking the Manchester Science Walk (Chapter 55) ending at the Museum of Science and Industry.

Practical Information

Visiting information can be found at *http://www.jb.man.ac.uk/*; details on the telescope can be found at *http://www.jb.man.ac.uk/aboutus/lovell/*.

052

Kelvedon Hatch Nuclear Bunker, Kelvedon Hatch, England

51° 40′ 18.5″ N, 0° 15′ 23.6″ E

Three Months of Underground Autonomy

The Kelvedon Hatch Nuclear Bunker looks like a fantasy hiding place straight out of a movie. Driving by, all you see is a bungalow tucked into a wooded area of the Essex countryside. But the bungalow is, in fact, the entrance to a secret underground facility that was designed to house up to 600 people.

The bunker was created in 1952 and used until 1994 as a safe place for central government to reside in the event of a nuclear war. Once inside the bungalow, the bunker is reached via a 120-meter-long tunnel leading to the lowest level of the facility. At the end of the tunnel are 1.5-tonne blast doors that protect the bunker from a nuclear explosion.

Inside, there is equipment for electricity generation, air filtration and pressurization (the bunker was kept at positive air pressure to prevent radioactive fallout from entering), and a water supply from a deep bore hole. Much of the bunker's equipment was built for communication with the outside world and to all the other military bunkers around the country. There's also a BBC broadcasting room for national broadcasts.

Above the equipment floor is a level dedicated to government operations, with amusing 1980s-era technology still sitting on the desks waiting to be used. Apparently, nuclear war was going to be managed with Apple IIs, Commodore PETs, and accompanying rotary dial telephones and teletype machines.

On the top floor there are dormitories with bunk beds, a small surgery, bathrooms, and a canteen, where today's visitors can enjoy a snack. At the time, there were sufficient supplies in the bunker for three months of operation, with no need to go to the surface.

Electromagnetic Pulse

The nuclear bunker is surrounded by a Faraday cage—a grounded wire mesh that completely covers the bunker. The cage prevents electrical fields from entering the bunker and vice versa. This both protects the bunker from an electromagnetic pulse (generated by a nuclear explosion) and prevents leakage of electrical signals that might be used for eavesdropping.

A Faraday cage relies on the fact that when an electric field is applied from the outside to a hollow metal cage, it causes a current in the cage. The electrons in the cage flow because of the current, and rearrange themselves until the charges on the inside and outside cancel each other out (see Figure 52-1). As long as the wavelength of the electromagnetic radiation hitting the cage is larger than the size of the holes in the mesh, the Faraday cage will completely dissipate the electricity and no electrical field will be created inside the cage.

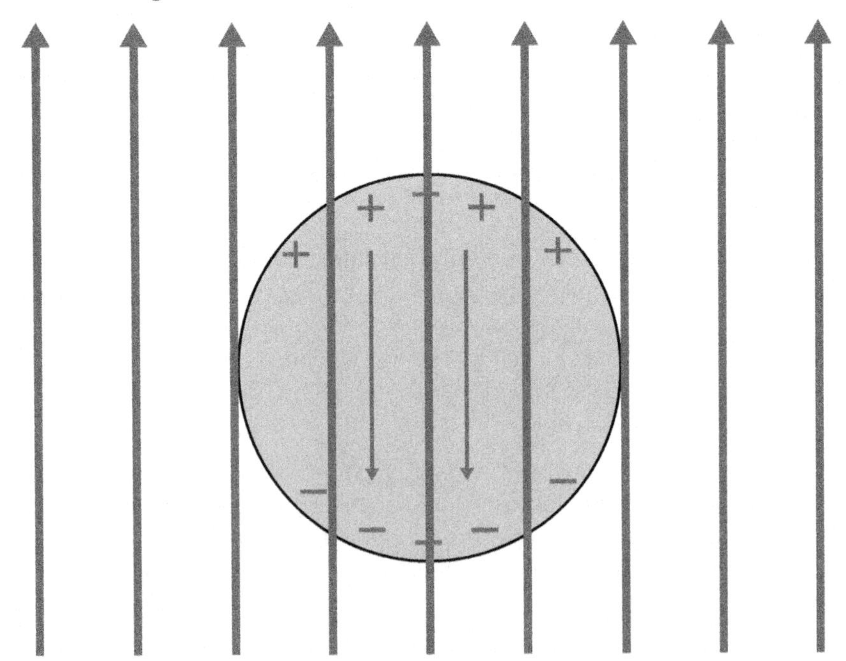

Figure 52-1. An external electric field creates an equal and opposite field inside the cage

Faraday cages are encountered frequently in everyday life. Cars and aircraft have metal skins that act as Faraday cages when struck by lightning; if a car is struck by lightning, the electricity flows around the outside of the car and jumps from the bottom to the ground. You can still use a cell phone in a car because the wavelength of the cell phone radio waves is short enough to pass through the gaps in the car's body caused by the windows.

Most homes also have a Faraday cage inside them—it's the fine mesh inside a microwave.

A nuclear bunker needs a Faraday cage because nuclear explosions create an intense electrical field. When a nuclear bomb explodes, it emits a large amount of gamma radiation. This radiation strikes air molecules at high energy, and the air becomes ionized with electrons taking energy from the gamma radiation. The electrons fly away from the positively charged air molecules, and a large electrical field is created.

This collision, called Compton scattering, was early evidence of the particle nature of electromagnetic radiation (including light). Physicist Arthur Compton showed, while studying the scattering of X-rays, that X-ray photons were colliding with electrons, causing the X-rays to be deflected and the electrons to gain energy. Compton subsequently won the Nobel Prize in Physics in 1927.

The electromagnetic pulse induces electrical currents in equipment in the vicinity of the explosion. For example, in 1962 a nuclear bomb test called Starfish Prime tested a nuclear explosion in space. The bomb was detonated at an altitude of 400 kilometers near Johnston Island in the Pacific Ocean.

The 1.4-megaton bomb created an electromagnetic pulse that knocked out street lights, set off car alarms, and damaged car ignition systems in Hawaii, over 1,300 kilometers away. It also disrupted radio communications between the U.S., Japan, and Australia for up to 20 minutes.

Many parts of the bunker have been mocked up to give a feeling of life underground. Mannequins sit ready to work on government business, and British Prime Minister John Major (who was in power when the bunker was closed) appears to be sleeping in his bed. Margaret Thatcher is down in the BBC room, preparing a broadcast.

Surrounding the bunker is a 3-meter-thick reinforced concrete wall, and above the bunker giant concrete slabs are buried in the ground to protect it from a nuclear blast. Surrounding the concrete wall is a Faraday cage to protect against an electromagnetic pulse (see sidebar) and waterproofing.

The topmost part of the bunker is 6 meters underground. Outside, the only conspicuous part of the complex is a 46-meter-tall broadcasting antenna and mast.

Practical Information

Details about the Kelvedon Hatch Nuclear Bunker are available from *http://www.secretnuclearbunker.com/*. The bunker is well sign-posted, with amusing road signs pointing to the "Secret Nuclear Bunker," and the admission price includes an hour-long, self-guided audio tour.

053

Kempton Park Waterworks, Kempton Park, England

51° 25′ 31.84″ N, 0° 24′ 13.20″ W

Triple Expansion Steam Power

It's a rare treat to see a working steam engine, and it's even rarer to see one in situ, right where it was meant to be. At Kempton Park, the huge Number 6 Worthington Triple Expansion Steam Engine was used to pump 72 million liters of water to North London each day. The engine (along with a twin) ran 24 hours per day, 7 days a week from the late 1920s until 1980.

Today, Kempton Park Waterworks is still supplying water to London (at a rate of 284 million liters per day) using electric pumps. Happily, the Number 6 engine has been restored to full working order—which required building a new steam boiler, because the old one had been demolished—and can be seen steaming seven weekends per year.

The engine stands almost 19 meters high and weighs 725 tonnes. It consists of three cylinders of different sizes. High-pressure steam enters the first, smallest cylinder at almost 14 times atmospheric pressure. The steam then passes into the second, larger cylinder at a lower pressure, and finally into the third and largest cylinder. Each cylinder has a piston connected to the same mechanism that drives the pumps.

The steam entering the first cylinder has been superheated, which means it has been heated beyond its boiling point. At Kempton, the steam is heated an additional 65°C. The superheated steam increases the efficiency of the engine because it condenses less than steam at a lower temperature.

Keeping the engine running required 12 tonnes of coal per day, which was delivered by a specially built railway. The building that houses the engine is a UK National Monument, and speaks of an age of municipal grandeur.

Triple Expansion Steam Engines were popular for water pumping, but they were also the mainstay of marine propulsion. The RMS *Titanic* was powered by two slightly larger, but similar, steam engines.

Measuring Efficiency

Part of the visit to Kempton is a close look at the instrumentation used to monitor the engine's performance. Two instruments are of particular interest: the flow meter and the combustion indicator.

The flow meter is used to measure the amount of water the engine is pumping. It does this using the Venturi Effect. As the water is leaving the pumping station, it passes through a restriction, called a Venturi. The water pressure is decreased by the restriction, and the difference in pressure is proportional to the rate at which the water is flowing (Figure 53-1).

The flow meter at Kempton works by connecting the pressure difference to a U tube filled with mercury. The mercury moves a pen, which records the flow rate on a cylinder of paper. The cylinder revolves once every seven days to create a hard-copy record of the flow. To get the total number of liters pumped, a mechanical computer integrates the rate of flow and displays the total on a series of dials.

The other interesting instrument is the combustion indicator. It is used to measure the efficiency of the coal combustion by sampling the gas in the boiler's flue and measuring the quantity of CO_2 present. The amount of CO_2 in the flue gas is a direct measure of the efficiency of coal burning—high percentages of CO_2 would indicate inefficient burning.

To measure the CO_2 content, a pressure differential is also used. Gas from the flue enters a pair of bellows: one bellow gets the gas directly from the flue, the other gets the gas after it has passed through a soda lime filter that absorbs the CO_2, creating a pressure difference. The pressure difference deflects a pen that traces a line on a rotating chart, recording one week's worth of combustion efficiency.

Soda lime is a mixture of calcium hydroxide ($Ca(OH)_2$), water, and sodium hydroxide (NaOH). In the presence of carbon dioxide, soda lime undergoes a sequence of chemical reactions that absorb the carbon dioxide and create calcium carbonate ($CaCO_3$) and water (Equation 53-1).

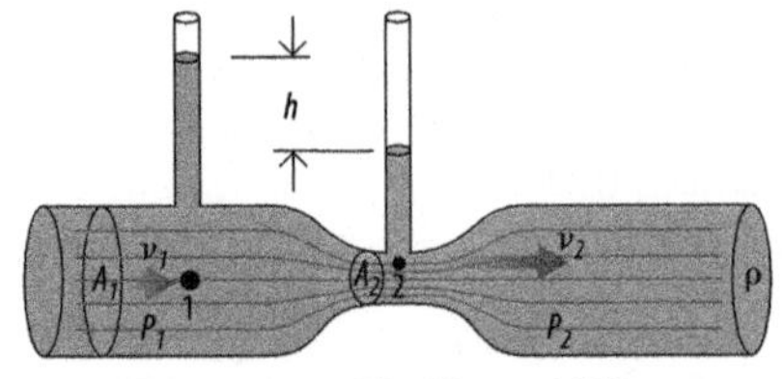

Figure 53-1. The Venturi Effect

$$CO_2 + H_2O = H_2CO_3$$

$$H_2CO_3 + NaOH + NaOH = Na_2CO_3 + H_2O + H_2O$$
(and heat)

$$Na_2CO_3 + Ca(OH)_2 = CaCO_3 + NaOH + NaOH$$

Equation 53-1. Soda lime reacts to carbon dioxide and absorbs it

Also on display at Kempton are steam turbines, which use steam at the same pressure and temperature as the steam engine. The turbines were installed in 1933 and weigh a mere 25 tonnes each. Between the steam engines and turbines, the pumping station was capable of supplying 326 million liters of water per day.

Adjacent to the pumping station are the filter beds where water was filtered and stored in the nearby reservoir.

Practical Information

The waterworks is about 35 minutes west of London and is accessible by train. Details can be found at *http://www.kemptonsteam.org/*. It's also possible to organize a special visit, where you can operate the engine by turning on the steam and keeping the speed at 18 rpm.

If you can't make it to Kempton, there's also a triple expansion steam engine at the Kew Bridge Steam Museum in London. The museum has a number of other historic engines that steam on many weekends. The oldest engine at the museum was built in 1820 by Boulton and Watt, and remains in working order. Details about the museum are at *http://www.kbsm.org/*.

054

Lacock Abbey, Wiltshire, England

51° 24′ 53.1″ N, 2° 7′ 1.85″ W

William Henry Fox Talbot

William Henry Fox Talbot invented the positive/negative photography technique while living at Lacock Abbey. He was a contemporary of Frenchman Louis Daguerre, who created the daguerreotype around the same time. Talbot's invention allowed multiple prints to be made from a negative; Daguerre's method created only a single image.

Initially, Talbot did not use a positive/negative: all his photographs were negatives. He made light-sensitive paper and used it to create contact prints: the object to be photographed (for example, a piece of lace) was placed in contact with the paper (or separated from it with a sheet of glass) and exposed to sunlight. This created a negative image that he termed a photogenic drawing.

Later, in 1841, he used these negatives to create positive prints by first taking a negative photograph, and then taking a photograph of the negative using the same technique. This he termed the calotype process. With one negative he could then make many positive prints.

In 1844, Talbot published the first book containing photographs. *The Pencil of Nature* contained just 24 photographs (including the image of Lacock Abbey shown in Figure 54-1) and a technical description of how his photographic technique worked. The book also discussed possible uses of photography (such as using photographs as evidence in court). Very few copies have survived, but pages from the book are on display in the museum at Lacock.

Figure 54-1. Lacock Abbey as photographed by William Henry Fox Talbot; courtesy of Glasgow University Library, Department of Special Collections

Lacock Abbey is a medieval cloistered abbey that was converted to a country house in the 16th century. Today, the abbey is owned by the National Trust and houses a museum of Talbot's life and work. The museum has many of his photographs, an explanation of the process, and the equipment he used.

Talbot was also interested in botany (many of his early photogenic drawings showed flowers, leaves, and plants) and the abbey is surrounded by woodlands and gardens, including his mother's rose garden (which has been restored based on Talbot's own photographs) and his own botanical garden (recently restored by the National Trust).

Talbot's first photograph was about the size of a postage stamp, and depicted the oriel window in Lacock Abbey (Figure 54-2). The oriel window is still there today; it's a popular snapshot for tourists.

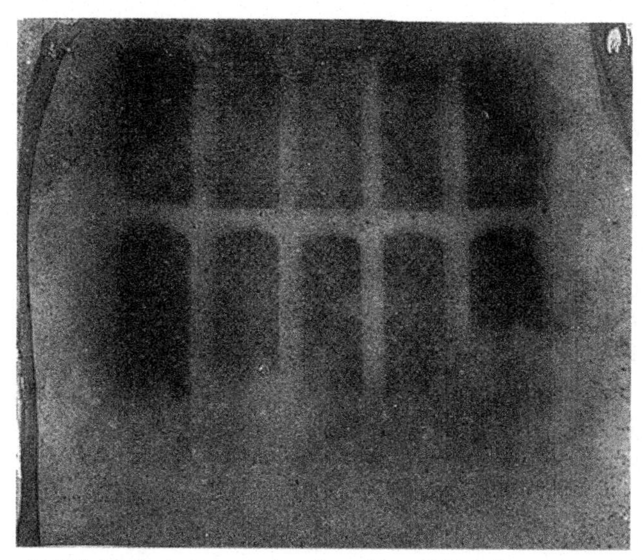

Figure 54-2. Talbot's photogenic drawing of the oriel window at Lacock Abbey

Close to Lacock Abbey is Lacock Village. The village seems to have been spared from modernization—dating back to the 13th century, it's filled with half-timbered stone houses and is a popular location for television and film. If you find yourself walking around the abbey with a sense of déjà vu, it's probably because it served as a filming location for two of the Harry Potter films.

Practical Information

Details about visiting the abbey and museum are best obtained from the National Trust: *http://www.nationaltrust.org.uk/lacock/*.

Salted Paper and Photogenic Drawings

The first step in Talbot's invention of photography was the making of his photogenic drawings. These contact prints were made using silver chloride soaked into drawing paper. To make this salted paper, Talbot first soaked the paper in salt (NaCl) solution, dried it, brushed it with silver nitrate ($AgNO_3$), and dried it again. This resulted in silver chloride (AgCl) and sodium nitrate ($NaNO_3$): $NaCl + AgNO_3 = AgCl + NaNO_3$.

Silver chloride is sensitive to both heat and light; when exposed to light it breaks down, releasing chlorine and leaving behind the silver. To make a print, the silver chloride–soaked paper was put in contact with the item to be photographed and then exposed to light for tens of minutes.

The silver was then oxidized and became silver oxide (Ag_2O), a dark-brown powder that is responsible for the brownish tint seen in old photographs. Finally, to prevent the image from changing with further exposure to light, the image was "fixed" by removing any unexposed silver chloride by washing the paper with potassium bromide.

This resulted in a single negative print, but the exposure time was too long for anything but a totally immobile subject.

Talbot proceeded to create more sensitive photographic paper to be used as a negative. The photogenic drawing paper was used to create a positive print from the negative.

To create the sensitive paper for the negative, Talbot washed drawing paper with silver nitrate and gently dried it. Then he soaked the paper in potassium iodide (KI), rinsed it, and dried it. This resulted in the creation of potassium nitrate (KNO_3) and the important light-sensitive component, silver iodide (AgI): $AgNO_3 + KI = KNO_3 + AgI$.

Just before taking the photograph, the paper was washed with a mixture of sodium nitrate and gallic acid. This final wash increased the sensitivity of the paper, reducing the exposure time to at most a minute. The paper was placed in a simple pinhole camera and exposed.

The paper was then removed and washed once more with sodium nitrate and gallic acid. During the washing, the negative image would appear on the paper. When the image had darkened enough, the paper was washed with potassium bromide to stop any further darkening.

With this negative image, Talbot was able to make as many prints as he wanted. He placed a piece of his salted paper under a sheet of glass, placed the negative on top, and exposed the combination to sunlight using the photogenic drawing process to make a positive print (the negative of the negative) on salted paper.

055

Manchester Science Walk, Manchester, England

53° 28′ 40.63″ N, 2° 14′ 40.48″ W

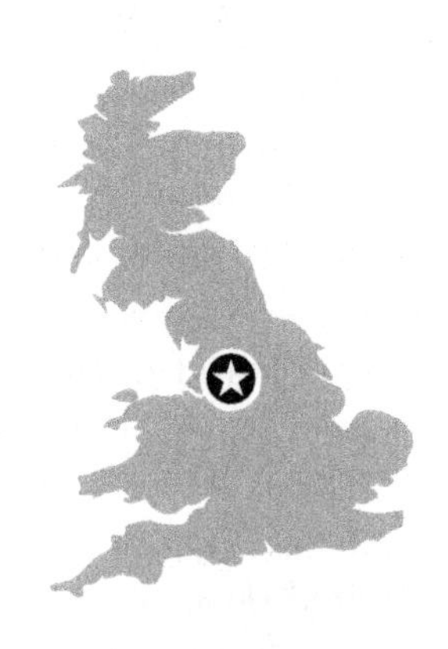

The Atomic Theory

John Dalton was a late-18th-century chemist who lived and worked in Manchester. His most important work was his atomic theory, which laid the groundwork for chemistry as we know it. Dalton's theory was that the elements are made from tiny particles called atoms, that the atoms that constitute an element are all the same, that atoms can be distinguished by their mass, that different elements have different atoms, and that chemical compounds are combinations of atoms from different elements. Only one part of his theory didn't stand the test of time: he thought that atoms were indivisible.

In 1803, Dalton published a table of atomic masses for a variety of elements in his book *A New System of Chemical Philosophy* (see Figure 55-1). He gave hydrogen the nominal mass of 1, and all the other elements were weighed relative to hydrogen. Dalton was also the source of the Law of Multiple Proportions, which says that the ratio of one atom to another in a chemical compound is a whole number—since atoms could not be divided, they had to combine in fixed, whole-numbered ratios.

Dalton also favored the idea that atoms combined in binary relationships, and erroneously thought that water was HO and not H_2O.

The best place to learn about John Dalton is at the Manchester Museum of Science and Industry, and a great way to get there is by following a free audio tour of scientific Manchester that ends at the museum.

The Manchester Science Walk is a 17-stop walk through central Manchester, narrated by an actor playing the part of John Dalton and available as a free download of MP3 files and a PDF map. The tour was created as part of the Manchester Science Festival, held each year in October.

The 2.5-kilometer walk starts at the central library, and it takes about 1.5 hours to reach the museum. The audio tour is largely historical, and explains Dalton's life and the life in Manchester at the time.

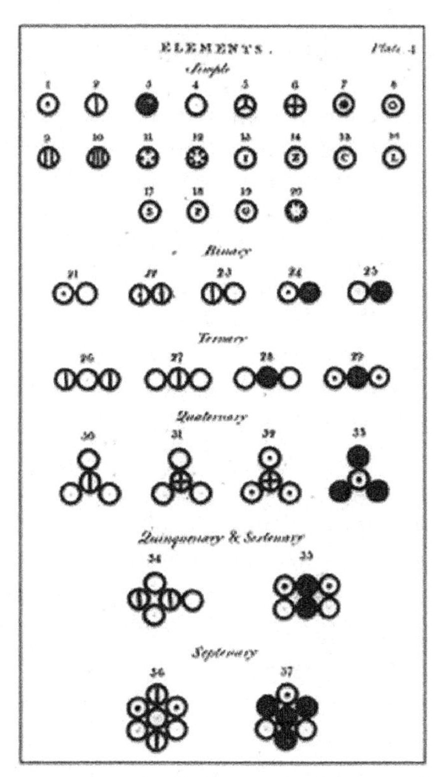

Figure 55-1. Dalton's elements and their weights

The museum's Manchester Science Gallery highlights those Manchester scientists who have had a large impact: John Dalton, James Joule, Ernest Rutherford, and Sir Bernard Lovell (see Chapter 51). There's also a working replica of the first stored-program computer (called the Baby), originally built in 1948.

Manchester was at the center of the textile industry and the early railway—in fact, the museum is housed inside the oldest surviving railway station in the world. It has an important collection of locomotives and demonstrations of machines used in cotton mills. Children will enjoy the tour though a Victorian sewer.

If you have time after the walk and museum, visit the Alan Turing statue in Sackville Street Gardens (see Chapter 66).

Practical Information

The website of the Manchester Museum of Science and Industry is at *http://www.msim.org.uk/*. Information about the Manchester Science Walk can be found at *http://www.verm.co.uk/Sciencetour/*, and about the Manchester Science Festival at *http://www.manchestersciencefestival.com/*.

The Changing Weight of Hydrogen

Although Dalton's theory was correct when he assigned hydrogen the atomic weight of 1, today the atomic weight of hydrogen stands at 1.00794. The weight of a hydrogen atom seems like a simple thing to agree upon, but it has been the subject of controversy, learned commissions, and inquiries ever since Dalton introduced the idea. Even the term "atomic weight" is contested—some scientists, notably physicists, prefer "atomic mass" because weight is gravity dependent (see Chapter 69).

During the 19th century, atomic weights were calculated (with varying degrees of accuracy) by using Dalton's assignment of 1 to hydrogen. Because there was disagreement about the actual weights of various elements, commissions were set up, culminating in the 1901 international agreement to calculate atomic weight by assigning oxygen the value 16. But following that there was international disagreement, and tables were published based on hydrogen with weight 1 and oxygen with weight 16.

A problem arose in 1929 with the discovery of two new isotopes of oxygen: ^{17}O and ^{18}O. These new isotopes created a problem because the atomic weight used by chemists is the weighted average mass of atoms of an element in their proportions as found in the Earth's atmosphere and crust. When new isotopes are discovered or measurements made, the atomic weights have to be updated. For this reason, atomic weight tables are updated regularly.

But there was still disagreement between physicists (who were interested in the mass of specific isotopes) and chemists (who were weighing naturally occurring elements). In 1961, the atomic weights were changed based on assigning the carbon isotope ^{12}C the weight 12. Because ^{12}C is the most abundant form of carbon on Earth, this change was acceptable to chemists as it did little to change the weights they were already using, and it was acceptable to physicists because ^{12}C is stable and has no free electrons: it has six protons, six neutrons, and six electrons.

Today, all atomic weights are based on ^{12}C. The atomic weight of carbon (which is still a weighted average) is 12.0107.

056

Museum of the History of Science, Oxford, England

51° 45′ 15.95″ N, 1° 15′ 18.68″ W

A Hidden Oxford Treasure

Tucked away on Broad Street in the center of Oxford is an ancient science museum. Since 1683, the Museum of the History of Science has been collecting scientific knowledge and apparatus and putting it on public display. The museum's collection covers science from antiquity to the 20th century, and from the simple (such as collections of platonic solids) to the intricate and beautiful (such as its collection of astrolabes dating back to the 9th century).

It has a collection of 15,000 items; those on display are arranged in pristine cabinets over three floors. There's a lot to see here—you should plan to spend at least a couple of hours visiting the museum.

The museum's iconic exhibit is a blackboard that was used by Einstein to give a lecture at Oxford in 1931 (see Figure 56-1). The blackboard shows Einstein's calculation of the age of the universe based on its expansion; he estimated it to be 10 billion years old. (The accepted value today is somewhere between 12 and 14 billion years old.)

The museum is a stop on the Oxford Science Walk, which passes through central Oxford. (You can buy a copy of the walk booklet at the museum.) The first stop on the walk is Oxford's Botanic Garden, which was established in 1648. It's still going strong, and even if the weather is poor, you can spend time in the garden's greenhouses, where it's warm enough to grow palm trees.

The next stop is the Penicillin Memorial and Rose Garden, which commemorate the isolation and purification of penicillin at Oxford by the Australian Howard Florey and the German Ernst Chain. A few stops later, there's a plaque on the High Street that marks the site of Robert Boyle and Robert Hooke's laboratory, where Boyle formulated his law and Hooke built an efficient air pump.

Figure 56-1. Einstein's blackboard; courtesy of Garrett Coakley (garrettc)

On New College Lane the walk stops at Edmond Halley's house. Halley is best known for his comet, but he was also interested in the Earth's magnetic field, how air pressure varies with altitude, ocean salinity, and optics. He also encouraged Newton to publish his *Principia Mathematica.*

The tour passes by Wadham College, where John Wilkins, Christopher Wren, Robert Boyle, John Wallis, and others met to discuss natural philosophy. They later went on to form the Royal Society in London.

Another stop is the University Museum (known formally as the Oxford University Museum of Natural History), which covers geology, mineralogy, zoology, and entomology. The front lawn of the museum has an amusing set of footprints from a Megalosaurus, the bones of which were found in a quarry near Oxford.

Practical Information

Admission to the Museum of the History of Science is free, and details of its location and collection can be found at *http://www.mhs.ox.ac.uk/*. Information about the Science Walk can be found at *http://www.mhs.ox.ac.uk/features/walk/*.

Boyle and the Gas Laws

Robert Boyle is best remembered for Boyle's Law, which states that if a fixed amount of gas is kept at a fixed temperature, then its pressure and volume are inversely proportional. This can be written as an equation combining the pressure, P, the volume, V, and an unknown constant value, k (Equation 56-1).

Boyle published this law in 1662, after experimenting with an air pump made by his assistant, Robert Hooke.

In 1787, French scientist Jacques Charles (who is better known by his experiments with the hydrogen balloon) discovered Charles's Law, which states that if a fixed amount of gas is kept under constant pressure, its volume is proportional to its temperature. This can be written as an equation combining the volume, V, the temperature, T, and another unknown constant, k (Equation 56-2).

$$PV = k$$

Equation 56-1. Boyle's Law

$$\frac{V}{T} = k$$

Equation 56-2. Charles's Law

Another Frenchman, Joseph Louis Gay-Lussac, published his gas law (called, inevitably, Gay-Lussac's Law) in 1802. This law states that if a fixed amount of gas has a constant volume, then its pressure is proportional to its temperature. This can be written as an equation combining the pressure, P, and volume, V, with an unknown constant, k (Equation 56-3).

With all three laws combined, a single law is possible: that the ratio between the pressure-volume product and the temperature of a fixed amount of gas is constant (see Equation 56-4).

The only missing piece is the unknown constant, k. In 1834 the Ideal Gas Law was formulated based on Avogadro's Law. An ideal gas has molecules of zero volume, no forces between the molecules, and random molecular motion. Avogadro's Law states that for ideal gases, equal volumes of gas at the same pressure and temperature have the same number of molecules. Using this law, it's possible to replace the unknown constant in the Combined Gas Law.

In the Ideal Gas Law (Equation 56-5), the constant, k, is replaced by nR, the product of the number of moles of gas, n, and the universal gas constant, R. With no unknowns, the Ideal Gas Law can be used for real-world calculations.

$$\frac{P}{T} = k$$

Equation 56-3. Gay-Lussac's Law

$$\frac{PV}{T} = k$$

Equation 56-4. The Combined Gas Law

$$PV = nRT$$

Equation 56-5. The Ideal Gas Law

Although the Ideal Gas Law is based on a fictional ideal gas, it is close enough to the behavior of real gases to be used frequently by engineers.

057

Napier University, Edinburgh, Scotland

55° 55′ 59.93″ N, 3° 12′ 50.21″ W

John Napier and Napier's Bones

Just as William Shakespeare was writing about the fictional Macbeth scheming to become King of Scotland, the real-life 8th Laird of Merchiston, the mathematician John Napier, was devising schemes for simplifying multiplication. His best-known inventions are Napier's Bones and logarithms (see sidebar). He also helped popularize the decimal point.

Napier was born in 1550 at Merchiston Tower in Edinburgh. The tower was the seat of the Clan Napier, which still exists today; John Napier was the clan's 8th laird. The tower is also still standing, but is no longer used by the clan; it forms part of Napier University's Merchiston campus. Outside the main entrance of the university's Craighouse campus stands a statue of a rather fearsome-looking John Napier, holding his "bones" in one hand.

Unfortunately, it's not possible to visit Merchiston Tower's interior, but standing in front of it you can still imagine John Napier working on his invention here, 400 years ago. And that invention has underpinned much of the rest of technological progress. The ability to easily do multiplication and division and the subsequent invention of the slide rule (page 138) made it possible to engineer buildings and machines with great accuracy before electronic computers came along.

Napier's Bones are a simple device for doing multiplication (Napier was obsessed with speeding up multiplication) and division, and for extracting square roots. They are a form of abacus, the history of which can be traced back to the House of Wisdom in Baghdad in the 9th century.

The bones consist of a rectangular board, with the numbers 1 through 9 along the side and a set of rods that fit into the board running vertically. Each rod corresponds to a single "times table" and has a single number at the top. For example, the rod numbered 7 at the top has entries 14, 21, 28, and so on, that line up with the numbers 1 through 9 on the board. The entries on the rod are arranged diagonally, with a line between them (Figure 57-1).

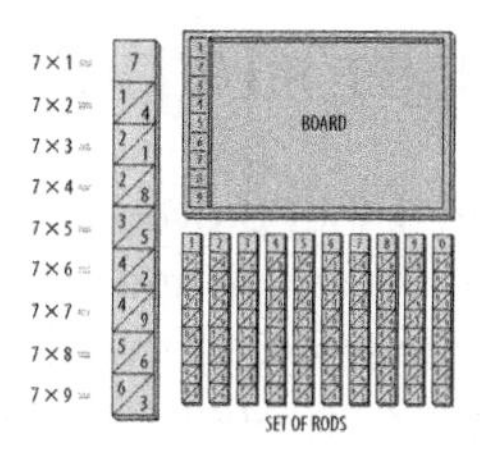

Figure 57-1. Set of Napier's Bones

Multiplication is performed by taking a set of rods that correspond to one of the multiplicands, placing them in the board in order, and then performing a long multiplication by looking up the correct answer for each digit of the other multiplicand. Because the digits are arranged in a slanting fashion, adding up and carrying numbers is easy (Figure 57-2).

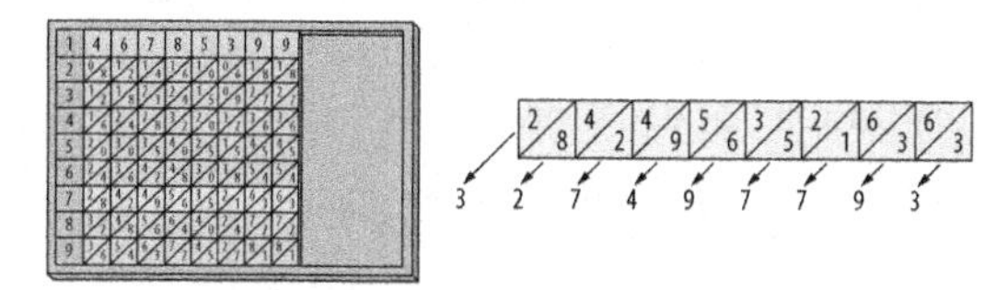

Figure 57-2. Multiplying by a single digit (7 × 46,785,399)

A long multiplication is performed by calculating the values for each digit and then adding them together (see Figure 57-3).

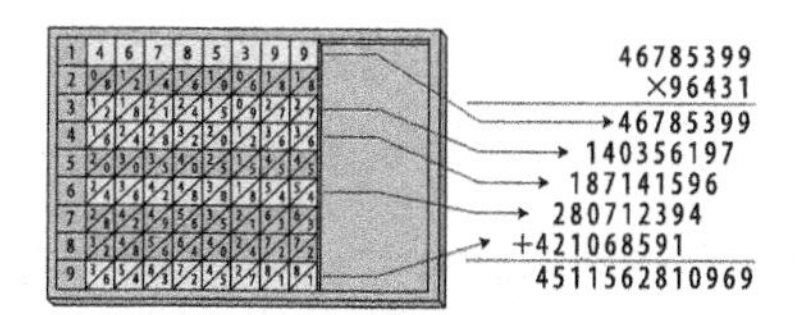

Figure 57-3. Long multiplication (96,431 × 46,785,399)

Since Napier's Bones are simple to build yourself, they make an excellent tool for teaching multiplication tables and long multiplication to children.

Because Napier lived so long ago, and because his work was mostly theoretical, the two places described here don't make a full day's worth of sightseeing. Happily, Edinburgh has lots to offer in the way of scientific tourism, at 14 India Street (the home of James Clerk Maxwell; see Chapter 35) and the National Museum of Scotland (Chapter 59).

Practical Information

Merchiston Tower is on the Merchiston Campus of Napier University on Colinton Road; it is easily reached from the city center by bus. The statue of John Napier is on the Craighouse Campus, which is off Craighouse Road; it can be reached by taking a number 17 bus after visiting Merchiston Tower.

Napier's Logarithm

Once logarithms were invented, they proved popular because they turned multiplication and division into addition and subtraction, and made the computation of square and cube roots simpler by turning them into multiplication. But to perform those calculations, it was necessary to have a table of precalculated logarithms; John Napier set out to define the logarithm and publish a book of tables, *Mirifici Logarithmorum Canonis Descriptio*.

Since Napier didn't have all the mathematical terminology available today (after all, it was he who helped to popularize the decimal point), he constructed his logarithms geometrically. Imagine a pair of lines with equally spaced ruled marks, as seen in the Figure 57-4. Both are number lines—the top line is laid out arithmetically (the number at each mark is one more than the previous number), and the bottom line is laid out geometrically (in this case, the numbers double for each mark).

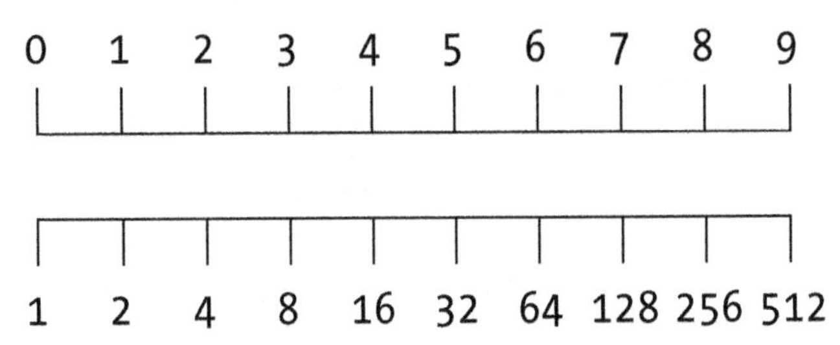

Figure 57-4. Arithmetic and geometric number lines

Napier coined the term *logarithm* from the Greek words *logos*, meaning proportion, and *arithmos*, meaning number. The logarithm of a number on the geometric line is just the corresponding number on the arithmetic line. So the logarithm of 4, for example, is 2. The logarithm of 33 is just past the logarithm of 32 (which is 5); if you could measure carefully enough, you'd discover that it's approximately 5.04.

Using modern notation, the number lines in Figure 57-4 are related by exponentiation. That is, if x is a number on the arithmetic line, then the corresponding number y on the geometric line satisfies the equation $y = 2^x$, and the logarithm is defined by saying that x is the logarithm of y, which is written $x = \log_2 y$.

The subscript 2 is important: it is known as the base, and the number lines in Figure 57-4 represent logarithms using the base 2. The base is just the factor used in the exponentiation. So, a more general logarithm for a base b would be defined as $x = \log_b y$, where $y = b^x$.

Logarithms are now commonly defined using the bases 2, e, or 10. $\log_2$ is used in computer science because of its relationship with the binary system. $\log_e$ is called the natural logarithm, where the base is e, and it is often written ln. It springs up in calculus because it is equal to the area under the curve 1/x (see Equation 57-1).

$$\ln(a) = \int_1^a \frac{1}{x}\,dx$$

Equation 57-1. Natural logarithm

$\log_{10}$ is used extensively in engineering, and is the logarithm used to lay out most slide rules.

Napier's logarithm didn't have any of these bases. He constructed it geometrically with a pair of number lines similar to those in Figure 57-4, but the geometric line didn't scale exponentially in the same way. He was actually calculating logarithms of sines of angles in a circle with a radius of 10,000,000 (he choose this large number so that he would get many digits of accuracy).

In modern terms, his logarithm had a base of 1/e, but the importance of his definition was that it nevertheless could be used for the simplification of multiplication, division, and the taking of roots.

Having defined his logarithm, he then published a description of the logarithm and a table of logarithms accurate to five decimal places in his book *Mirifici Logarithmorum Canonis Descriptio* (The Canon of Marvellous Logarithms) in 1614.

058

National Museum of Computing, Bletchley, England

51° 59′ 54.6″ N, 0° 44′ 36.6″ W

Colossus

It's hard for the British National Museum of Computing to rival the Computer History Museum in the U.S. (Chapter 86), as the history of computing has largely been written by U.S. companies. Still, Britain's role in the early history of computing is important (not least because of the influence of the greatest computer scientist of all, Alan Turing; see Chapter 66).

The star exhibit at the museum is the reconstructed Colossus computer created and used during the Second World War. Colossus was one of the first computers that used vacuum tubes (instead of mechanical relays), thus making it an electronic computer. Like modern computers, it also used the binary system (see sidebar) and was programmable—albeit for the limited task of breaking the Nazi German Lorenz code.

The machine read an intercepted message from a rapidly moving paper tape, and then worked out the settings of the Lorenz machine used to transmit the message. The Lorenz code, unlike Enigma, was based on the binary system. Each character to be transmitted was first converted to a binary number consisting of 5 bits (1s or 0s) using the standard Baudot code used for telegraphy. The letter A might have been transmitted as 00011, B as 11001, C as yet another combination of 1s and 0s, and so on.

The Lorenz machine produced a pattern of apparently random sets of 5 bits, with a new pattern appearing for each letter to be transmitted. The machine was not actually producing random patterns, but was following a difficult-to-decipher sequence based on the machine's settings.

For each letter to be transmitted, its 5 bits were matched up with the next 5 bits coming from the Lorenz machine, and they were combined using an operation called XOR (which is read as exclusive-or). The XOR worked bit by bit, taking the first bit of the letter and the first letter of the Lorenz code and combining them to produce a new first bit; it then proceeded to the second bits of the letter and the Lorenz code to produce the second bit, and so on (see Figure 58-1).

Transmission		***Reception***	
0 0 0 1 1	Letter to encrypt (e.g., A)	0 1 0 1 0	The bits that are received
0 1 0 0 1	Random bits from the Lorenz machine	0 1 0 0 1	Same random bits from the Lorenz machine
XOR		XOR	
0 1 0 1 0	Bits that are transmitted	0 0 0 1 1	Decoded letter (e.g., A)

Figure 58-1. The Lorenz code

The XOR operation does the following: if the bits are the same (either two 1s or two 0s), then its output is 0, but if the bits are different (one is 1 and the other is 0), then its output is 1. A nice property of XOR is that if it's used twice with the same Lorenz code, the original letter would appear; since both the receiving and transmitting Lorenz machines had the same settings, they produced the same sequence of apparently random bits.

With a binary code being transmitted by the Nazis, the code breakers at Bletchley built a binary computer to do the decryption. The original Colossus computers were destroyed, along with their plans, by order of Winston Churchill, who asked that they be broken into pieces "no larger than a man's fist." With painstaking work based on surviving plans and the memories of those who built the machines, a reconstructed Colossus now sits at Bletchley and breaks codes once more.

At the museum, Colossus is joined by displays on predigital computing, pocket calculators, personal computers, air traffic systems, and the beginnings of the electronic office with massive mainframe computers. The aim is to show working machines and explain their significance.

For a real treat, book a ticket for one of the museum's evening events, where the workings of Colossus are explained and the machine is fired up to break the Lorenz cipher once more.

Practical Information

The National Museum of Computing is part of Bletchley Park (see Chapter 40), but has its own website at *http://www.tnmoc.org/*.

The Binary System

Modern computers, like the Second World War's Colossus, use the binary system to represent numbers (and use numbers to represent everything else). To a computer, pictures, music, documents, and anything else are just lists of numbers, and those numbers are stored as binary 1s and 0s.

Binary is used because it is the simplest number system, it is easy to handle electrically (a 1 can be represented by the presence of a voltage, and a 0 by no voltage), and any number can be written in binary. Not using binary would mean finding a way to represent more complex numbers electrically—for example, before binary computers became common, analog computers used voltage levels to represent numbers (0.5 might have been 0.5V, 2 might have been 2V, etc.). Binary has the great advantage of simplicity: 1 is on, 0 is off. This simplicity means that binary can be also be used on hard disks to store information magnetically (1 could be the north pole of a magnetic portion of the disk, 0 the south pole) and optically (CDs and DVDs are read by shining a laser into the tracks of the discs and detecting the presence or absence of pits in the disc's material representing 1s and 0s).

Binary is also known as the base-2 system. Everyday arithmetic uses the base-10 system. Take the number 128, for example: going back to primary-school days, this can be thought of as a 1 in the 100s column, a 2 in the 10s column, and an 8 in the 1s column. As you move from right to left reading a number, the columns are numbered 1, 10, 100, 1,000, and so on: each column is 10 times the previous column. This multiplying factor is called the base.

The binary system, on the other hand, has columns that are numbered 1, 2, 4, 8, 16, 32, 64, 128, and so on. Each column is twice the previous one, so the number that is written 13 in base 10 is written 1101 in base 2. That is, there's a 1 in the 8s column, a 1 in the 4s column, and a 1 in the 1s column. You can quickly verify this by adding up the columns where there's a 1: that is, 8 + 4 + 1 = 13.

The base also determines the digits that can be used in each column. For base 10, the digits are 0, 1, …, 9; for binary, the only digits that make sense are 0 and 1 (just as in the decimal system it wouldn't make sense to try to put 11 in the 10s column, in binary it's not possible to put 3 in the 2s column).

Mathematically, a number in the decimal system with n digits is represented by its digits (d_{n-1}, d_{n-2}, ..., d_0). Each digit is multiplied by the power of 10 corresponding to its column (remembering that $10^0 = 1$) and the result is summed (see Equation 58-1).

$$\sum_{i=0}^{n-1} d_i 10^i$$

Equation 58-1. A decimal number

Binary is the same, except that 10 is replaced by 2 and the digits (b_{n-1}, b_{n-2}, ..., b_0) are restricted to being 0 or 1 (see Equation 58-2).

$$\sum_{i=0}^{n-1} b_i 2^i$$

Equation 58-2. A binary number

Binary makes computers particularly good at dealing with certain operations—just as multiplying and dividing by 10 is easy in the decimal system (just add a 0 or remove the rightmost digit), multiplying or dividing by 2 is easy in binary (once again, either add a 0 or remove the rightmost digit). It also allows a computer to deal with numbers and logic at the same time—binary is just as good at representing numbers as it is the concepts of true and false. By assigning the value 1 to the concept of true and 0 to false, a computer can be used to make logical decisions (more on computer logic on page 333) such as "IF X > 10 AND X < 20, THEN LAUNCH MISSILE." This logic consists of two comparisons of numbers done using binary numbers and a logical AND decision.

To a computer, the round binary numbers 1, 10, 100, 1000, and 10000 correspond to the decimal numbers 1, 2, 4, 8, and 16. It's for that reason that computers have 4, 8, 16, or 32 gigabytes of memory (as opposed to 5, 10, 20, or 30 gigabytes).

It also explains the title of this book. The 128 in the title is a round binary number (10000000), and is meant as a signal to any computer geek seeing this book on a shelf. If this book had been written by a non-nerd, it probably would have had 100 places, and you'd have been shortchanged 28 places.

059

National Museum of Scotland, Edinburgh, Scotland

55° 56′ 50″ N, 3° 11′ 23″ W

Dolly the Sheep

The National Museum of Scotland is an ideal excuse to visit Edinburgh, the capital of Scotland. Edinburgh was the home of many great scientists, including Alexander Graham Bell (see Chapter 4), James Clerk Maxwell (see Chapter 35), and John Napier (inventor of the logarithm; see Chapter 57). It is also the final resting place of the first animal cloned from an adult cell: Dolly the Sheep (Figure 59-1).

Figure 59-1. Dolly at the National Museum of Scotland

Dolly was born in Scotland on July 5, 1996, and she was euthanized on February

14, 2003, after suffering from lung disease. During her life at the Roslin Institute near Edinburgh, she bred—the natural way—with a ram and had six lambs.

Dolly was cloned by taking cells from the udder of a six-year-old adult sheep. The nucleus of the udder cells was removed and placed in an egg cell. The egg was allowed to divide for six days and was then implanted in a ewe, Dolly's mother. The egg then grew normally, and Dolly's mother gave birth naturally to the first cloned sheep.

Because Dolly was grown from the udder (or mammary gland) of a sheep, the scientists named her after country music star Dolly Parton.

The team that cloned Dolly went through 277 failed attempts before successfully creating an egg with a cloned nucleus that would divide and go on to be born. Although animals had been cloned previously, Dolly was a major step because the nucleus containing the DNA came from an adult sheep. Prior to Dolly it was not known whether using an adult nucleus would work, as it was feared that parts of the DNA would have been switched off as the donor sheep had aged.

Dolly (now stuffed) is surrounded by an exhibit that explores the technology and history of cloning.

The museum's other exhibits include a 1,000-year-old Islamic astrolabe, electrical apparatus used by Lord Kelvin, chemist Joseph Black's glassware, and a barometer standing almost 9 meters tall. If you tire of scientific exhibits, the museum also traces the history of Scotland, from the Picts to the present day.

Practical Information

Practical details about visiting the National Museum of Scotland are available at *http://www.nms.ac.uk/nationalmuseumhomepage.aspx*.

Cloning by Somatic Nuclear Cell Transfer

Dolly was cloned by a process called somatic nuclear cell transfer (depicted in Figure 59-2), in which the nucleus of a somatic cell (any cell from the body of an organism; i.e., not a cell coming from a sperm or egg) is transferred into an egg. First, cells (in this case, from a sheep's udder) are collected and the nucleus removed. This is done by holding the cell in place under a microscope using a suction pipette (a hollow tube) and extracting the nucleus using a very thin glass needle.

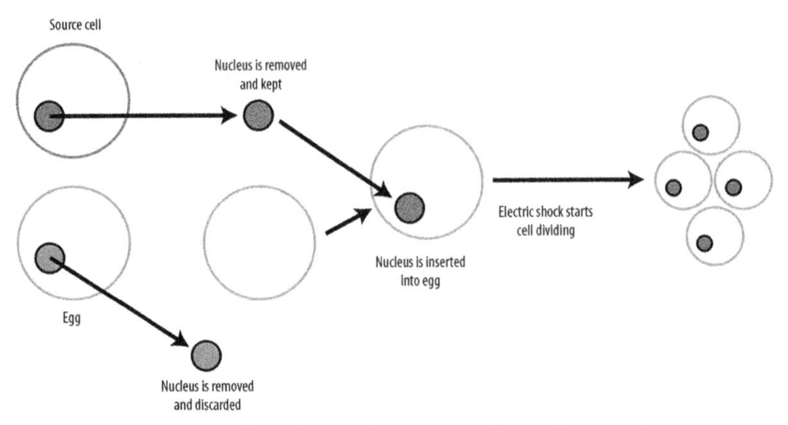

Figure 59-2. Transferring the nucleus of one cell to another

The rest of the cell is discarded. At the same time, a similar procedure is used to remove the nucleus of an egg cell, but this time the cell is kept and the nucleus discarded. Then the reverse procedure is performed, placing the nucleus inside the egg cell.

The cell is then given a small electric shock to start the division process. After five or six days of division in a test tube, the group of cells has become a blastocyst: a cluster of cells whose inside will grow into an embryo and the outside into the placenta. The blastocyst is implanted in the womb of the future mother and allowed to grow normally.

The key to Dolly's creation, where other nuclear cell transfers had failed, was the use of a quiescent cell. After the somatic cells were harvested, they were grown in a test tube and fed a rich diet to keep them dividing and growing. Before implanting the cells, their diet was severely cut back. The cells stopped dividing, but stayed alive, and became quiescent.

Once the nucleus had been replaced and an electric shock provided, the quiescent cell starting dividing again.

060

National Railway Museum, York, England

53° 57′ 38.76″ N, 1° 5′ 47.58″ W

Chariots of Fire

You don't have to be a trainspotter to appreciate the power, romance, and beauty of rail travel, and the National Railway Museum in York is the place to understand the development of the railway from the earliest days of the Industrial Revolution to the present.

No British railway museum would be complete without the early steam locomotive Stephenson's Rocket, and the National Railway Museum has two. Both are replicas (the original is in the Science Museum in London; see Chapter 77). One of the replicas is in working order, and the other is cut into sections so that visitors can understand the design and operation of the first successful steam locomotive. The Rocket traveled at about 20 kph in 1829.

The fastest steam train ever, the Mallard, is on display and has been completely restored. On July 3, 1938, it set the world record for steam locomotives, traveling at over 202 kph. The locomotive is one of the most important exhibits at the National Railway Museum and is in complete working order.

There's also an enormous KF7 steam train, which was built for the Chinese railway and ran between Canton and Hankou, and Nanking and Shanghai. Because of the steep inclines and weak bridges along the routes, the train had to be both powerful and light, and it also had to be capable of burning the poor-quality coal available in China at the time.

Perhaps the most famous steam train of all, the Flying Scotsman, is also preserved at the National Railway Museum. It ran from 1923 to 1963 along the East Coast line from King's Cross station in London to Edinburgh. It was the first train to pass 100 mph (160 kph), and could make the trip from London to Scotland on a single tender full of coal, with no need to stop along the way.

Double-Acting Steam and the Kylchap Blastpipe

A typical steam engine works by burning coal to heat water, and then using the steam to drive a piston or pistons to turn the wheels. To get the maximum amount of power, many steam engines are double-acting—that is, the steam drives the piston in both directions (Figure 60-1).

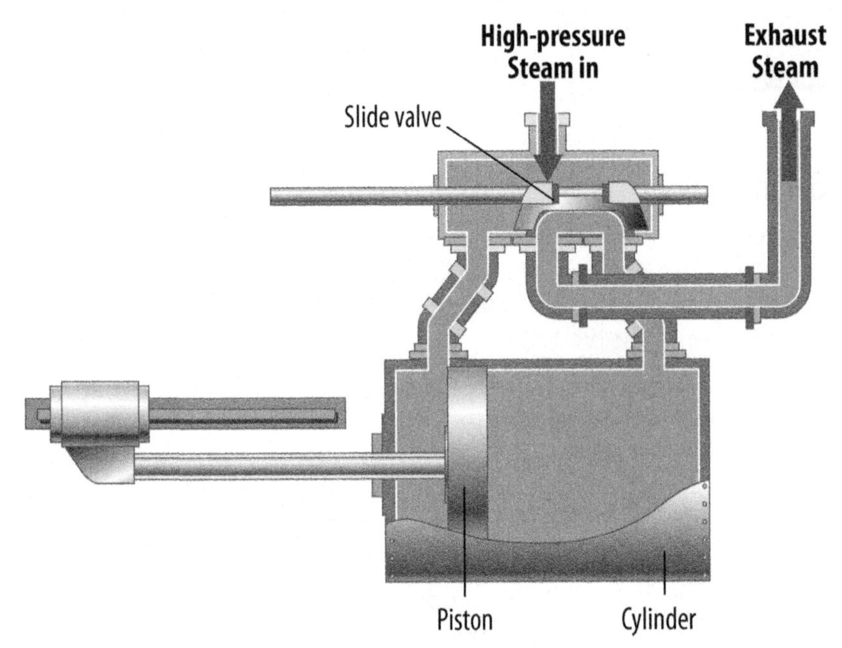

Figure 60-1. Double-acting steam engine cylinder

With the piston at one end of the cylinder, high-pressure steam from the locomotive's boiler is directed into the cylinder via a valve. The piston is forced down the cylinder, creating movement of the piston rod that eventually drives the wheels. Once the piston reaches the other end of the cylinder, the valve closes, reopening at the other end. High-pressure steam enters the cylinder again and pushes the piston in the other direction.

As the piston is being pushed by the steam, it is pushing out the cooled steam from the other side of the cylinder. This cooled steam exits via a pipe. In steam engines, the entire process can be achieved with a single valve that covers and uncovers the appropriate hole into which steam is injected, while at the same time connecting the other side of the cylinder to the exhaust.

The steam leaving the cylinder is not wasted. In steam engines, the exiting steam is mixed with the smoke coming from the fire that powers the boiler. By injecting the steam into the smoke, it's possible to increase the draught up the locomotive's chimney and draw more air in through the boiler, increasing its heat. The waste steam is thus used to make even more steam.

Maximizing the draught was a major part of steam engine design. The Mallard had two chimneys and a double Kylchap blastpipe. The blastpipe is the part of the locomotive that mixes the used steam and smoke before it enters the chimney.

In the Kylchap design, the used steam enters a splitter that consists of four vertical funnels. On top of the funnels is a stack of two or more tubes that open at the top and bottom. The steam shoots out of the funnels into the first tube, drawing in smoke, and then continues up into the next tube, drawing in more and more smoke, before exiting through the chimney.

But the Museum doesn't just have steam engines. It also has a current Eurostar train of the type that runs through the Channel Tunnel, a 1976 Japanese 0 Series Shinkansen bullet train capable of 220 kph (the only one outside Japan), a 1972 prototype of the British Rail High Speed Train (top speed 230 kph), and diesel and electric trains of all types that were used by British Rail. There's also a large collection of carriages and wagons that include Queen Victoria's special railway carriage.

In addition to all these trains, there are models, track, signal equipment, signage, and even crockery. In fact, there's everything you'd need to create your own railway system, including maintenance facilities and an extensive library of artifacts, technical manuals, and historic photographs.

Conspicuously absent from the museum is the French TGV, the current world record holder for train speed at 574.8 kph, as the museum focuses on the development of railway from a British perspective.

The museum offers daily talks and demonstrations; a full list of events is available from the website. Twice a day there's a special talk about the Shinkansen, and there are demonstrations of the enormous train turntable used to move trains from track to track. And if all that isn't enough, admission is free!

Practical Information

Full information about the National Railway Museum, its collection, and its location is available at *http://www.nrm.org.uk/*.

061

Natural History Museum, London, England

51° 29′ 45.54″ N, 0° 10′ 34.94″ W

The Stuffed Museum

There are natural history museums, and then there is the Natural History Museum in London. Set in the grandest of buildings and containing a truly extensive collection going back over 250 years, the Natural History Museum is simply the best museum explaining the natural world. And best of all, like the neighboring Science Museum (Chapter 77), entry is free.

The Natural History Museum traces its beginnings to the collection of natural curiosities made by Sir Hans Sloane during the late 17th and early 18th centuries. Sloane was a practicing doctor (mostly to the upper classes, making him a wealthy man) but also made a number of voyages, including one to Jamaica where he catalogued 800 then-unknown species of plant. While in Jamaica he found the local drink of cocoa mixed with water to be nauseating, but invented a drink made of milk and cocoa that went on to great success when manufactured by the Cadburys.

Upon his death Sloane left his collection to the British Museum. Over the years the museum added other collections, and eventually became an independent museum. In 1881, it moved into its current home, a specially built building of Victorian grandeur. Entering the museum, visitors find themselves in the imposing Central Hall, where a complete skeleton of a Diplodocus greets new arrivals and people passing from one section of the museum to another. While in the Central Hall, be sure to tear your eyes away from the Diplodocus (it's been standing there for the last 100 years and will still be there when you look back) and look up at the richly decorated ceiling with its frescoes of plants and animals.

To help guide visitors, the museum has been divided into four color-coded zones: red (which covers the planet itself), green (which features the original collection of the Natural History Museum), blue (where you'll find everything from tiny invertebrates to dinosaurs), and orange (which includes the wildlife garden and is only open between April and October).

The Miller-Urey Experiment

Although Darwin proposed the mechanism through which the variety of species seen at the Natural History Museum came about, he didn't offer an explanation of how the process got started. Nevertheless, he did write, in a letter, about the possibility of a "warm little pond" filled with a mixture of basic chemicals and exposed to light, heat, and electricity, creating proteins that could have been the building blocks of life.

In 1953, a pair of scientists at the University of Chicago, Stanley Miller and Harold Urey, decided to conduct an experiment bringing together the ingredients that were likely in the warm little pond, to see if the building blocks of life could have been created spontaneously. The Miller-Urey experiment was a success, and is the most celebrated attempt to recreate the conditions of the earliest days on Earth.

The experiment worked by using an enclosed apparatus in which water was heated to produce water vapor. The water vapor traveled to another part of the apparatus, where it joined the gases ammonia (NH_3), methane (CH_4), and hydrogen (H_2), all believed to have been present in the Earth's early atmosphere.

The gaseous mixture was then subjected to simulated lightning, with a spark created from an external source of electricity. The resulting mixture was cooled and sampled to test for complex molecules. The cooled liquid mixed back with the water to be heated, and allowed to cycle through the same lightning simulator over and over again. The complete equipment is shown in Figure 61-1.

Miller analyzed the resulting mixture after a week of operation, and found five amino acids: glycine, alpha-alanine, beta-alanine, aspartic acid, and alpha-amino-n-butyric acid. Since amino acids are the basic building blocks of proteins, and proteins are the building blocks of living organisms, the Miller-Urey experiment showed at least one way that the complex chemicals needed for life could have occurred.

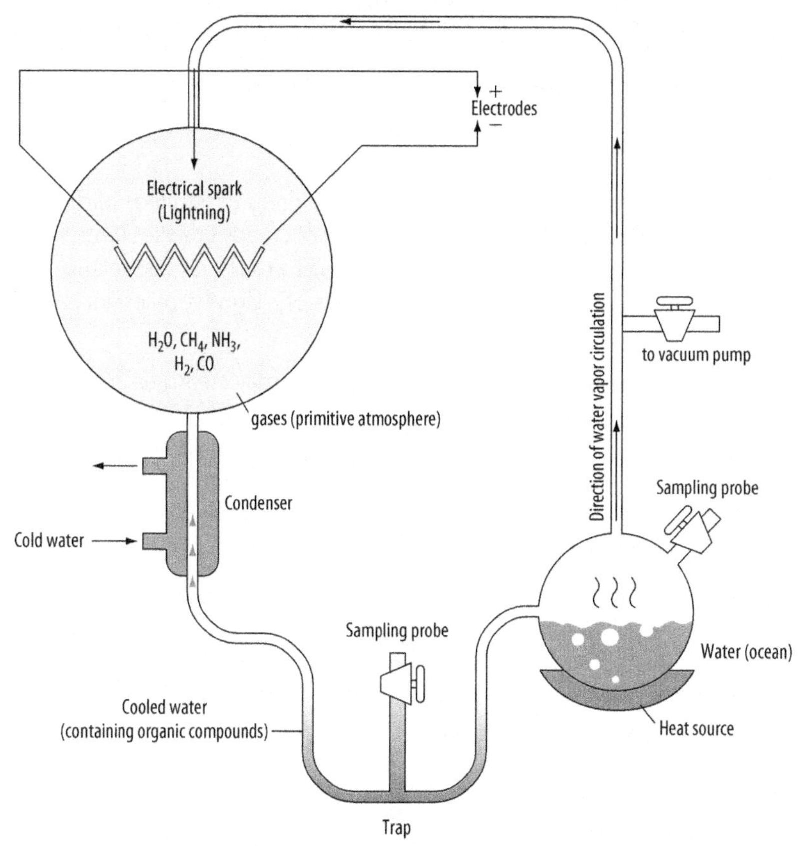

Figure 61-1. The Miller-Urey experiment

An analysis of vials of liquid from Miller's other experiments, using techniques available in 2008, showed that he had succeeded in creating a total of 22 different amino acids.

The Red Zone explains the process of evolution, and displays the museum's incredible collection of gems, rocks, and minerals. Earth science is covered by a section on earthquakes and volcanoes, another section on the forces that have forged the current surface of the planet, and a section on sustainable use of the Earth's resources.

The Green Zone has the museum's historic collection of (stuffed) birds, fossils from the British Isles, a spine-tingling collection of creepy crawlies (ants, moths, termites, crabs, and more), an exhibit dedicated to man's closest relatives (the primates), and the wonderfully displayed fossils of marine reptiles, including a pregnant ichthyosaur.

The Blue Zone is filled with the fossils of large and small creatures and includes an animatronic T. Rex, the fearsome-looking skeleton of a Triceratops, a large collection of stuffed mammals, a life-sized model of a blue whale, a special section on human biology, and a collection of marine invertebrates (such as coral, squid, and worms). The fish, amphibians, and reptiles collection is equally impressive—there are giant snakes, tortoises, and fish of all sizes.

If the main museum isn't exhausting enough, then book a tour of the Darwin Centre. This newly constructed building houses the museum's collection of bottled specimens. The most stunning item on display is an 8.62-meter-long giant squid that was caught off the coast of the Falkland Islands in 2004. It now sits in a specially constructed transparent tank, where it is preserved in a solution of formol-saline.

Practical Information

Everything you need to know about arranging a visit to the Natural History Museum can be found at *http://www.nhm.ac.uk/*. For visitors bringing children, there's also a Parents' Survival Guide with tips on what to see and how to avoid being exhausted by the size and scope of the museum (*http://www.nhm.ac.uk/visit-us/parents-survival-guide/*).

062

Poldhu, Cornwall, England

50° 1′ 46.30″ N, 5° 15′ 47.80″ W

The First Transatlantic Radio Transmission

In the southwest corner of the UK lies the county of Cornwall. It was there that the first transatlantic radio transmission was made, from the remote and windswept Poldhu Cove on December 12, 1901.

The rugged Lizard Peninsula, which contains Poldhu Cove and Poldhu Point, is the southernmost part of Great Britain and was chosen by Italian radio pioneer Guglielmo Marconi as a site for experimental radio communication with ships at sea. Geologists will enjoy the presence of green serpentine rock, and it was here that titanium was first discovered in 1791.

The wireless station built by Marconi and his company, Marconi's Wireless Telegraph Company, was not only the site of the first transatlantic transmission but also the place where, in 1910, the first SOS (dot dot dot, dash dash dash, dot dot dot in Morse code) was received in Britain from a ship at sea. And in 1962, the Goonhilly Downs (see Chapter 48) area of the peninsula was chosen as a ground station for the first transatlantic satellite, Telstar.

But 61 years before Telstar's ability to send television pictures, Marconi's message was just the letter S sent in Morse code (dot dot dot). It was transmitted continuously between 1500 and 1900 local time on December 12. Marconi (with assistant George Kemp) had traveled to St. John's, Newfoundland, to listen for the signal using a kite trailing a 155-meter-long wire as an antenna. He had originally planned to use antennas installed on the site, but these were destroyed in a storm. The same fate had befallen the antennas installed at Poldhu, and the first transatlantic transmission was achieved with makeshift equipment for transmission and reception.

Marconi's Transmitting Equipment

In early wireless experiments by the German physicist Heinrich Rudolf Hertz, there was no antenna at all: the entire transmission was achieved by a spark jumping across an air gap. This sent out a wide range of frequencies and could be detected using another spark gap connected to a coil of wire. You can still hear these types of transmissions by listening to the radio during a lightning storm: the cracks coming from the radio are "transmissions" from the lightning bolts.

To achieve more reliable transmission, Marconi and others used a simple method of tuning the output of the spark to certain frequencies and adding an antenna. In his earliest experiments, Marconi's transmitters were nothing more than an antenna plus a spark gap connected to the ground (i.e., the Earth).

But by 1901 he had improved the circuit to include the equivalent of an LC resonant circuit. That type of circuit consists of an inductor (a coil of wire wrapped around some ferromagnetic material, such as an iron bar) and a capacitor (called a condenser in Marconi's day, and typically consisting of a pair of metal plates separated by some non-conducting material called the dielectric). The capacitor C and inductor L are connected as shown in Figure 62-1.

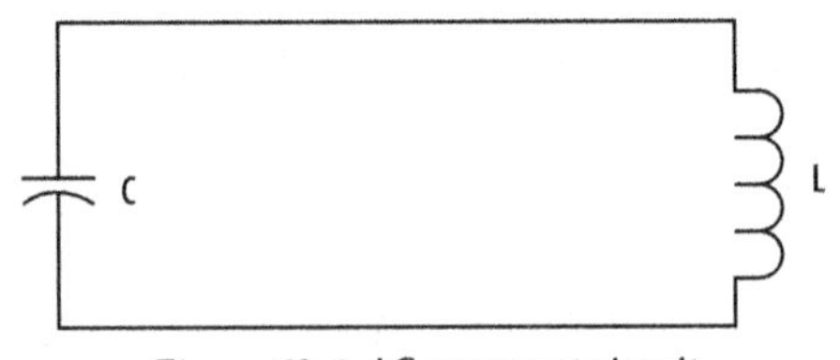

Figure 62-1. LC resonant circuit

When electricity is applied to this circuit, the capacitor charges up, acting like a battery, until it cannot hold further charge and starts to discharge through the inductor. The inductor turns the current into a magnetic field, since the current flowing through the coil induces a magnetic field in the bar it is wrapped around. Once the capacitor is empty (completely discharged), the inductor will start to recharge the capacitor using the energy stored in its magnetic field (the magnetic field now induces a current in the coil of wire). This circuit continues to oscillate between capacitor and inductor at a frequency determined by the two components.

Using such a circuit meant that Marconi could coarsely tune his transmitter. The circuit diagram in Figure 62-2 shows a simplified version of the transmitter he used in 1901.

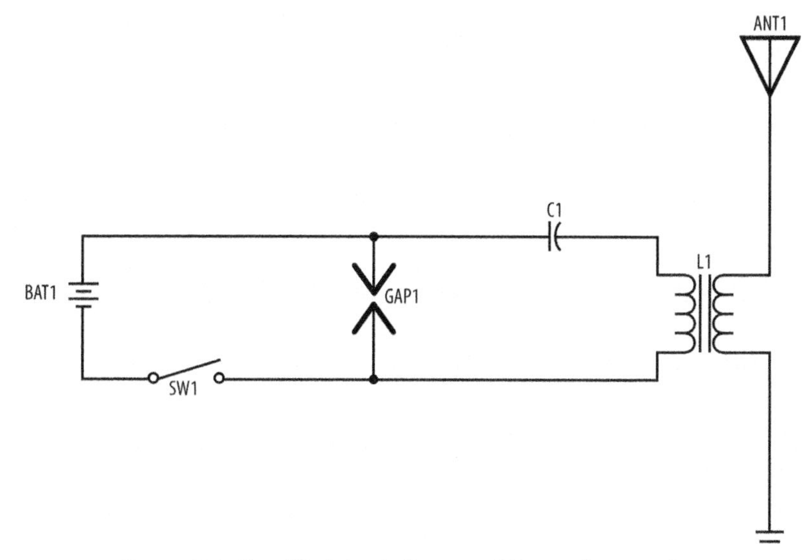

Figure 62-2. Simplified circuit diagram of Marconi's transmitter

Marconi's circuit worked by charging the spark gap until the voltage became large enough to cause a spark to jump across it. The jumping spark had the effect of closing the LC circuit consisting of capacitor C and inductor L and providing a jolt of power to it; the circuit then started to oscillate. L actually consisted of a pair of coils with one coil connected to the antenna. The magnetic field of L induced a current in the second coil, which passed into the antenna and created the radio signal.

The transmitting equipment consisted of an enormous condenser made of metal plates and capable of providing 1500V after being charged up by a steam-powered generator. This large voltage was applied to a "spark gap" (literally a pair of electrodes with an air gap between them). The condenser charged up until a spark was able to jump between the electrodes, creating the radio signal.

Listening through the background static, Kemp and Marconi claimed to hear the S signal transmitted from the UK some 2,100 miles away on three occasions (at 1230, 1330, and 1430 local time). A few months later, Marconi repeated the experiment, sailing west on the SS *Philadelphia* and recording signals from the Poldhu station up to 2,099 miles away. By the end of 1902, Marconi had sent transmissions in both directions between Poldhu and Nova Scotia. For his many wireless achievements, Marconi received the Nobel Prize in 1909.

Today the entire Poldhu area is owned by Britain's National Trust, and you can have a superb day out walking along the cliff tops and visiting the restored Lizard Wireless Station and the Marconi Centre. The 800-year-old town of Helston is nearby, and is an ideal starting point for wireless explorers. A suitable end to the day is a traditional cream tea—consisting of warm scones, clotted cream, and strawberry jam—in one of Helston's many tea rooms.

For the complete wireless experience, and because it's a good base for exploration, stay at the Housel Bay Hotel, where Marconi himself stayed in 1900 while scouting for a location to set up his wireless telegraphy station. The hotel is just 200 meters from the spot where Marconi first received signals from the Isle of Wight (some 186 miles away). The hotel is also close to Poldhu.

Practical Information

To visit the Poldhu site, check in with the National Trust; its website can be found at *http://www.nationaltrust.org.uk/main/w-vh/w-visits/w-findaplace/w-thelizardandkynancecove.htm.*

063

Porthcurno Telegraph Museum, Porthcurno, England

50° 2′ 34.8″ N, 5° 39′ 14.4″ W

Nerve Center of the Empire

In 1870 a submarine telegraph cable came ashore at Porthcurno in Cornwall, linking Britain to India via Portugal. By relaying between telegraph stations along the route, a message could be sent from London to Bombay in four minutes.

By 1885 there were more than 48,000 kilometers of submarine telegraph cables linking Britain to all parts of its Empire (Figure 63-1). By the start of the Second World War, there were over 560,000 kilometers of submarine cables. At the center of this vast network was the small village of Porthcurno, on the southwestern tip of England.

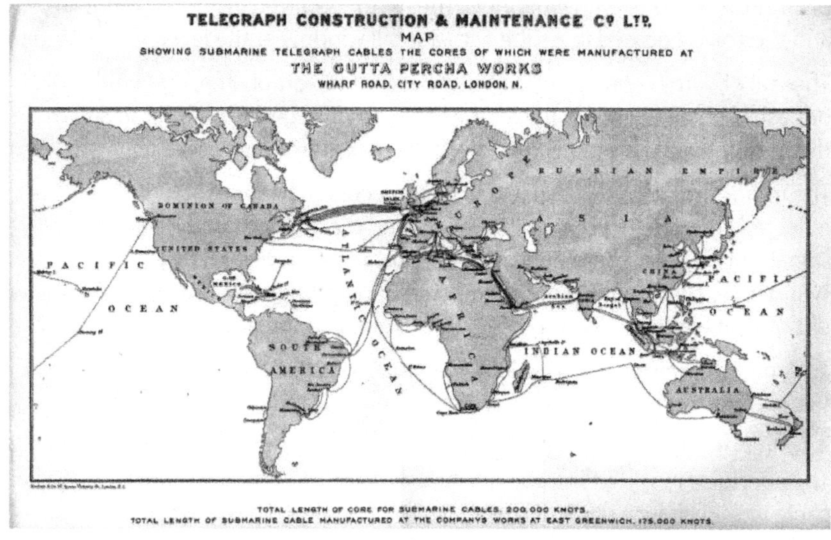

Figure 63-1. 1903 Map of undersea telegraph cables

The Mirror Galvanometer

A typical 19th-century transatlantic telegraph cable was powered by a 30-volt battery, and its total resistance was around 4 kΩ. A quick application of Ohm's Law (V = IR; i.e., voltage = current × resistance) shows that the current flowing in the cable was tiny—somewhere between 4 and 8 mA. Detecting such a tiny current was a major problem for long-distance telegraphy.

On short, overland runs of telegraph cable, it was possible to connect a simple sounder containing an electromagnet that would cause an audible click when a current was received. But with the tiny currents coming from an undersea cable, a much more sensitive instrument was needed. The mirror galvanometer, invented by Lord Kelvin, solved the problem (see Figure 63-2).

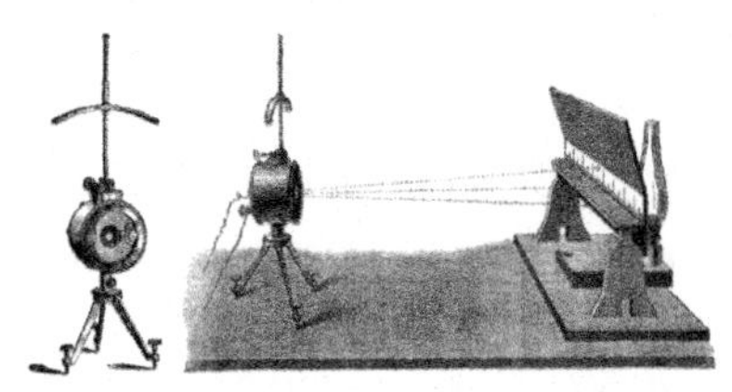

Figure 63-2. A mirror galvanometer

In a mirror galvanometer, a small mirror is suspended by a silk thread. On the back of the mirror are tiny magnets. The mirror is placed inside a coil of wire, and the coil is encased in a tube with a small window at the level of the mirror.

The coil of wire is connected to the incoming telegraph line, and when a dot or dash is received, the coil becomes an electromagnet that interacts with the magnets on the back of the mirror to make it rotate.

The mirror's rotation can be viewed by projecting a beam of light onto it. The reflected light falls on a graduated scale, and the strength of the telegraph signal can be seen. The polarity of the signal is also indicated by the mirror swinging to the left or right. The telegraph operator would watch the swinging dot of light—a swing in one direction was a dot, in the other direction a dash.

To center the mirror, a movable curved magnet was raised or lowered above the galvanometer.

Another problem affecting long-distance telegraph lines was capacitance. The conductor inside the cable formed a capacitor with the sea water; the cable would charge up, and then have to discharge. The effect of this capacitance was that telegraph signals were slowed and distorted, reducing the effective bandwidth.

Capacitance was eventually overcome by inserting repeaters into the cables, and by reversing the polarity of the current each time a Morse signal was sent.

During the Second World War, the Porthcurno station needed to be protected. Tunnels were dug into the granite rock, and the entire facility was moved underground. Today, fiber optic cables come ashore at Porthcurno and lead inland to a new connection point at Skewjack. Telegraphy stopped in 1970, but the underground station lives on as the Porthcurno Telegraph Museum.

Visits to the museum start with a talk and a demonstration of the telegraph equipment, and visitors are then free to walk through the tunnels. The museum is split into seven major areas.

The Maritime Room explains the process of laying submarine telegraph cables from cable ships. Cable laying was a major 19th-century challenge, and cables had to be waterproofed using a natural latex called Gutta Percha because plastics had yet to be invented.

The Main Showcase has a large collection of telegraphy equipment, including part of the original cable linking Cornwall to the Scilly Islands, parts of original Atlantic cables, and a Kelvin siphon recorder. The siphon recorder drew faint telegraph signals on a reel of paper tape by siphoning electrostatically charged ink from a bowl. The siphon tube swung back and forth with the incoming signal, without touching the paper. The paper was also charged and attracted the ink from the siphon tube.

The Instrument Room has a complete display of operational telegraphy equipment, including regeneration equipment that amplified weak telegraph signals so that they could be relayed to another station. Here you'll also find early semi-automatic telegraph equipment that used manually punched tape to send a signal to a siphon recorder.

The Cable Testing Room explains how breaks and faults in undersea cables were detected so that they could be repaired. Cable testing involved measuring the resistance and capacitance of a cable to determine the location of a break, and sending a cable ship out to repair it.

The Vintage Workshop is used by museum volunteers to maintain the museum's equipment. The Generator Room contains the diesel generator that was installed to provide back-up power in case the main current was cut off.

Finally, the Escape Stairs lead to the surface, and were originally built as an emergency exit in the case of an attack.

Practical Information

Visiting information can be found at *http://www.porthcurno.org.uk/*. Porthcurno is close to Poldhu, where Marconi conducted wireless experiments (see Chapter 62), and both sites can easily be visited in a day.

064

Royal College of Surgeons Hunterian Museum, London, England

51° 30′ 55″ N, 0° 6′ 57″ W

Surgery Laid Bare

The squeamish should avoid this museum—its centerpiece is 18th-century Scottish surgeon John Hunter's collection of over 3,000 anatomical and pathological specimens (Figure 64-1). Hunter's collection includes gems like P 1051 ("Colon from a patient with dysentery") and P 1056 ("Portion of a rectum with anus showing effects of a tuberculous infection"). There's also a wide range of animal specimens, human skulls, and human (adult and child) skeletons.

Figure 64-1. Specimens in the Hunterian Museum; courtesy of Joanna Ebenstein/www.astropop.com

In fact, the museum has such an extensive collection of pathological items that it's probably got an example of any body part you can think of, showing the after-effects of a disease. Serious students of the body can access information about every single item in the collection via a computer in the museum (or on the Web, from the comfort of their own homes).

The Lymphatic System

John Hunter, despite having no formal training as a doctor, made significant contributions to the medical field through the study of teeth, inflammation, venereal diseases (he is rumored to have infected himself with gonorrhea so that he could study it), the digestive system, the placenta, and the lymphatic system.

As blood flows around the body, its fluid leaks through the walls of the capillaries into the surrounding tissue. This leakage provides oxygen and nutrients to the body and is caused by the pressure created by the heart. The fluid, called interstitial fluid because it resides between the cells, also contains the waste products of the cells.

Most of the fluid returns to the capillary by osmosis. As the water leaves the capillary, the remaining fluid becomes more concentrated. Eventually this results in a pressure imbalance (of osmotic pressure caused by the imbalance of liquid concentration), and the water returns to the capillary.

Nevertheless, some of the fluid remains in the tissue and must then be drained away. The drainage is achieved by the lymphatic system, which removes the interstitial fluid and filters it. The lymphatic system consists of tubular vessels similar to the blood vessels throughout the body (Figure 64-2). There are lymphatic capillaries almost everywhere, which connect to a network of ever-larger vessels.

The fluid in the lymphatic vessels (the *lymph*) is not pumped by the heart, but by a process called peristalsis. Peristalsis is the contraction of smooth muscles surrounding the lymphatic vessels that drives the lymph along. The vessels contain valves that prevent the lymph from flowing in the wrong direction.

The lymph finally reaches a lymph node, where it is filtered to remove materials foreign to the body such as micro-organisms. Each bean-sized lymph node captures these foreign organisms and exposes them to cells from the body's immune system for destruction.

Eventually the lymph makes it way back to two major vessels—the right lymphatic duct (through which lymph from the righthand side of the body and head is drained) and the thoracic duct (which drains the rest from the lefthand side). These two ducts reconnect with the cardiovascular system in the subclavian veins behind the shoulder blades, where the filtered lymph reenters the bloodstream.

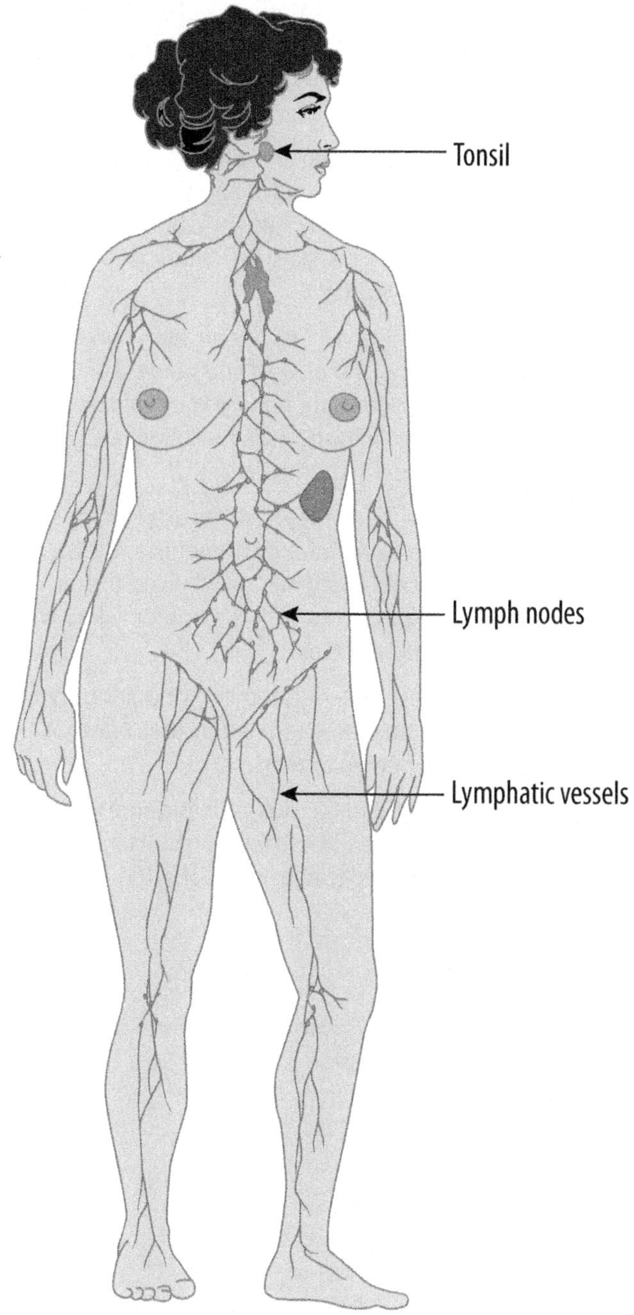

Figure 64-2. The lymphatic system

Not on display, but hidden away in the museum collection and accessible by special request, is half a human brain—that of 19th-century computing pioneer Charles Babbage, no less (see page 294). However, plenty of other brains (human and otherwise) are on display.

If the museum's Crystal Gallery specimen collection doesn't send you running for the door, there's lots more to see. The Silver and Steel collection is an entire gallery filled with the instruments used by surgeons, both modern and ancient. And the Science of Surgery gallery explains the techniques and technologies used by surgeons in the past and today.

Among the interesting instruments is a 1957 heart-lung bypass machine that is used to take over the functions of the heart and lungs during heart surgery. There's a display of apparatus for anesthesia (using ether, chloroform, and nitrous oxide), early equipment for infusing saline solution, and a big collection of scalpels and other sharp-edged instruments.

The museum holds a large collection of instruments used by Joseph Lister, who was best known for insisting on sterilizing instruments and cleaning wounds. On display are his antiseptic spray machines, which filled the air with carbolic acid. Lister used carbolic acid (or phenol, as it is commonly known) to sterilize instruments and wounds as well as the air.

A good way to get the most out of the museum is by inquiring about their talks and tours. Each Wednesday there's a free tour led by the curator at 1 p.m. There are also regular free talks by expert volunteers.

William Hunter, brother of John Hunter, was a physician and anatomist who worked in Scotland and specialized in obstetrics. The Hunterian Museum in Glasgow (see Chapter 73) commemorates his life and the lives of other great Scots.

Practical Information

Visiting information for the Royal College of Surgeons is available at *http://www.rcseng.ac.uk/museums/*. Admission is free.

065

Royal Gunpowder Mills, Waltham Abbey, England

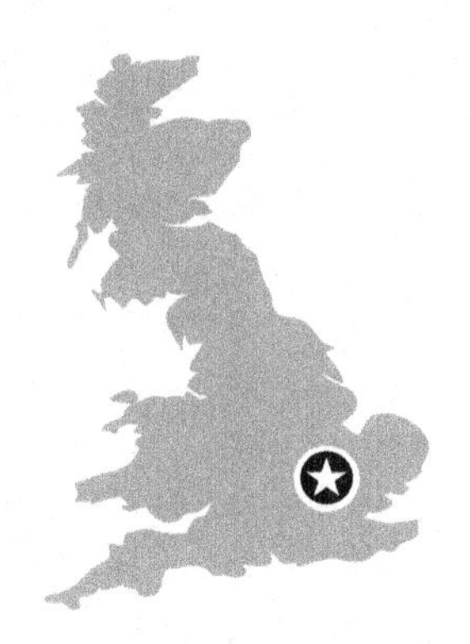

51° 41′ 34.8″ N, 0° 0′ 32.4″ W

The Royal Gunpowder Mills are set in 70 hectares of countryside. Such a bucolic location might seem unusual for a factory, but if your factory has a habit of exploding, it's best to have it spread out over a wide area to start with. The Waltham Abbey mills have blown up at least 12 times in their 300-year history.

Luckily for visitors, the mills are no longer likely to explode, as they stopped operating in 1991. The site has been open to the public since 2001.

The site's history goes back to 17th century, when it started life as a mill creating gunpowder by grinding up potassium nitrate, charcoal, and sulphur into black powder. As technology progressed, the site was used to create guncotton and Cordite (see sidebar) and then TNT and RDX (the power behind the well-known plastic explosive C-4). After the end of the Second World War, when the demand for explosives was drastically reduced, the site was used to do research and development of new explosive compounds and propellants for rockets.

There are multiple exhibitions on the site, and beautiful woody parkland to enjoy if weather permits. A number of the exhibitions are purely historic (such as the 1940s Experience, the Gunpowder Plot exhibit, and the Farewell to Arms exhibit of firearms), but there are three items of scientific interest.

First is an interactive exhibition that explains the process of making gunpowder and the differences between propellants (used to fire bullets) and explosives. There's also a rockets exhibit, which shows the connection between explosives and rocket fuel. It has a number of rockets on display and contains information about the rocket fuels used. The exhibit is a good place to explore the British contribution to rocket and missile science.

Gunpowder, Guncotton, and Cordite

Gunpowder was the first explosive, and was first discovered in the 9th century in China. It then made its way through the Arab world and into Europe. It consists of a mixture of carbon, sulphur, and potassium nitrate.

The three components of gunpowder perform three different jobs. Potassium nitrate (KNO_3, which is also called saltpeter) provides a supply of oxygen; carbon (in the form of charcoal) fuels the explosion; and sulphur acts as a catalyst.

When the mixture is ignited, the potassium nitrate releases oxygen, which allows the charcoal to burn vigorously, producing carbon dioxide. The carbon dioxide gas is produced so rapidly that in a confined space, the ignition of gunpowder forms an explosion (which can be used for driving a bullet down the barrel of a gun, for example).

A typical gunpowder mixture consists of 75% potassium nitrate, 15% charcoal, and 10% sulphur. When ignited, the mixture undergoes a complex reaction that creates heat and pressure and converts the gunpowder to a mixture of potassium carbonate, potassium sulphate, and nitrogen and carbon dioxide (Equation 65-1).

$$10KNO_3 + 8C + 3S \rightarrow 2K_2CO_3 + 3K_2SO_4 + 6CO_2 + 5N_2$$

Equation 65-1. The gunpowder reaction

Gunpowder gets its reputation as an explosive from the speed with which it burns—so quickly that it can propel a bullet. The actual speed of burning is determined by how finely the powder is ground; the finer the powder, the faster the burning. Gunpowder can also be used as a rocket propellant or even a fuse; fireworks use gunpowder (sometimes mixed with chalk to slow the burning) to propel rockets, and a thin hollow tube filled with gunpowder makes an effective fuse.

But gunpowder is not a high-explosive, and is completely ruined it if gets wet. But it's stable—dropping or rubbing gunpowder won't set it off.

In contrast, guncotton, or nitrocellulose, has a tendency to explode all by itself. It was first discovered when a spill of nitric acid (HNO_3) was cleaned up with a cotton cloth. When the cloth had dried, it suddenly exploded. The cellulose ($C_6H_{10}O_5$) in the cotton had reacted with the nitric acid to create nitrocellulose (and some water)—see Equation 65-2.

$$3HNO_3 + C_6H_{10}O_5 \rightarrow C_6H_7(NO_2)_3O_5 + 3H_2O$$

Equation 65-2. Guncotton production reaction

Nitrocellulose is a much more powerful explosive than gunpowder, but it is relatively unstable, and was mostly used for motion-picture film stock by treating it with camphor to make celluloid. Celluloid is still very flammable, but doesn't explode.

Nevertheless, guncotton did find an explosive use when used in conjunction with the high-explosive nitroglycerine. Alfred Nobel manufactured nitroglycerine (which is made from nitric acid, sulphuric acid, and glycerol) for use as a liquid high-explosive, and although it did work, it proved to be very unstable (liquid nitroglycerine can explode spontaneously if knocked). Nobel went on to create dynamite by mixing nitroglycerine with diatomaceous earth (a soft, chalky rock that crumbles into a powder mostly made of silica). This made nitroglycerine relatively safe to use and transport.

The British military used nitroglycerine in conjunction with guncotton to make a military propellant for bullets and shells that had more power than simple gunpowder and was smokeless (unlike gunpowder, which produces characteristic white smoke when ignited). The mixture, known as Cordite, consisted of nitroglycerine, guncotton, and vaseline that was extruded into long cords and then packed into shells.

Finally, there are large exhibits that include the canal system used to transport explosives (canals have the advantage of being calm, and canal boats are slow moving); the fire alarm system for the mills (which used Morse code driven by a clockwork mechanism to signal the location of any fire on the large site); and the narrow gauge railway system that was used for transport within the site.

The mills are about a one-hour drive from Bletchley Park (see Chapter 40), making it possible to enjoy a full day's worth of scientific sightseeing.

Practical Information

Information on the Royal Gunpowder Mills can be found on its website: *http://www.royalgunpowdermills.com/*.

066

Sackville Street Gardens, Manchester, England

53° 28′ 36″ N, 2° 14′ 9″ W

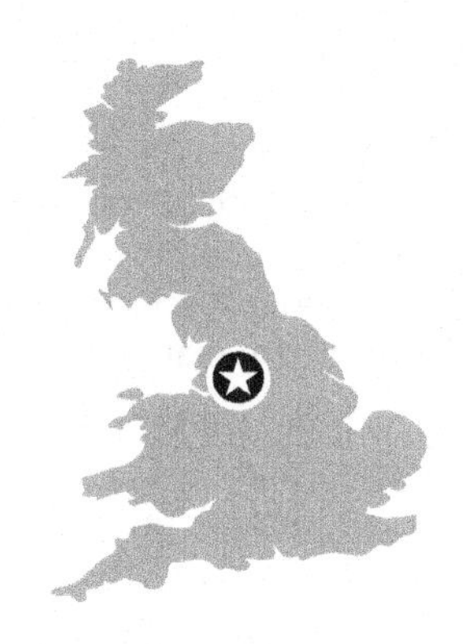

"IEKYF ROMSI ADXUO KVKZC GUBJ"

In this small green park, close to the University of Manchester Institute of Science and Technology, sits a bronze statue of the British thinker Alan Turing.

Turing is sitting on a bronze park bench holding an apple in one hand, his shirt collar undone and his tie loosened. At his feet sits a plaque reading "Father of Computer Science, Mathematician, Logician, Wartime Codebreaker, Victim of Prejudice." If Turing is known at all, it is because of his work on the decryption of the Nazi Enigma cipher during the Second World War (see Chapter 40). But Turing's legacy is far greater than that—he is responsible for large contributions to computer science as we know it.

Turing invented a theoretical computer known today as a Turing Machine. A Turing Machine is a very simple computer with four important components. The first is a tape divided into squares (called cells); the Turing Machine can store one symbol from its alphabet of characters in each square. Symbols are written and read from the tape using the Turing Machine's head; either the head or the tape moves, so the head can read and write from any cell (Figure 66-1). The Turing Machine uses a table of instructions to decide what to do—it can perform actions like moving the head one cell, reading the symbol under the head, writing a new symbol, or erasing a symbol from the tape. Finally, there's a "state register" that tells the machine where it is in the table of instructions.

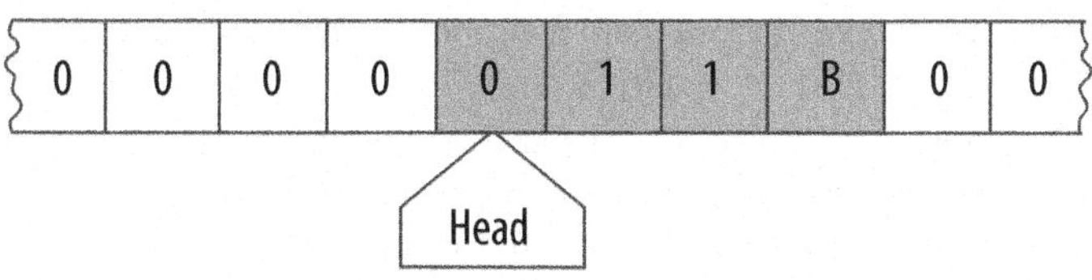

Figure 66-1. A Turing Machine's tape and head

The Halting Problem

Turing's Halting Problem asks, "Is it possible to write a computer program that determines whether another computer program will run forever?" Turing showed that the answer is no—it is not possible. To prove this, Turing used a technique common in mathematics called diagonalization (see the sidebar in Chapter 47).

The proof starts by imagining every possible computer program and numbering them 1, 2, 3, and so on. Using this numbering, it's possible to refer to any computer program by its number. The same trick is done for the input to the program: all possible inputs are numbered 1, 2, 3, and so on.

The Halting Problem itself can then be written as a mathematical function that is 1 if a program halts for a given input, and 0 if it does not (Equation 66-1).

The Halting Problem can then be rephrased as, "Is there a program to compute the halting function $h(p,i)$?" To establish that, Turing came up with a second mathematical function that is undefined (which is the same thing as nonhalting for a computer program) for some values, and 0 for others (Equation 66-2).

$$h(p, i) = \begin{cases} 1, \text{ if program } p \text{ halts with input } i \\ 0, \text{ if not} \end{cases}$$

Equation 66-1. The Halting Function

$$g(i) = \begin{cases} 0, \text{ if } h(i,i) = 0 \\ \text{undefined, if not} \end{cases}$$

Equation 66-2. The Diagonal Function

$g(i)$ only has one parameter, i, which is just a number. It is used by $h(i,i)$ to find both the program numbered i and the input numbered i. And $g(i)$ is 0 if the program numbered i does not halt with input i; otherwise, $g(i)$ is undefined. Now for the mind-bending bit. Suppose that there is a computer program for $h(p,i)$. It's fairly easy to see two things:

1. *There must be a program for $g(i)$ because it is defined in terms of $h(p,i)$.*
2. $g(i)$ is computed by the program number j for some number j.

So what happens if you try to calculate $g(j)$? What happens if we feed $g(j)$ to itself?

1. If $g(j)$ is 0, then it means that $h(j,j)$ is 0, which means that program j does not halt with input j. But program j is the program for $g(j)$, which we've just said is 0. This is contradictory, and hence $g(j)$ cannot ever be 0.
2. If $g(j)$ is undefined, then it means that $h(j,j)$ is 1, which means that program j does halt with input j. But program j is the program for $g(j)$, which we've just said is undefined (doesn't halt). Another contradiction.

This reasoning leads to contradictions and means that the original assumption was incorrect: there is no program to compute $h(p,i)$. The Halting Problem was significant because it showed that Turing's Machines could not be used to solve important unsolved mathematical problems.

It also tells us that no computer is going to be able to predict whether a program will stop responding. So at least we humans aren't totally out of a job when it comes to debugging.

The tape is like the memory of a modern computer; the head has been replaced by electronic access to the memory, and the table of instructions is just the program. Turing described this Turing Machine while he was at Cambridge University in the late 1930s, and it remains the theoretical design of the computers we use today.

Turing came to invent this computer because of his interest in a mathematical puzzle called the Entscheidungsproblem (or Decision Problem). In the early 20th century (and even earlier in the minds of some great mathematicians like Leibniz), many mathematicians were interested in the possibility of creating a machine to do mathematics. The Decision Problem asked whether it would be possible to find an algorithm that could be fed some mathematical problem (written in a suitable mathematical language) and output either True or False. If such an algorithm did exist, then difficult unsolved mathematical problems could be fed to it to determine whether they were true or not. Turing's contribution was to show that the Halting Problem could not be solved on a Turing Machine (see sidebar).

After his codebreaking work during the Second World War, Turing worked on some of the first electronic computers including ACE (the UK's first computer with a program stored in it) and later on the Manchester Mark I at the University of Manchester. In 1950 Turing wrote about artificial intelligence and proposed a test for determining whether a computer could really think. What we now call the Turing Test involves two people, A and B, and a computer, C. Suppose that A is able to communicate with B (the person) and C (the computer) in writing only. A's job is to determine which of B and C is human; if he cannot, then C is considered to be thinking.

The park is also close to Canal Street (running alongside the Rochdale Canal), which has been the center of Manchester's gay community since the 1960s. Alan Turing was gay, and was prosecuted for "gross indecency" in 1952. To avoid prison, Turing accepted treatment for homosexuality in the form of estrogen injections, but in doing so he lost his security clearance and was no longer allowed to work on cryptography for the British government. Two years later, Turing was found dead in his home. He had apparently taken his own life by eating an apple laced with cyanide.

One Turing mystery remains. Set into the bench is the string of characters IEKYF ROMSI ADXUO KVKZC GUBJ, which is alleged to be the encryption of FOUNDER OF COMPUTER SCIENCE using an Enigma machine. But professional codebreakers have pointed out that that would mean the U in COMPUTER had been encrypted as the U in ADXUO. That's impossible, since Enigma would never encrypt a letter as itself. The exact decoding is still unknown.

Practical Information

The park is easy to find because of its proximity to Canal Street (and Sackville Street), the University, and the canal. (Canal Street itself is a lively location for lunch or a drink.) Scientific visitors may want to coordinate their visits with an evening at Café Scientifique, where the University of Manchester hosts public talks and demonstrations of science in the café of the Manchester Museum; for details, see *http://www.cafescientifique.manchester.ac.uk/*.

067

Sound Mirrors, Dungeness, England

50° 57′ 22″ N, 0° 57′ 14″ E

Echoes from the Sky

Dungeness is a peculiar bit of British landscape: it looks like there's almost nothing there. It's a vast area of shingle, swept by salty sea winds, and yet it has a very dry microclimate. It was designated as a National Nature Reserve and is full of wildlife, including the largest medicinal leeches in Britain.

But this empty landscape also has a number of non-living attractions: there are two nuclear power stations, two lighthouses, a lifeboat station, and a military firing range. But the most fascinating of all are three concrete acoustic mirrors built in the 1920s and 1930s as a simple early-warning system against an airborne attack from across the English Channel.

This desolate place was chosen as the site for the mirrors because there was little noise, and the seaward-facing mirrors could listen for an attack without interference. Sound mirrors were built in other places around the UK, but the three mirrors at Dungeness are well preserved and show three different mirror forms: a pair of dishes and a 60-meter-long curved wall.

The mirrors were built as a response to the bombing of Britain during the First World War, when Germany had used Zeppelin air ships and conventional aircraft to bomb various cities around Britain. The mirrors were designed to detect approaching enemy aircraft by listening for their engine noise.

Radar had not yet been invented, and early warning was initially achieved by people listening for aircraft approaching the coast. However, by the time the aircraft were audible to the human ear, it was too late for an effective early warning, and the concrete sound mirrors were proposed as a solution because they could detect approaching aircraft from further away and give an approximate bearing of the sound. Once detected, the early-warning stations could warn the bombers' targets by telephone.

The first mirror (shown in the middle of the photograph in Figure 67-1) was built in 1928. It measures 6 meters in diameter and was operated by a person standing in front of it and listening for the sound of aircraft using a large stethoscope. The operator would move the stethoscope to find the loudest sound, and would then read the bearing of the sound from a scale on the mirror itself.

Figure 67-1. Dungeness sound mirrors; courtesy of Dauvit Alexander (the justified sinner)

The second mirror is 9 meters in diameter and points toward the sky. The operator of this dish was spared from standing for hours exposed to the elements, and instead got to sit in a specially constructed booth. The operator still moved the stethoscope around to find the sound and its bearing, but used a handle and foot-pedal combination for horizontal and vertical movement.

The large mirror wall dispensed with the stethoscope altogether, and instead used 20 fixed microphones positioned in front of the wall and fed to an operator. The small mirrors were capable of detecting sounds effectively at around 370 Hz; the much larger mirror wall was designed to pick up the low frequency and large wavelength sounds from aircraft.

When their powers were combined, the large mirror could give an indication of the direction of an approaching aircraft when it was very distant, and the dishes could be used to pinpoint more accurate locations once the general direction was known.

The mirrors were never used in wartime, however, because they were doomed by two inventions: faster aircraft (which made the early warning possible with the dishes much too short) and radar (which was able to quickly detect aircraft at a much greater distance than was possible by listening for engine noise).

Sound Mirror Operation

The sound mirrors operated by simply reflecting sounds hitting the curved surface back toward the operator. By moving the listening device (a stethoscope with a horn attached), the operator could find the point where the sound was loudest. This point corresponded to the focal point of the reflected sound waves (see Figure 67-2).

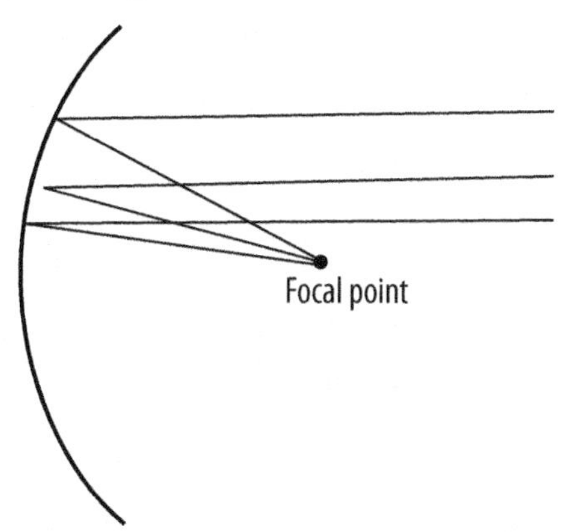

Figure 67-2. Sound mirror operation

Sound arriving at the mirror was reflected by the curved surface; parallel lines of sound arriving from the same distant aircraft were reflected at slightly different positions on the dish at slightly different angles. These reflected waves converged at a focal point; the operator's job was to find that focal point. Knowing the shape of the dish and the simple rule that the angle of incidence is equal to the angle of reflection, the focal point could be used to determine the direction of the sound.

The sound mirrors had a major drawback compared to radar: they couldn't judge the distance to the aircraft. In radar, the radio waves are sent from the radar station and bounce off the aircraft; by timing this reflection, it's possible to get the distance of the aircraft in addition to the bearing.

Sound mirrors relied on sound waves coming from the aircraft, so no timing was possible. Therefore, the sound mirror could not distinguish two aircraft at the same bearing but at different distances.

Practical Information

General information about the National Nature Reserve is available at *http://www.dungeness-nnr.co.uk/*. A book called *Echoes from the Sky* by Richard Scarth details the history of the dishes. The book is unfortunately out of print, but Dr. Scarth himself provides guided walks to the mirrors; details can be found at *http://www.rmcp.co.uk/*. The walk is the best way to see the dishes, and the only way to see them close up.

If you're driving, don't go straight to Dungeness itself—the mirrors are actually located about 5 kilometers to the north.

068

SS Great Britain, Bristol, England

51° 26′ 56.99″ N, 2° 36′ 30.12″ W

Iron Hulled and Propeller Powered

The ship SS *Great Britain* (Figure 68-1) was launched in 1843 and was, at the time, the largest ship ever built. Designed primarily by Isambard Kingdom Brunel (see Chapter 70), she had two major innovations: the hull was made of iron, and the ship was powered by a propeller instead of paddle wheels.

Figure 68-1. The SS Great Britain today; courtesy of the SS Great Britain Trust/Mandy Reynolds

Cavitation and Supercavitation

Brunel's propeller was unlikely to have been spinning fast enough to suffer the damage that plagues marine propellers today—cavitation. Cavitation causes pits and erosion of propellers, turbines, and pumps. Anywhere a liquid is accelerated to high speed, cavitation can be a problem.

All liquids boil at some combination of temperature and pressure—increasing the temperature of water can cause boiling, as can decreasing its pressure. When a propeller moves through sea water, it can cause some of the water to boil (i.e., to turn to water vapor) because of localized areas of low water pressure. The low pressure comes about because the moving propeller blade separates the water through which it is rotating (a little like on an aircraft wing), which can create a pressure difference.

This underwater boiling causes small bubbles (cavities) to be formed, typically on the surface of the propeller. But the low pressure inside the bubble is unable to resist the higher surrounding water pressure (which may increase as the propeller moves), and the bubbles collapse with an inrush of water. The collapse damages the surface of the propeller blade because of the sudden change in pressure. This process is called cavitation.

Cavitation is a major problem for propellers, which must be designed to either avoid cavitation completely, or to avoid having the bubbles implode on the propeller surface. And cavitation is noisy—in a submarine, the telltale sound of the collapsing bubbles on the propellers could give away the craft's location.

Some watercraft, most notably torpedoes, actually exploit the cavitation effect to move very fast underwater by essentially flying through water inside a bubble of vapor. If a torpedo is correctly shaped, its nose will create cavitation vapor as it moves through the water. For a small projectile (such as an underwater bullet), this movement may create enough cavitation all around the projectile that it flies through the water without actually coming into contact with it. This process is called supercavitation (see Figure 68-2).

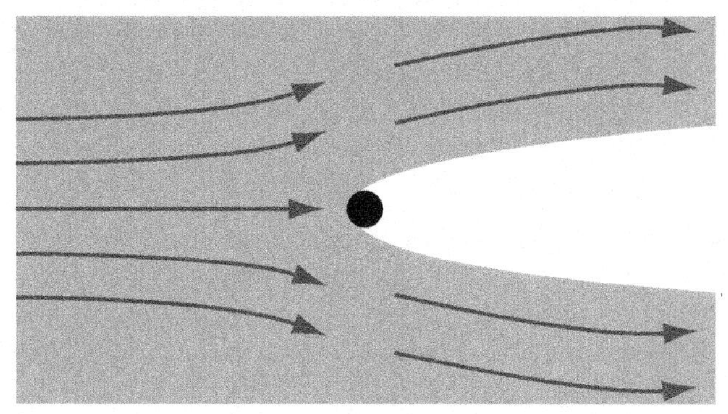

Figure 68-2. Supercavitation

The Russian Shkval torpedo (Figure 68-3) exploits supercavitation to fly underwater at very high speed (rumored to be as high as 200 knots). The torpedo has a special nose cone that creates cavitation vapor at the nose. It also has a rocket motor for propulsion, and some of the rocket gases are fed to the nose to expand the cavitation bubble all the way along the torpedo for minimum resistance underwater.

Figure 68-3. Russian Shkval torpedo

The SS *Great Britain* sailed from 1843 until 1886, crossing the Atlantic in record time on her maiden voyage in 1845 and later serving as an emigration ship taking people from Great Britain to Australia. In 1886, she was abandoned in the Falkland Islands after an on-board fire, and was used to store coal until she was scuttled in 1937.

In 1970, the ship was refloated on a pontoon and returned to Bristol, to the very dry dock where she had been built. After extensive work to repair, preserve, and restore the ship, the SS *Great Britain* is now a superb tourist attraction, with a wonderful illusion that keeps the ship dry yet makes it appear to be afloat.

The iron hull corroded badly in the Falkland Islands, and would have continued to corrode in the high humidity found in Bristol (especially in the old, slightly leaky dry dock). To prevent further degradation, the dry dock has been covered over, and the lower part of the ship is kept desert-dry by large dehumidifiers. The dock cover is filled with water, so that to visitors above the water line the ship appears to be afloat.

At the same time, visitors can enter the dehumidified section and view the ship from below. So visiting the SS *Great Britain* means seeing the entire ship—it's a unique experience, covering the interior and decks, the exterior from above and below water, the propeller, and the engines.

The ship's original engine was destroyed, so the restorers built a full-scale replica in its place. The replica doesn't steam, but it does move, giving an idea of the power of the ship's original 1,000-horsepower, four-cylinder, V-formation engine. Each cylinder is 2.2 meters across, and originally drove a crankshaft that was the largest forged object in the world in 1843.

Even the ship's boiler was grandiose. It contained 180 tonnes of sea water that would be boiled to power the engine. Heat came from 1,000 tonnes of coal aboard the ship.

Today, free audio tours are available with a range of different subjects; the most interesting for the science-minded is the "Maritime Archaeologist" tour. The interior of the ship has been staged to give an idea of life aboard a luxury ocean liner in 1843, but for technologists the real excitement lies in the engine room and below the water line.

Isambard Kingdom Brunel also built the Clifton Suspension Bridge in Bristol, which is also well worth a visit. Guided tours are available.

Practical Information

The SS *Great Britain* has its own website at *http://www.ssgreatbritain.org/*, as does the Clifton Suspension Bridge at *http://www.clifton-suspension-bridge.org.uk/*.

069

The Apple Tree, Trinity College, Cambridge, England

52° 12′ 26.22″ N, 0° 7′ 3.98″ E

The System of the World

Sir Isaac Newton, arguably one of the greatest scientists of all time, lived and worked at Trinity College at the University of Cambridge for much of his life. Just outside Newton's old rooms is an apple tree believed to be descended from a tree at Newton's home in Woolsthorpe, Lincolnshire: the tree that inspired Newton's theory of gravitation (Figure 69-1).

Figure 69-1. Newton's apple tree at Trinity College, Cambridge; courtesy of Fred Parkins

Newton is not the only famous scientific alumnus of Trinity College; Trinity was the one-time home of Lord Rayleigh, J. J. Thomson, Ernest Rutherford, Niels Bohr, Sir William Henry Bragg, and 21 other Nobel Prize winners. Newton was Cambridge University's Lucasian Professor of Mathematics (a post currently held by Stephen Hawking and previously held by Charles Babbage and Paul Dirac, among others).

Weight Versus Mass and Apollo 15

The mass and weight of an object are easily confused. The mass of an apple, measured in kilograms, is a measure of its inertia: it's a measure of how hard it would be to move it. Its weight, on the other hand, is entirely dependent on gravity: it's a measure of the gravitational force pulling on the apple.

On Earth, an apple might have a mass of 100 grams; on the Moon, the apple would have the same mass, but would weigh much less. The Moon has 1/6 of the gravity of the Earth, so the apple would weigh 1/6 as much as it does on Earth.

In scientific terms, mass is measured in kilograms and weight in newtons. One newton is the amount of force needed to accelerate a mass of one kilogram at a rate of one meter per second per second. On Earth, where the force of gravity is about 9.81 ms^{-2}, an object has a weight of one newton if its mass is about 102 grams. So the apple Newton saw fall to the ground could well have had a weight of one newton!

On the Moon, the apple weighs about 1/6 of a newton, but it still has the same mass. So an astronaut trying to throw an apple on the Earth or the Moon will use the same amount of energy.

In 1971, the astronaut Commander David Scott performed an experiment on the Moon that demonstrated that objects of different masses undergo the same amount of acceleration. Live on television, Scott dropped a 1.3-kilogram geological hammer and a 30-gram falcon's feather from the same height. They hit the lunar surface at the same moment (see Figure 69-2).

This showed that Galileo Galilei's theory—that objects undergo the same acceleration in a vacuum (and that it's the presence of air that slows down the feather on Earth)—was correct. And Newton later showed that the acceleration was due to gravity through his law of universal gravitation.

Although the feather and the hammer remained on the Moon, visitors to the Science Museum in London (see Chapter 77) can see the apparatus used in 1761 to demonstrate to King George III that Galileo was correct. In this apparatus, a feather and a guinea coin are suspended in a tube that is evacuated; when released, they fall at the same speed.

Figure 69-2. The hammer and feather on the Moon's surface; courtesy of NASA

Newton's great contribution, published in 1687 in his *Philosophiæ Naturalis Principia Mathematica*, is the law of universal gravitation and his three laws of motion. These showed that Kepler's laws about the motion of planets—that the Earth did revolve around the Sun—were correct, and gave the reason: gravity.

The law of universal gravitation is that two bodies (such as planets, but also apples) exert a gravitational force on each other. The force is proportional to the product of the masses of the bodies (the larger the body, the more force is exerted) and inversely proportional to the square of the distance between them.

This inverse-square law shows that the further the bodies are apart, the weaker the force, and that doubling the distance between two bodies results in a force 1/4 of the strength. Inverse-square laws abound: electromagnetic radiation (including light and ordinary radio transmissions) diminishes with the square of the distance between the source and the receiver; the force of attraction between two electrically charged particles does the same; and even the intensity of sounds diminishes with the square of the distance.

Newton's three laws of motion govern how forces act on bodies:

1. If no force is applied to a body, it either stays where it was, or keeps moving at a constant speed.
2. If a force is applied to a body, then its acceleration is proportional to the force, and inversely proportional to the body's mass. The acceleration occurs in the direction of the force.
3. If a body A exerts a force on body B, then the opposite is true; in other words, every action has an equal and opposite reaction.

Newton's own copy of *Philosophiæ Naturalis Principia Mathematica* is kept in the Trinity College library, as well as many other books that he wrote or owned.

A visit to Trinity starts with Newton's apple tree just outside the Great Gate. Once inside, you're standing in the Great Court, the largest courtyard of any Oxford or Cambridge college. Many of the university's colleges are built around courtyards, and this one is well known to filmgoers as the setting for the race in *Chariots of Fire*. The college clock used in the film (and in real life) to time races around the courtyard has been in use since the 1700s.

Surrounding the courtyard you'll see the college chapel, the Master's Lodge, and the Hall (where students eat). In the middle is a fountain that until recently was fed from a spring; the fountain is believed to have been used by students for washing. Standing in this famous courtyard, it's not hard to imagine Newton working here: little seems to have changed.

In the second courtyard you'll find the Wren Library, designed by Sir Christopher Wren. The library is open to the public and includes a bust of Newton. Behind the library is the River Cam, where students and visitors alike can try their hand at punting: navigating a flat-bottomed boat by pushing it with a 5-meter-long pole. To propel the punt, stand at the back of the boat, drop the pole straight down into the water (under the force of gravity, or with a little push), and push on the pole to drive the punt forward. Once you've got the punt moving, let the pole float to the surface and use it as a rudder until the next push is needed. The greatest danger in punting is pushing too hard, getting the pole stuck in the muddy river bottom, and realizing that you can either fall in trying to pull it out, or let the pole go. (Oars are usually provided for those who choose to not fall in.)

If you do make it down the river, you'll come to Queens' College and a pretty wooden bridge spanning the Cam. Don't believe the stories about this being a "mathematical bridge" built by Newton. It was built after his death, has been rebuilt twice since, and is called simply the Wooden Bridge.

Practical Information

Trinity College is open to the public. For information on reaching the college and a proposed tour of the buildings and grounds, see its website at *http://www.trin.cam.ac.uk/*.

070

The Brunel Museum, London, England

51° 30′ 5.76″ N, 0° 3′ 10.8″ W

The Thames Tunnel

In 1843, the first tunnel passing under a body of water opened beneath the Thames River, between Rotherhithe and Wapping in London. The tunnel, which is known simply as the Thames Tunnel, was built by Marc Isambard Brunel and his son Isambard Kingdom Brunel. It opened as a pedestrian tunnel featuring underground shops and entertainment, but within 30 years it had been purchased by a railway company and to this day is used to run trains.

On the Rotherhithe side, there's a museum dedicated to the life of Isambard Kingdom Brunel and to the construction of the Thames Tunnel. Despite being the son of a Frenchman, Brunel is the most celebrated British engineer of all because he built tunnels, bridges, and steam ships that are still around today. He was also the most audacious engineer of the 19th century (and perhaps of any modern century).

His best-known achievements are the Clifton Suspension Bridge in Bristol (which opened in 1864 and was the longest bridge in the world) and the iron-hulled (and propeller-driven) SS *Great Britain* (see Chapter 68), but he was also the chief engineer of the Great Western Railway that linked London to western England and southern Wales. As chief engineer, Brunel supervised the construction of bridges, viaducts, and tunnels, and he also designed the London terminus of the Great Western Railway, Paddington Railway Station, which opened in 1854 and is still in use.

His last project was the steamer SS *Great Eastern*, which was initially intended to be a passenger ship but was the only vessel capable of carrying the enormous weight of cable needed to lay the first telegraph link between the UK and U.S. The SS *Great Eastern* went on to lay cables between France and Canada, and Yemen and India. When Brunel began planning the ship, he sketched an outline of a vessel much larger than any other ship afloat at the time—and the final ship was even bigger.

The Davy Lamp

In constructing the Thames Tunnel, the Brunels had to contend with digging into soft ground that was constantly flooded with water from the river above. There were also eruptions of marsh gas (methane) that would have caused terrible explosions had it not been for the 1815 invention of the Davy Lamp.

Since mining was being performed well before the invention of electric light, it was necessary to light mines with naked flames. But mining inevitably led to the release of pockets of methane (which miners termed firedamp) that would ignite on the candles and lamps used underground.

The British chemist Sir Humphrey Davy invented a lamp that was safe for mining use by enclosing the flame in a fine metal gauze (Figure 70-1). If methane was present, the gas would enter the gauze and the flame would turn blue and lengthen, but it would not cause an explosion. Miners could keep an eye on the flame to know when to leave.

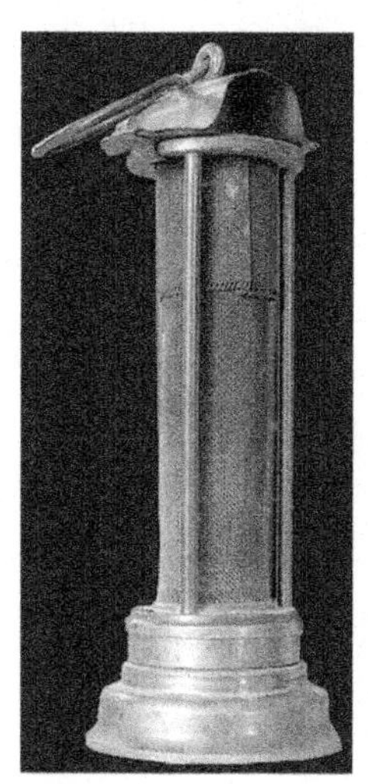

Figure 70-1. A Davy Lamp; courtesy of David Greenhalgh

The gauze acted as a flame arrester—the heat of the burning methane was dispersed across it and the temperature dropped below methane's ignition temperature (around 540°C), preventing the flame from passing through the gauze.

Similar technology is used today to prevent flames from progressing down gas pipelines. Car and boat engines contain flame arresters to prevent flames from the engine traveling backward down the fuel pipe to the fuel tank, to prevent flames from leaving the exhaust pipe, and to prevent flames from hitting the air filter.

Today many of Brunel's bridges and tunnels are still in use, and the British landscape is literally dotted with his handiwork. The Brunel Museum sits atop his Thames Tunnel, in a building that originally held a steam pump used to remove water from the tunnel.

The museum offers trips through the tunnel, sometimes on foot in the dead of night, and sometimes in trains that creep along while the walls are floodlit. From the train, you can clearly see the Brunels' carefully constructed tunnel, with its columns and arches forming spaces that were once occupied by stall holders in the first underground shopping center.

Above ground, the museum has a detailed exhibition of the construction and use of the Thames Tunnel, as well as information about Brunel's many other innovative projects.

Practical Information

The Brunel Museum's website is at *http://www.brunel-museum.org.uk/*. At the time of this writing, the tunnel is closed while the train line is renovated as part of the creation of a new overland railway due to be completed in mid-2010. The museum remains open, though, and is working to restore its tunnel trips as soon as possible.

071

The Eagle Pub, Cambridge, England

52° 12′ 14.4″ N, 0° 7′ 5.52″ E

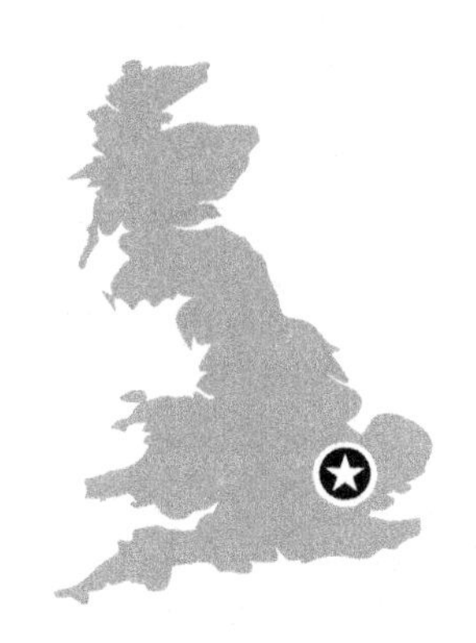

"We Have Found the Secret of Life"

At lunchtime on February 28, 1953, James Watson and Francis Crick announced to the patrons of the Eagle Pub in central Cambridge that they had found the secret of life. Watson and Crick were finally convinced (after a week of verification) that they understood the structure of DNA. Nine years later, they were awarded the Nobel Prize.

Watson and Crick conducted their work at Cavendish Laboratory, which at the time was located on Free School Lane; the Eagle Pub was a short walk away, and a common eating place for laboratory scientists. The laboratory was set up in 1874 as part of Cambridge University's physics department. Watson and Crick were working on biological applications of X-ray crystallography: the study of the structure of molecules using X-rays to determine the placement of atoms.

Their study of DNA, along with that of others around the world, led to the theory that DNA was helical. After playing with cardboard models of the chemical components of DNA, Watson and Crick realized that two helices were involved (not three, as others had suggested), and that the four chemicals, called bases, that held the genetic code were on the inside of a double helix, holding it together. This structural insight brought together much that was known about DNA prior to 1953, and suggested how DNA was able to replicate when cells divide (see sidebar).

A blue plaque outside the Eagle Pub commemorates Watson and Crick's discovery and announcement. The pub itself is large yet cozy, and is divided into five rooms with two bars. One of the bars, known as the RAF bar, has the names and squadrons of Second World War RAF and USAF airmen burned into the ceiling using cigarette lighters. The pub serves Greene King beers from a brewery (founded in 1799) in nearby Bury St. Edmunds. Greene King's great ales include Abbot Ale and IPA (good on a hot day); there's also Old Speckled Hen if you're in the mood for something stronger.

Replication

Crick and Watson's paper describing the double-helix structure of DNA appeared in the journal *Nature* in April 1953. It contains a single line describing one of the most important aspects of DNA: the ability to replicate. The line is a masterstroke of understatement:

> It has not escaped our notice that the specific pairing we have postulated immediately suggests a possible copying mechanism for the genetic material.

The DNA double helix (seeFigure 71-1) is held together by four chemicals, each with a single-letter abbreviation: thymine (T), adenine (A), guanine (G), and cytosine (C). These four chemicals appear in only two possible combinations: T joined to A (and vice versa) and G joined to C (and vice versa).

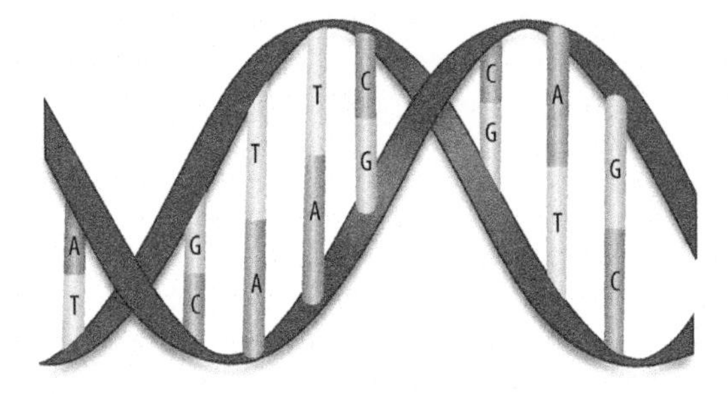

Figure 71-1. The DNA double helix

What Watson and Crick realized was that because the links between the two helices were composed of four chemicals in only two pairings, it would be possible to split the double helix in two, with each part keeping one half of each chemical pair. Thus the double helix could split into complementary chains of chemicals, similar to a photograph and its negative.

Once split in two, the double helix could be used to create two new complete double helices. If the split DNA was sitting in a pool of the ACGT chemicals, each chemical would bond only with the chemical that complemented it, thus building a new double helix (Figure 71-2). By splitting and copying itself, DNA is able to replicate when cells divide.

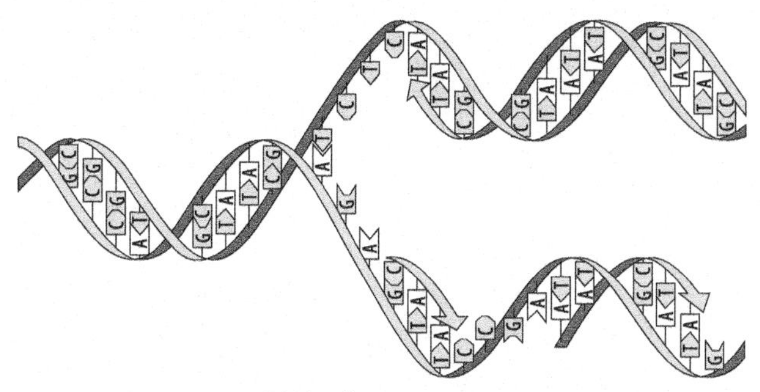

Figure 71-2. DNA splitting to create two copies

With this newfound understanding of the replication mechanism, Crick and many other scientists went on to discover how the four possible combinations (AT, TA, GC, and CG) are decoded to build cells. It was discovered that the pairs were grouped together into groups of 3 (called codons), giving 64 (4×4×4) possible codes.

All but three of the codons correspond to the creation of a single amino acid, and a set of codons (called a gene) causes the creation of a chain of amino acids that come together to form a protein. The three non-amino acid codons are actually instructions meaning "the protein ends here," and are used to separate proteins being made from the DNA. These proteins are the building blocks for living things.

If you've ever programmed a computer, you'll find it striking how the underlying code looks like a computer program (complete with HALT instructions to tell the system to stop operating). Even more striking is that the number of codons is a power of 2, just like the number in the title of this book.

The pub serves typical English pub food such as fish and chips. Start your tour of Cambridge at the nearby Whipple Museum of the History of Science, and pop into the pub for lunch. On a fine day, you can sit outside in the beer garden, which is overlooked by the student rooms in nearby colleges.

When you've finished lunch at the pub, head over to the new Cavendish Laboratory on J J Thomson Avenue off Madingley Road. The laboratory has a small museum filled with fascinating artifacts, including the original model of DNA made by Crick and Watson, X-ray diffraction photographs that gave the clue to the helical structure, and models of the base pairs.

The museum also has many instruments used by James Clerk Maxwell (who was the first Cavendish Professor of Physics, and who oversaw creation of the laboratory). There's the cathode ray tube used by J. J. Thomson in the discovery of the electron, and parts of the equipment used by J. D. Cockcroft and Ernest Walton to create a particle accelerator and disintegrate atomic nuclei (or "splitting the atom").

Other sections of the museum are dedicated to X-ray crystallography, electron microscopes, radio astronomy, and low-temperature physics. The museum is small and free of charge, and won't take more than an afternoon to visit.

Be sure not to confuse the Eagle Pub (which was originally called the Eagle and Child) in Cambridge with the Eagle and Child in Oxford, which is famed as the watering hole of writers such as J. R. R. Tolkien and C. S. Lewis.

Practical Information

The Eagle Pub is at 8 Benet Street, in the city center of Cambridge. The old Cavendish Laboratory is right around the corner on Free School Lane, where you'll also find the Whipple Museum of the History of Science. The new Cavendish Laboratory has visiting information and directions on its website: *http://www-outreach.phy.cam.ac.uk/*.

072

The Falkirk Wheel, Falkirk, Scotland

56° 0′ 1.08″ N, 3°50′ 30.02″ W

Raising and Lowering Canal Boats Using Just 1.5 Kilowatts

The Union Canal starts in Edinburgh and heads roughly westward, following the contours of the land until it reaches Falkirk in central Scotland. The Forth and Clyde Canal starts outside Glasgow at the Firth of Clyde (which links to the Atlantic Ocean) and heads roughly eastward, ending at the Firth of Forth (which links to the North Sea). But at Falkirk the two canals meet, and are linked by a stunning and unique boat lift.

The boat lift is needed because there's a 35-meter drop from the Union Canal to the Forth and Clyde Canal. The link was originally navigated by traversing 11 locks over 1.5 kilometers. But in 1933 the locks were dismantled, disconnecting Edinburgh from its historic water link to Glasgow.

The link was reestablished in 2002 with the spectacular Falkirk Wheel. The locks have been replaced by a rotating boat lift that uses a tiny amount of power to lift 300 tonnes of boats and water from the Forth and Clyde Canal up to the Union Canal, while lowering another 300 tonnes of boats and water in the opposite direction.

Boats arriving from the Union Canal enter the Falkirk Wheel from an aqueduct—the canal comes to an abrupt end inside an open tunnel that connects to the boat lift. Boats then enter a caisson (a watertight box) that is mounted inside one of the arms of the wheel. As the wheel turns, the caisson rotates in the opposite direction, keeping the boats level until they are delivered to the water level of the Forth and Clyde Canal. The caisson then opens, and the boats can leave.

Exactly the same procedure happens in reverse on the other side of the wheel, where boats in another rotating caisson are raised to the level of the Union Canal. The process of rotating through 180° takes about five minutes. Amazingly, the procedure is almost silent—since the two caissons counterbalance each other, only a tiny amount of power (about 1.5 kilowatts) is needed to start the wheel turning and overcome friction.

Archimedes's Principle

Archimedes's Principle is often explained in terms of the story of King Hiero II's crown. As the story goes, the king wanted to know whether his crown was made of solid gold. Archimedes weighed the crown, and then prepared a block of gold and a block of silver, both having the same weight as the crown.

The crown was submerged in a bowl filled to the brim with water, and the overflowing water was collected. The process was repeated with the block of gold and the block of silver. Archimedes realized that the volume of overflowing water (which is called the volume of water displaced) equaled the volume of the object under the water.

He could then compare the volumes of the crown and of the gold and silver blocks. He quickly discovered that the crown displaced more water than the gold block and less than the silver. This indicated that the crown was less dense (had greater volume) than the gold block and was therefore impure.

The simplest form of Archimedes's Principle is thus "the volume of water displaced by a submerged object is equal to the volume of the object." More accurately, the principle states that a body completely or partially submerged in water experiences an upward force equal to the weight of the water displaced.

For example, a boat in one of the Falkirk Wheel's caissons floats because the upward force acting on it equals the weight of the boat balancing the force of gravity. The volume of water displaced by the boat weighs the same amount as the boat, and is determined by the volume of the boat below the water line.

An object will sink in water if the volume of water equal to its volume weighs less than the object (that is, if the object is denser than water).

A skin diver displaces water when submerged and feels lighter—he experiences an upward force equal to the volume of water that his body displaces. When the skin diver tries to climb out of the water, it becomes more and more difficult as less and less water is displaced and the diver's apparent weight increases.

The weight of water displaced depends on its density and volume, so a swimmer experiences different upward forces depending on the composition of the water. Sea water is denser than pure water, and in the Dead Sea the water is so dense (because of the high salinity) that swimmers literally float.

Key to keeping the system in balance, and using as little power as possible, is the fact that the two caissons weigh exactly the same, regardless of the number of boats on each side of the wheel (a caisson can even be empty if there's not enough traffic). Equal weight is achieved by a simple application of Archimedes's Principle—the weight of water displaced by a boat in the caisson is determined by the weight of the boat (as long as it floats). By allowing the water displaced by the boat to flow back into the canals, the wheel's operators can ensure that both caissons have the same weight.

Keeping the two caissons level is achieved by a very simple mechanism consisting of five cogs (Figure 72-1). A central cog is fixed to the non-moving aqueduct support. Each of the caissons is inside a cog of the same diameter as the central cog. Each arm of the wheel has a small cog that joins the central cog to the one containing the caisson. As the wheel rotates, the small cogs are rotated by the fixed central cog, causing the outer cogs to rotate and thus keep the caissons level.

Figure 72-1. Cog mechanism; courtesy of Jacqui Napier (Jax60)

You don't need to rent a canal boat to experience the Falkirk Wheel. You can view the wheel turning from the visitor center, which explains the history of the canals and the construction of the wheel itself. You can also book a one-hour trip on the wheel and canals—visitors start in a boat at the bottom and are taken up to the Union Canal. The boat trip continues through the 180-meter

Roughcastle Tunnel, which was constructed to link the canal to the wheel; the boat then returns to the wheel and descends to the Forth and Clyde Canal next to the visitor center.

In addition to enjoying the canal on the water, you can also walk or cycle along the canal paths. There's even a portable audio guide of a trail around the basin in which the wheel sits.

The Falkirk Wheel is just outside Falkirk itself, which is roughly halfway between Glasgow (where you can visit the Hunterian Museum; see Chapter 73) and Edinburgh (site of the National Museum of Scotland; see Chapter 59). One hour's drive takes you to either of these destinations.

Practical Information

Visiting information for the Falkirk Wheel is available from its website: *http://www.thefalkirkwheel.co.uk/*.

073

The Hunterian Museum, Glasgow, Scotland

55° 52′ 19″ N, 4° 17′ 19″ W

William Hunter and Lord Kelvin

The 18th-century anatomist and obstetrician William Hunter (brother of John Hunter; see Chapter 64) bequeathed the money for the creation of the Hunterian Museum in Glasgow. He also provided the core of its collection—his medical equipment and specimens, and his wide-ranging collection of art works, coins, books, and minerals. In the 200 years since its creation, the museum has vastly expanded its collection of scientific and artistic works, and also contains the most important public exhibition of the work of another great Glaswegian, William Thomson (who became Lord Kelvin).

William Hunter introduced the practice of using cadavers to teach medical students. He eventually specialized in obstetrics, publishing his illustrated *The Anatomy of the Human Gravid Uterus* in 1774 (see Figure 73-1).

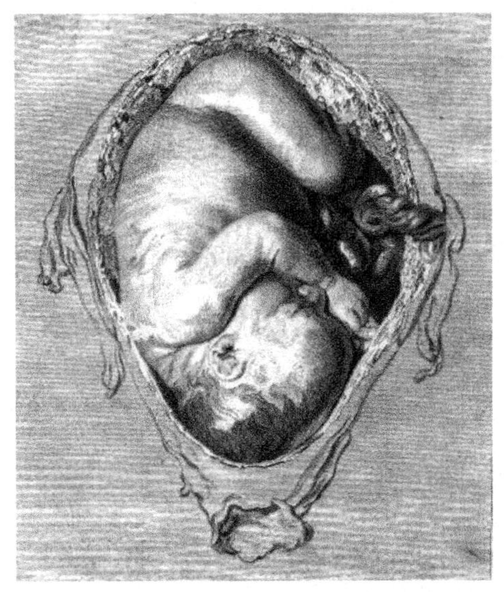

Figure 73-1. Illustration from The Anatomy of the Human Gravid Uterus

Measuring Temperature

Lord Kelvin's name is so well known partly because it is used as the standard international unit of temperature, the kelvin. For most scientific work, the kelvin has displaced Celsius and Fahrenheit.

The Fahrenheit scale was proposed in 1724 by the German physicist Daniel Gabriel Fahrenheit. His scale was developed, at least in one telling of the story, by fixing three interesting points: 0°F, 32°F, and 96°F. Fahrenheit set 0°F as the temperature of a mixture of water, ice, and ammonium chloride. This curious mixture was actually a common way of getting a fixed, low temperature at the time, since the mixture will automatically find and stay at a specific temperature.

32°F was the temperature of water with ice melting in it, and 96°F was the temperature of a healthy person taken under the arm. Later recalibration of the scale led to the standard healthy human temperature being 98.6°F.

Around 1742, the Swede Anders Celsius proposed a different scale, where 100°C was the freezing point of water and 0°C the boiling point of water. He was careful to point out that the boiling point of water is affected by air pressure, and that fixing 0°C should be done at mean pressure at mean sea level. Not long afterward, the scale was reversed to form the familiar Celsius scale used today.

In 1848, Lord Kelvin wrote the paper *On an Absolute Thermometric Scale* and proposed a radically different way of looking at temperature based on the actual nature of temperature itself. Kelvin's paper proposed that a temperature scale be based on the lowest temperature possible (the infinitely cold); today we call this Absolute Zero.

Temperature arises from the motion of particles. For example, the temperature of the air arises from the random motion of air molecules—the higher the temperature, the faster the molecules move about and bounce off each other. Absolute Zero is the temperature at which motion is at a minimum and can go no lower.

In defining his unit of temperature, Kelvin used Absolute Zero as the zero value and the same degrees as Celsius. He calculated Absolute Zero to be –273°C; today, the accepted value is –273.15°C. Thus the freezing point of water is 273.15 K and the boiling point is 373.15 K.

In accurately defining the kelvin scale, today two values are used—Absolute Zero and the triple point of Vienna Standard Mean Ocean Water. The official definition is that the kelvin "is the fraction 1/273.16 of the thermodynamic temperature of the triple point of water."

The triple point of water (Figure 73-2) is a specific temperature and pressure at which all three forms of water (liquid, solid, and gas) coexist. This occurs at 273.16 K (0.01°C) with a pressure of 611.73 pascals. At that temperature and pressure, it is possible to cause all the water to turn into any one of the forms (liquid, ice, or vapor) by changing the pressure or temperature.

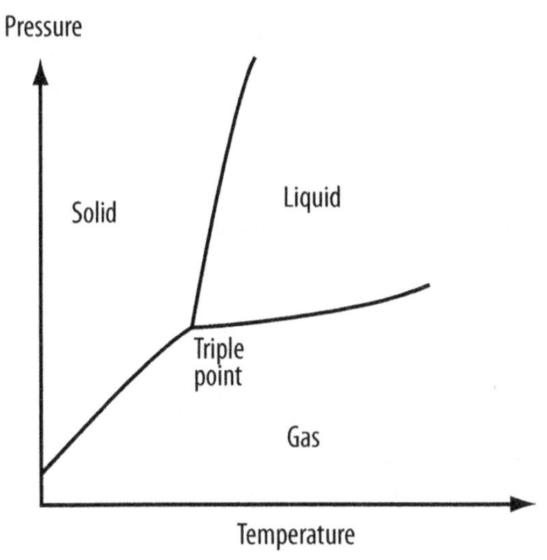

Figure 73-2. The triple point

Of course, the actual composition of the water makes a difference to its triple point, thus the specification of Vienna Standard Mean Ocean Water (VSMOR). The VSMOR is a precise definition of the composition of pure water that specifies the proportions in which isotopes of hydrogen and oxygen are found (and is intended to reflect real average ocean water). Despite the "ocean" in its name, the water is pure—there are no salts or minerals present.

Because the triple point of water is fixed, it is commonly used as a calibration point for thermometers. This is achieved by placing the thermometer inside a triple point cell—an enclosed tube nearly filled with water, leaving a gap containing just water vapor.

The cell contains a slot for the insertion of a thermometer. To get the water to the triple point, the entire cell is cooled down to near freezing. Then dry ice is placed in the thermometer slot, which causes the water around the slot to freeze. Once this mantle of ice extends for a few millimeters, the dry ice is removed and a warm metal probe briefly inserted in the slot. The ice mantle melts slightly and floats off into the water. At that point, water, ice, and vapor are all present in the tube, and it can be used to calibrate a thermometer very precisely.

The museum celebrates Hunter's life through his collection, with a special emphasis for scientists on his medical equipment and pathological specimens. The collection also includes early X-ray and ultrasound equipment. The most up-to-date part of the exhibition examines the MRSA (multiple-resistant Staphylococcus aureus) bacterium, which is hard to treat because of its drug resistance, and the Glasgow Coma Scale (a scale of 3 to 15 indicating the degree of responsiveness of a coma patient).

William Thomson was born in Belfast, and came to Glasgow when his father was appointed professor of mathematics at the University. Thomson went on to Cambridge University and a spectacular career that spanned both theoretical and practical work. The Hunterian Museum describes him as the greatest Glaswegian scientist, which is undoubtedly true, but his impact was felt far from Glasgow and to this day.

Lord Kelvin's practical inventions included the mirror galvanometer, which extended the telegraph to all parts of the British Empire (see page 236). Kelvin also gave his name to the standard unit of temperature (see sidebar), coined the term "kinetic energy," built a highly accurate machine for predicting tides, and was a mentor for James Clerk Maxwell (see Chapter 35). He stands among the world's greatest scientists, along with Newton, Einstein, and Faraday. He is buried in Westminster Abbey next to Newton.

One of Kelvin's often-overlooked inventions is his compass (Figure 73-3). In 1874, he wrote and lectured extensively about magnetic compasses, the Magnetic North Pole (see Chapter 128), and problems with navigation. His compass was both very sensitive and able to cope with the ravages of a moving ship at sea, and it could be tuned to eliminate magnetic deviation caused by the use of iron in the ship's construction. That was achieved by the placement of a small magnet close to the compass to eliminate the ship's magnetic field.

The museum's special Kelvin exhibition contains his original equipment and a range of hands-on experiments ideal for schoolchildren. Apart from an oddly fearsome-looking Kelvin dummy, which seems a little out of place, the exhibition does a good job of explaining Kelvin's theoretical and practical work.

The museum also makes use of multimedia to add depth to Kelvin's life, and offers the ability to further explore the museum's collection of Kelvin's equipment, and computer-based animations explaining his work.

Figure 73-3. Kelvin's compass; courtesy of Don Turnbull (http://donturn.com)

Practical Information

The Hunterian Museum's website is at *http://www.hunterian.gla.ac.uk/*. Admission is free. A useful online resource for Kelvin research is Scran (*http://www.scran.ac.uk/*), a fully searchable database of items held in Scottish museums—just do a search for "lord kelvin".

074

The Iron Bridge, Ironbridge, England

52° 37′ 38.08″ N, 2° 29′ 7.92″ W

Birthplace of the Industrial Revolution

The Iron Bridge crossing the river Severn is the first bridge ever made from cast iron (see Figure 74-1). Its 30-meter span, created from 379 tons of cast iron, has been standing since 1779. Today, as in the 18th century, the bridge is a major tourist attraction. Though closed to cars, it is open to foot traffic.

Figure 74-1. The Iron Bridge

Prior to its construction, the only way to cross the river was by ferry, and the surrounding area had rich coal deposits supplying industries in local towns. The businesses needed a more reliable crossing, and shares were issued to raise the money necessary to build a bridge.

The bridge was constructed by Abraham Darby III, whose grandfather, Abraham Darby, had perfected coke smelting to produce iron goods. Coke smelting greatly reduced the cost of making cast iron, and led the surrounding area to be called "the Birthplace of the Industrial Revolution."

Once the bridge was open, the town of Ironbridge grew beside it, and with advent of steam power, cheap cast iron, and spinning wheels, the Industrial Revolution was underway. Today the entire Ironbridge area is a UNESCO World Heritage site, and features the Blists Hill Victorian Town (with costumed staff portraying daily 19th-century life and a weekly demonstration of ironwork at The Foundry), a number of museums covering the area and local industry, and the bridge itself.

The Coalbrookdale Museum of Iron contains the excavated remains of Abraham Darby's first coke-burning blast furnace and an explanation of the cast iron production process. For children, the interactive design and technology center Enginuity has hands-on exhibits and demonstrations for all ages.

Since the entire village of Ironbridge is an open-air museum, visitors should plan to spend the entire day in the area. If you're staying overnight, try to get a room at the Tontine Hotel. The hotel was built in 1784 to profit from the popularity of the bridge—it sits at the end of the Iron Bridge, and some rooms have a view directly over the bridge itself.

The hotel gets its unusual name from a form of financing called a tontine. Investors buy shares in a business, such as the hotel, and receive dividends. When a shareholder dies, their shares are divided among the remaining living shareholders, providing an enormous motivation for shareholders to benefit from each other's deaths (accidental or otherwise). For that reason, tontines are illegal in many countries.

Practical Information

Extensive information about the Iron Bridge can be found at its website: *http://www.ironbridge.org.uk/*.

Iron Smelting

Prior to Abraham Darby's invention of the coke blast furnace, charcoal was used to extract iron from iron ore. In a blast furnace, a mixture of fuel (charcoal or coke) and iron ore is continuously added to the top of the furnace. Air is pumped into the bottom of the furnace, the fuel burns, and the ore reacts with carbon in the fuel to extract iron. The molten iron and the residue (called slag) are extracted from the bottom of the furnace.

Charcoal-based blast furnaces required an enormous amount of wood to operate. The wood was first burned to create charcoal, and the charcoal then used to extract iron. Darby's use of coke was more efficient because coke is made from coal instead of wood (which was in critically short supply as the demand for iron rose), and coke is physically stronger than charcoal, which meant that much larger blast furnaces were possible. Coke blast furnaces were capable of producing 5 to 10 tons of iron per week.

Coke is produced by burning coal at up to 2000°C in an airless oven. This high-temperature baking leaves just the carbon (and a little ash), extracting water, gases, and tar from the coal.

In the blast furnace, the carbon from the burning coal reacts with the oxygen in the air, creating carbon dioxide and additional heat. Further up the furnace, the carbon dioxide reacts with more carbon in the coke and turns into carbon monoxide (see Equation 74-1).

$$\begin{aligned} C + O_2 &= CO_2 \\ C + CO_2 &= CO + CO \end{aligned}$$

Equation 74-1. Coke turns into carbon dioxide, then carbon monoxide

Near the top of the furnace, the carbon monoxide reacts with the iron ore (such as naturally occurring haematite, Fe_2O_3) to separate the iron and produce carbon dioxide gas (Equation 74-2).

$$\begin{aligned} &Fe_2O_3 + CO + CO + CO = \\ &\quad Fe + Fe + CO_2 + CO_2 + CO_2 \end{aligned}$$

Equation 74-2. Carbon monoxide extracts the iron

075

The Royal Institution of Great Britain, London, England

51° 30′ 35.28″ N, 0° 8′ 33″ W

Furthering the Public Understanding of Science

When it comes to the public understanding of science, one ancient institution stands out—the Royal Institution of Great Britain. It was founded in 1799 with a specific goal: "diffusing the knowledge, and facilitating the general introduction, of useful mechanical inventions and improvements; and for teaching, by courses of philosophical lectures and experiments, the application of science to the common purposes of life." To this day, the Royal Institution hosts public lectures including the popular Christmas Lectures (specifically aimed at children), which were started in 1825 by Michael Faraday and only took a break for the Second World War.

Since its establishment, many great scientists have worked and even lived at the Royal Institution. Faraday lived for many years in a small apartment in the Institution, where he worked (mainly in the basement) on chemistry and electricity. A self-taught man and a highly religious Christian, he discovered electromagnetic induction (see sidebar), worked out the laws of electrolysis (which show the relationship between the amount of electricity passing through a solution and the mass of chemicals produced), and demonstrated that magnetism could bend light. He discovered benzene, and coined the terms *anode*, *cathode*, *electrode*, and *ion*.

Faraday was also the first to make dynamos and electric motors, and invented the Faraday cage (see page 196).

Faraday initially became interested in science in part because he was able to attend public lectures at the Royal Institution given by Sir Humphrey Davy, who had helped get the Royal Institution going. Davy is best known for the Davy Lamp (page 265), but also discovered a list of elements including sodium, potassium, calcium, magnesium, barium, and boron. In total, the Royal Institution has seen 10 elements discovered on its premises.

Electromagnetic Induction

Michael Faraday was not the first person to discover electromagnetism, but he did bring together the work done by others to create the theory of electromagnetism. And Faraday discovered electromagnetic induction, which is fundamental to the operation of motors, dynamos, generators, and transformers.

Electricity and magnetism are inextricably linked—electricity passing through a wire generates a magnetic field around the wire, and a wire moving through a magnetic field has electricity created in it. In fact, wherever there's a changing magnetic field there's an electric field, and wherever there's a changing electric field there's a magnetic field.

Faraday discovered that if he wrapped two coils of wire around an iron ring, then a current passed through one coil would create a current in the other. His experiment is on display at the Royal Institution and is the first example of a transformer. It shows electromagnetic induction—a changing current in one coil creates a magnetic field, and the changing magnetic field then creates a changing current in the other coil.

He also showed that moving a coil of wire over a permanent magnet caused a current to be created in the wire—the current is said to be *induced* in the wire. And since a coil of wire creates a magnetic field when electricity passes through it, it's possible to create movement because the coil will be repelled or attracted to a permanent magnet. With these experiments, Faraday was able to create practical dynamos and motors.

He is also responsible for Faraday's Law of Induction (Equation 75-1). This states that the electromotive force, E, induced in a circuit by a changing magnetic field is equal to the rate at which the flux of the magnetic field, Φ, changes.

$$E = -\frac{\Delta\Phi}{\Delta t}$$

Equation 75-1. Faraday's Law of Induction

The electromotive force expresses the amount of energy that a device (such as a battery) will add to electrons passing through it. Strictly speaking, this is the energy added per unit of charge passing through the device and is measured in volts (in fact, Volta—see Chapter 26—was the first person to speak about the electromotive force). When talking about a battery or a generator, its electromotive force is more commonly called its voltage.

The flux of a magnetic field is the quantity of magnetism in a magnetic field over some area. When a magnetic field is inducing a current in a wire, the flux is based on the strength of the magnetic field and the area of the wire the magnetic field passes through. The flux of a magnetic field over some area is the strength of the field times the area.

Thus, increasing the strength of the magnetic field, or the amount of wire the field passes through (which is why coils of wire are typically used) increases the amount of electromotive force induced.

And Faraday's Law of Induction explains how to construct transformers. A simple transformer is exactly what Faraday originally built: a pair of coils of wire (of the same gauge) wrapped around an iron ring. When a changing electric current is passed through one coil, a changing electric current of a different voltage is created in the other coil (see Figure 75-1). The ratio of the two voltages is explained by Faraday's Law.

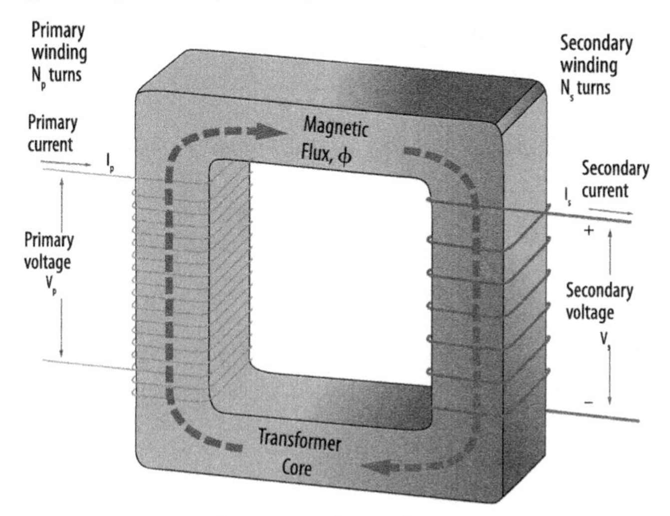

Figure 75-1. A transformer

The coils of wire in a transformer are referred to as the primary winding (where the voltage to be transformed is applied) and the secondary winding (where the transformed voltage is present). When a voltage of V_P is applied across the primary winding, it generates a magnetic field with a flux that follows Faraday's Law. Because the coil is wound around the iron many times, the effective area is proportional to the number of turns, N_P, meaning that the resulting magnetic flux depends also on the number of turns (Equation 75-2).

$$V_P = -N_P \frac{\Delta\Phi}{\Delta t}$$

Equation 75-2. Faraday's Law applied to the primary winding

The induced magnetic field passes through the secondary winding, inducing a current with a voltage, V_S, that is once again proportional to the number of turns, N_S (Equation 75-3) in the secondary.

$$V_S = -N_S \frac{\Delta\Phi}{\Delta t}$$

Equation 75-3. Faraday's Law applied to the secondary winding

Since the magnetic flux is the same in both cases (assuming an ideal transformer and no loss of flux), it can be eliminated to show that the ratio of voltages in a transformer is equal to the ratio of turns in the two coils (see Equation 75-4). To make a simple transformer capable of converting European 220V to U.S. 110V, a ratio of 2:1 (primary:secondary) is needed for the two coils. A cell phone charger that takes 110V and converts it to 5V would needed a transformer with a winding ratio of 22:1.

$$\frac{V_S}{V_P} = \frac{N_S}{N_P}$$

Equation 75-4. Voltage ratio in a transformer

The Australian physicist Sir William Lawrence Bragg (who worked on X-ray diffraction and won the Nobel Prize) was Fullerian Professor of Chemistry at the Royal Institution (as was Faraday).

The history and science behind the Royal Institution is explored in a completely renovated museum (formerly known as the Faraday Museum) split over three floors. In the basement where Faraday worked is an area entirely dedicated to experimentation. There's a reproduction of Faraday's own laboratory made with his actual equipment, and clear and varied displays of equipment used from the earliest days of the Royal Institution to the present.

The basement exhibition covers topics as varied as electrolysis and electromagnetism, nanotechnology and glass making, the creation of X-rays and the experiment that explained why the sky is blue.

The other floors are devoted to the people associated with the Royal Institution, and the institution's communication and education work.

Practical Information

Visiting information for the Royal Institution is available at *http://www.rigb.org/*. There's a small café called Time and Space inside the building, if you find yourself in need of a science-filled place for lunch; it too is filled with historical objects.

076

The Science Museum, Swindon, England

51° 30′ 44.1″ N, 1° 48′ 46″ W

The Little-Known Outpost

Even aficionados of the Science Museum in London (Chapter 77) are frequently unaware that the museum has an enormous outpost on the disused Second World War airfield at Wroughton near Swindon. The Swindon museum consists of six hangars (each 0.4 hectares in size) filled with the largest objects in the museum's collection (only about 8% of the overall Science Museum collection is visible in London).

Even though the outpost has been in use since the 1980s, it is not well known because it's only open to the public on prearranged tours during the summer months. To take a tour of the Science Museum in Swindon is to enter a secret world that few have had the privilege of seeing.

The Swindon museum contains 18,000 large objects including aircraft, cars, motorbikes, tractors, fire engines, large machines and scientific instruments, and equipment used in printing and telecommunications. There's the only Lockheed Constellation aircraft in the UK, a complete De Havilland Comet 4B airliner (the version with round windows that didn't tend to pop out in flight), two deactivated nuclear warheads, a machine used to try to dig a Channel Tunnel in 1921, lifeboats, literally tonnes of 1960s computer equipment, a Tucker Snocat (the first vehicle to cross the Antarctic), a truly enormous printing press (the size of two houses) that was used to print the *Daily Mail*, MRI machines (see sidebar), industrial cold stores, a British Blue Steel missile, the SRN1 Hovercraft, the submersible made famous by the James Bond film *For Your Eyes Only*, the world's smallest jet plane (the Bede BD-5), and racks and racks of other gear.

The site also has the Science Museum Library, which contains the museum's collection of books, maps, scientific journals, patents, catalogs, personal papers, and manuscripts. With prior arrangement, individuals can do research free of charge at the library.

Magnetic Resonance Imaging

Quantum mechanics may seem like an abstract theory with little bearing on everyday life, until you find yourself needing to have an MRI scan at a hospital. Magnetic Resonance Imaging (MRI) uses the quantum mechanical magnetic properties of hydrogen nuclei to create images inside the body through a process called nuclear magnetic resonance.

In quantum mechanics, all atomic nuclei, elementary particles, and well-known particles like protons and neutrons have a property known as spin. The spin can be imagined as the rotation of a particle or nucleus about its own axis, and for elementary particles (like quarks) it is an intrinsic property (just like electrical charge and mass).

The spin is related to the angular momentum of the particle, and since the angular momentum in quantum mechanics can only take on certain discrete (quantized) values, the spin can be represented by a spin quantum number. A photon has a spin quantum number of 1 (shortened to spin-1), an electron has a spin of 1/2 (as do quarks), and the Higgs boson (see page 119) has no spin at all.

The spin also has a direction, called up or down (or positive or negative), depending on whether the particle is spinning clockwise or counterclockwise around its axis.

Both protons and neutrons have a spin of 1/2. The spin of an atomic nuclei depends on the balance of protons and neutrons. For example, the most commonly occurring isotope of hydrogen (hydrogen-1) has a nucleus with a single proton and no neutrons, giving it an overall spin of 1/2.

Because hydrogen-1 has a non-zero spin, it also has a magnetic moment and hence has a changing magnetic field (the analogy with classical physics is that a spinning charged body would generate a magnetic field). It's this magnetism that MRI machines exploit.

The human body is filled with water, and the water is partly made of hydrogen (in particular, hydrogen-1). When that hydrogen is placed in a very strong magnetic field (such as from the enormous magnet in an MRI machine), the protons gain different energies from the field depending on whether their 1/2 spin is negative or positive. (This can be thought of as the protons being aligned with or against the magnetic field coming from the magnet.)

This overall difference in energy level is key to getting an image. When an electromagnetic radiation is applied to the hydrogen-1 nuclei at the right frequency (which corresponds to a radio signal), it is absorbed. This is called the resonant frequency.

To actually create an image of the body, an MRI machine uses a second set of magnets that are turned on so that another magnetic field is created in a slice through the patient. It's this "slice" that will be excited by the electromagnetic radiation. The combination of the radio frequency used, the strength of the magnetic field, and the nature of the tissue to be imaged is taken into account.

By pulsing the secondary magnetic field, it's possible to cause the electromagnetic radiation to be absorbed by specific nuclei. When the magnetic field is turned off, the nuclei return to the energy level corresponding to the large magnetic field of the main magnet, giving off electromagnetic radiation in the form of a radio signal.

The radio signal received from the nuclei as they return to their normal state in the fixed magnetic field is picked up by the MRI machine, and then transformed into a picture of the density of the material in the body.

Practical Information

Go to *http://www.sciencemuseumswindon.org.uk/* for details of the outpost at Swindon. To tour the museum, a group booking must be made by telephoning the Swindon office; a charge does apply. The museum is currently fundraising to open a public museum at the site, and public donations are welcome.

077

The Science Museum, London, England

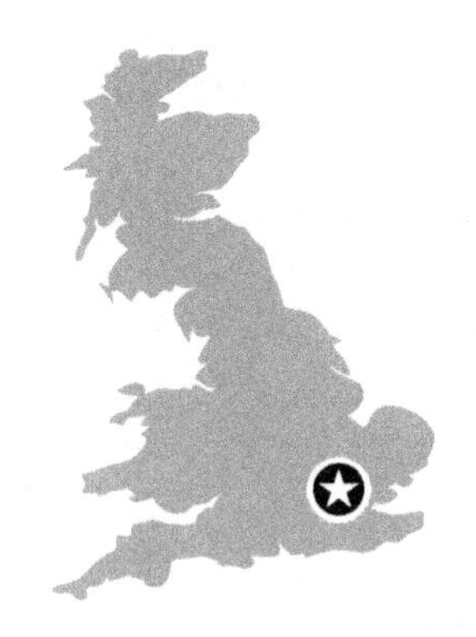

51° 29′ 51″ N, 0° 10′ 29″ W

A National Treasure

The Science Museum in London opened its doors in 1857, and is the oldest member of Britain's family of science and industry museums (the others are the Science Museum in Swindon, Chapter 76; the National Railway Museum, Chapter 60; and the National Media Museum). Its collection of historical artifacts is simply outstanding, and it's free. It's also right next door to the Natural History Museum (Chapter 61) and the Victoria and Albert Museum (where non-scientists can cool their heels while exploring its collection of art and design).

There are many famous objects on display at the Science Museum, including Stephenson's Rocket, the first atomic clock, the command module from Apollo 10, and Charles Babbage's Difference Engine. The ground floor is home to three exhibitions: Making the Modern World, the Energy Hall, and Exploring Space.

The Making the Modern World section has many of the museum's iconic objects (including Crick and Watson's model of DNA; see also Chapter 71), presented as a chronology of science and technology from 1750 to today. There are unusual objects such as Sir Henry Bessemer's converter used for turning pig iron into steel, a pair of freeze-dried transgenic mice (whose DNA has been modified), an early incandescent light bulb made by Edison, and a Rolls Royce vertical take-off engine.

The Energy Hall is all about steam, and has a collection of large, working steam engines. There's an enormous engine dating from 1903 that was used to power a mill that gets up steam regularly.

The small Exploring Space exhibition tells the story of Britain's small space program, including a Black Arrow rocket that made Britain a space power (albeit a minor one) by putting a satellite in orbit.

But the museum is not limited to the ground floor—there are seven levels of science to discover.

The first floor has exhibits related to weather forecasting, land surveying, agriculture, telecommunications, and material science.

The second floor has mathematics (and includes a fine collection of slide rules; page 138) and computing (where you'll find the Difference Engine; see sidebar), as well as shipping, docks, and diving.

On the third floor is King George III's collection of 18th-century scientific equipment and a gallery on the technology of flight. The third floor also houses the popular Launchpad gallery, with lots of hands-on exhibits. This area is very good for older children; younger children will enjoy the Garden in the basement, which has lots of things to explore using water, light, and sound.

The history of medicine is split between the third, fourth, and fifth floors.

Unfortunately, the museum is not immune to the "Science Is Fun!!" wave, and has installed an IMAX 3D theater and various "rides" throughout the building.

Practical Information

The Science Museum's website is at *http://www.sciencemuseum.org.uk/*. Plan to spend an entire day there (or if you're short on time, spend half a day there and half a day at the Natural History Museum).

The Difference Engine

One of the key exhibits at the Science Museum is a reproduction, not an original. The Difference Engine was originally conceived by the British polymath Charles Babbage in 1822. He planned to build a mechanical device capable of computing tables of numbers (such as logarithms, see page 214) with great accuracy.

Between Babbage's changing ideas and his constant demands for extra money from the British government, the Difference Engine was never completely constructed and only small sample parts remain. Even if he had managed to acquire the money, it was commonly believed that he could not have completed the proposed machine because of limitations of the Victorian technology to make the thousands of small parts required.

But between 1989 and 1991, the Science Museum decided to use Babbage's plans, and Victorian-era techniques to see if the Difference Engine could actually have been built. The attempt was a success, and the working machine now sits in the museum. It has 4,000 parts and weighs over 2.5 tonnes, but it works.

In 2000, the museum completed the associated printer—Babbage planned to eliminate transcription errors in tables of numbers by first calculating them mechanically and then printing them directly. The number of columns used for printing can be varied, and the printer produces a paper record of the results and a plate suitable for use in a printing press. The printer even does line wrapping of numbers automatically.

To calculate tables of numbers, the Difference Engine uses the method of finite differences to calculate values for a complex function such as a logarithm from simple addition.

For example, to calculate the exponential function e^x to build a book of tables, it's possible to approximate e^x using a Taylor series (Equation 77-1). A Taylor series is a way of expressing a mathematical function as a sum.

$$e^x = \sum_{n=0}^{\infty} \frac{x^n}{n!} = 1 + x + \frac{x^2}{2} + \frac{x^3}{3 \times 2} + \frac{x^4}{4 \times 3 \times 2} + \cdots$$

Equation 77-1. Taylor series of e^x

By cutting off the sum, it's possible to get an approximation to the function with a specific amount of accuracy (the more terms that are kept on the righthand side, the more accurate the approximation). For example, e^x could be approximated by cutting off the Taylor series after the first four terms; this simplified form is called a Taylor polynomial for e^x (see Equation 77-2).

$$e^x = 1 + x + \frac{x^2}{2} + \frac{x^3}{3 \times 2}$$

Equation 77-2. Taylor polynomial that approximates e^x

This gives an approximation to e^x that is close to the actual value. Table 77-1 shows the value of e^x calculated using the approximation compared to the actual value. In this case the approximation is accurate for two decimal places (the table shows three decimal places for clarity).

Table 77-1. Approximation of e^x using a Taylor polynomial

X	e^x	$1+x+x^2/2+x^3/6$
0.1	1.105	1.105
0.2	1.221	1.221
0.3	1.350	1.350
0.4	1.492	1.491
0.5	1.649	1.646
0.6	1.822	1.816

But calculating the Taylor polynomial is relatively difficult—it requires doing many multiplications, which in Babbage's time required the use of logarithm tables and was slow and laborious. And building a machine that could do multiplication was also relatively difficult. So Babbage set out to make a machine that would rely on addition instead.

Writing out the differences between the values of e^x given by the Taylor polynomial and continuing to work out the differences of differences shows an interesting property of all polynomials—eventually, the differences disappear. The second column of Table 77-2 shows the differences between successive pairs of numbers (values of x) in the first column, the third column shows the difference between pairs in the second column, and so on.

As you can see, by the fourth difference, there are no longer any differences.

Table 77-2. Table of differences

X	$1+x+x^2/2+x^3/6$	First difference	Second difference	Third difference	Fourth difference
0.1	**1.105**				
0.2	1.221	**0.116**			
0.3	1.350	0.128	**0.012**		
0.4	1.491	0.141	0.013	**0.001**	
0.5	1.646	0.155	0.014	0.001	0
0.6	1.816	0.170	0.015	0.001	0

The Difference Engine used this process in reverse. It started with the differences and performed repeated addition to get the results. For example, by setting up the machine with just the numbers shown in bold, it's possible to output as many values as necessary.

The value for x = 0.2 is calculated by adding 1.105 and 0.116, which yields 1.221. To get the value for x = 0.3, the first difference 0.128 is needed (so that it can be added to 1.221); it can be obtained by adding the previous first difference (0.116) and the second difference (0.012). And so on.

For each step, the Difference Engine can add one result and one first difference to get a new result, then add one first difference and one second difference to get the next first difference, and one second difference and one third difference to get the next second difference, and so on. So the Difference Engine proceeds down the table just doing additions to get the results, and keeping a diagonal of differences waiting to calculate the next result. The actual Difference Engine can calculate up to 31 digits using seven levels of differences for high accuracy.

078

Thinktank, Birmingham, England

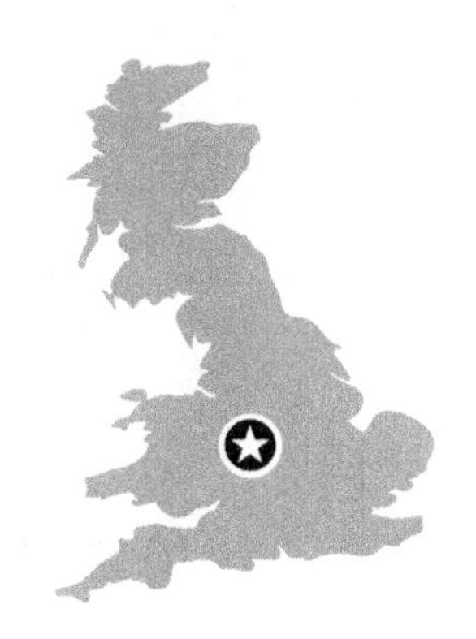

52° 28′ 58.47″ N, 1° 53′ 9.81″ W

The Smethwick Engine

The city of Birmingham in central England was once the powerhouse of the world—it was there that Matthew Boulton and James Watt formed a partnership to build Watt's efficient steam engines. Today Birmingham is still an important industrial city (although services such as banking now dominate the local economy), and the city sits at the heart of the West Midlands conurbation, the second-largest conurbation in the UK.

And now, in honor of Boulton and Watt and the city's industrial heritage, Birmingham has a brand-new science museum, called Thinktank, in the city center.

The museum is divided into four levels with the (perhaps too cute) names of thinkback, thinkhere, thinknow, and thinkahead. The first level (thinkback) is where you'll find the industrial history of Birmingham split into three parts called Move It, Making Things, and Power Up.

The Move It space covers road, canal, rail, and suburban transport with items including an early 20th-century electric tram car, the Foden Steam Wagon (a steam-powered truck), and the Railton Special. The Railton Special beat the land speed record by traveling at over 568 kph in 1938.

The Making Things space recounts the manufacturing history of Birmingham, showing the machines and goods produced in the 19th and 20th centuries in the area. These include Frederick Wolseley's sheep-shearing machine, and a modern chocolate-wrapping machine (in action) used by Cadbury's, another local business.

But the star exhibit is in Power Up—Boulton and Watt's Smethwick Engine, which operated from 1779 to 1892 (Figure 78-1). It was used to pump water back up the Birmingham canal at nearby Smethwick. Because the canal descended almost 150 meters through a flight of locks, a great deal of water was used in moving boats. The Smethwick Engine pumped water back up again to keep the canal filled. Today it sits at Thinktank and steams once more.

Figure 78-1. The Smethwick Engine; courtesy of Thinktank Trust

Thinktank has a number of other working engines that steam regularly, but to reduce wear and tear they do not all steam every day. Check with the museum for operating times.

The thinkhere area is focused on the history of Birmingham, from its beginnings as the tiny hamlet recorded in the Domesday Book through the Industrial Revolution to today. The thinknow exhibit covers modern technology and is particularly good for children. There's a gallery called "Things about Me" that helps children understand the body's inner workings and covers digestion, the muscles and skeleton, the brain and the senses, breathing, and the process of growing up.

Finally, thinkahead contains a planetarium that uses digital projection for its many different presentations, including shows explaining what you can see in the skies above Birmingham at night. There's an extra charge for the planetarium, and you'll need to book ahead.

Practical Information

Information about the Thinktank science museum is available from *http://www.thinktank.ac/*.

The Boulton and Watt Engine

The first practical steam engines were designed by Newcomen in the early 1700s (see Chapter 45), but they were relatively inefficient because of their method of cooling the steam in the cylinder. James Watt made a major alteration to the design of the steam engine by moving the condensing of the used steam into a separate cylinder. He also introduced double action (Figure 78-2), where the piston is powered by steam in both directions; automatic control of the engine's speed; and a method of taking an up/down motion and turning it into rotation.

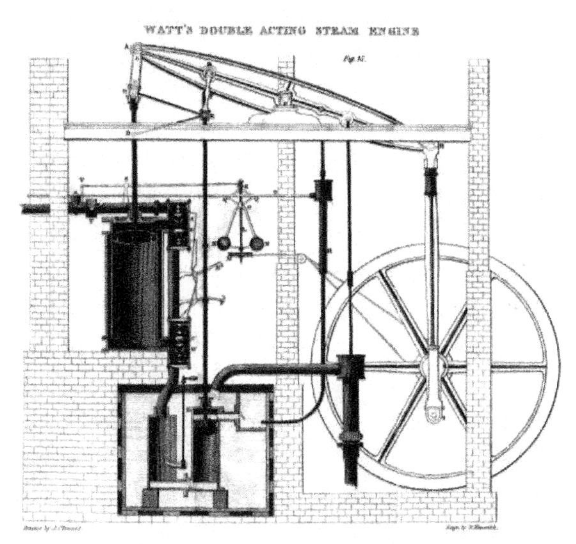

Figure 78-2. Watt's double-acting steam engine, with secondary condenser and vacuum pump at the bottom

Newcomen's engine steam worked by pushing a piston up inside a cylinder. Then water was sprayed inside to condense the steam, creating a partial vacuum and enabling atmospheric pressure to drive the piston down again. Watt's big innovation was to separate the condensation of steam into its own cylinder. This removed a major inefficiency from Newcomen's engine—the main cylinder never had to be cooled, and therefore never had to be reheated when new steam was allowed in.

Watt's secondary condenser was evacuated by a small pump connected to the main engine, so that when an appropriate valve was opened, steam from the cylinder would enter. The entire secondary condenser was kept cool (by immersing it in water) and so the steam entering would condense, creating a partial vacuum in the main cylinder that allowed atmospheric pressure to push the piston down again.

Next, Watt made the piston double-acting. Instead of relying on atmospheric pressure to push the piston in one direction, Watt used steam in both directions, with an appropriate system of valves to direct steam into either end of the cylinder. This simple idea introduced a major problem—in the Newcomen engine, the piston was connected to a chain attached to the beam. But if the piston generated power in both directions a fixed linkage was needed, and that fixed linkage needed to cope with the absolutely straight motion of the piston.

To solve that problem, Watt invented his parallel motion linkage, which translated the up-and-down motion of the piston rod into an identical up-and-down motion of a parallel rod that could be used to do work. In Figure 78-2, the parallel motion linkage takes the motion of the piston and turns it into the motion of the rod that descends into the vacuum pump.

Figure 78-3 illustrates parallel motion. As the piston is raised and lowered, it moves the corners of the parallelogram BCDE. Each corner of the parallelogram is a flexible joint, and BC is extended into the rod KC, which pivots around the point A. So the angle of the parallelogram changes with the piston's movement.

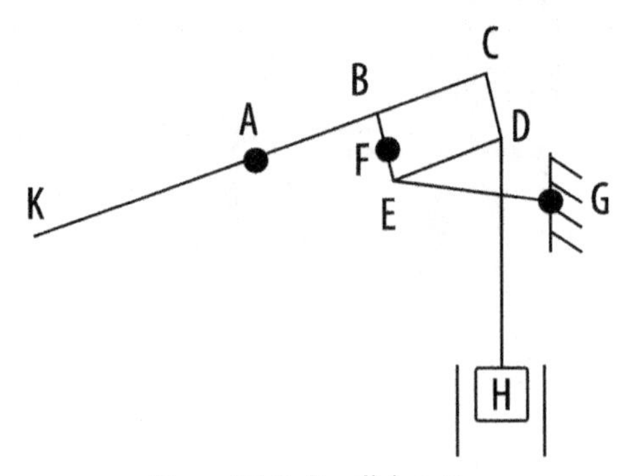

Figure 78-3. Parallel motion

To constrain the point F (where the additional parallel rod could be connected to extract power), the parallelogram is connected to a fixed point G from the corner E. As the piston moves, the resulting motion of F is an elongated figure eight that approximates a straight line.

To control the speed of the engine, Watt used a centrifugal governor (Figure 78-4), which consisted of a pair of metal balls connected to a linkage that allowed them to fly outward as the linkage rotated. By connecting the governor to the engine and to a valve controlling the steam entering the piston, the engine's speed was automatically controlled. As the engine sped up, the balls flew outward, reducing the incoming steam; when the engine slowed down, the balls fell inward and allowed more steam to enter.

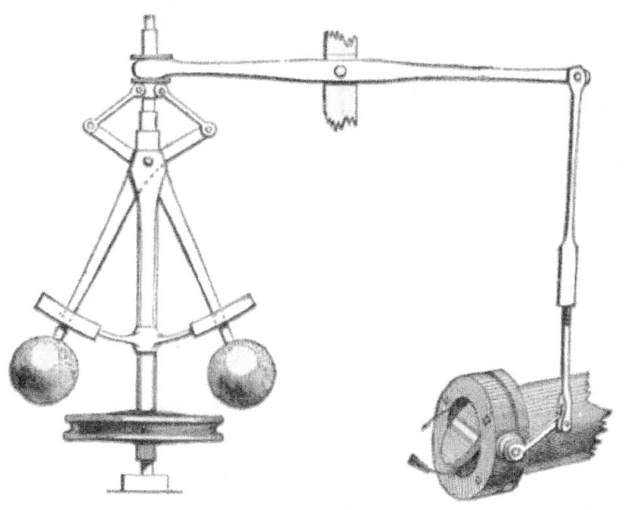

Figure 78-4. Centrifugal governor

Finally, the up-and-down motion needed to be turned into rotation of a flywheel so that power could be easily transmitted via belts and used to operate machinery. For this, Watt invented the sun and planet gear (Figure 78-5). The up-and-down motion of the piston is turned into rotation by a small gear that rotates around a gear fixed directly to the flywheel. This small "planet" gear does not rotate around its own axis; it is fixed to the rod, conveying motion from the piston.

Figure 78-5. Sun and planet gear

With these major innovations (and others such as the steam jacket, which enclosed the exterior of the cylinder in a wall of steam to keep it warm), Boulton and Watt's steam engines were very efficient, and became the standard for steam power during the Industrial Revolution.

079

Westminster Abbey, London, England

51° 29′ 58″ N, 0° 7′ 39″ W

The Collegiate Church of St Peter at Westminster

Most people think of Westminster Abbey as the church where British monarchs are crowned, married, and buried, and it is one of a few churches and chapels directly controlled by the British Queen. But Westminster Abbey is also a burial place where great Britons are laid to rest with honor, making it the most important scientific burial ground in the world.

Westminster Abbey honors the following great scientists: Charles Darwin (see Chapter 6), Paul Dirac, James Clerk Maxwell (see Chapter 35), William Herschel, Sir Isaac Newton (see Chapter 69), Ernest Rutherford, J.J. Thomson (who discovered the electron), and Lord Kelvin (see Chapter 73) in the Nave. Newton's grave is ornately decorated (Figure 79-1); Darwin's is marked simply with the words "Charles Robert Darwin, Born 12 February 1809, Died 19 April 1882". One notable British scientist who is honored by Westminster Abbey but not actually buried there is Michael Faraday. His grave can be found in London's Highgate Cemetery.

Two names that may be unfamiliar are Thomas Tompion and George Graham. Tompion and Graham were famous clockmakers who built accurate clocks and watches in the 17th and 18th centuries. Tompion built clocks for the Royal Observatory at Greenwich (see Chapter 49) that only needed winding once per year. He was also a friend of Robert Hooke, who developed the fine springs that Tompion used to make his watches.

The abbey itself sits in the City of Westminster, close to the British seat of government at the Palace of Westminster. The oldest parts of the abbey date to the 11th century and are now home to a small museum. The abbey has undergone extensive construction work throughout its lifetime. In the 19th century, rebuilding and restoration work was performed by the architect Sir George Gilbert Scott.

Figure 79-1. Newton's grave; courtesy of Flickr user ccr_358

Although a free audio tour is available, the best way to understand the importance of the abbey is to take a Verger-led tour. For a small fee, knowledgeable Vergers take visitors on a 90-minute tour that covers the history of the abbey and those who are buried there.

Because the abbey is centrally located in London, it makes an ideal starting point for a day of exploring London's scientific places, such as the Science Museum (Chapter 77), The Hunterian Museum (Chapter 64), or the Faraday Museum (Chapter 75).

Practical Information

Since Westminster Abbey is used for official ceremonies, it is important to check that it is open to the public on a given day. Details are at *http://www.westminster-abbey.org/*.

Hooke's Law and Clocks

Three of the notables buried in Westminster Abbey were clockmakers (Harrison, Tompion, and Graham), attesting to the importance of keeping time. Two of the basic mechanisms for timekeeping are pendulums and springs. The Dutch scientist Christiaan Huygens built clocks using a pendulum (see page 126) because a pendulum's period is dependent only on its length; that meant that a pendulum could be nudged by a mechanism to keep it swinging without affecting the period.

The other method of timekeeping, commonly used in small clocks and watches, is a combination of a balance wheel and a balance spring.

Mechanical clocks consist of three important pieces: a means of obtaining energy, an escapement for controlling the use of the energy, and a mechanism for timing. Energy is typically obtained from a falling weight or a tightly wound spring. The escapement consists of a toothed wheel (called the escape wheel), and a curved bar with a pair of teeth (called pallets) that connect with the toothed wheel.

Robert Hooke invented one form of escapement called an anchor escapement (so named because of the shape of the curved bar). In this mechanism, the escape wheel is driven by the clock's energy (say from a large spring) and wants to rotate. It is prevented from rotating by the anchor (seeFigure 79-2).

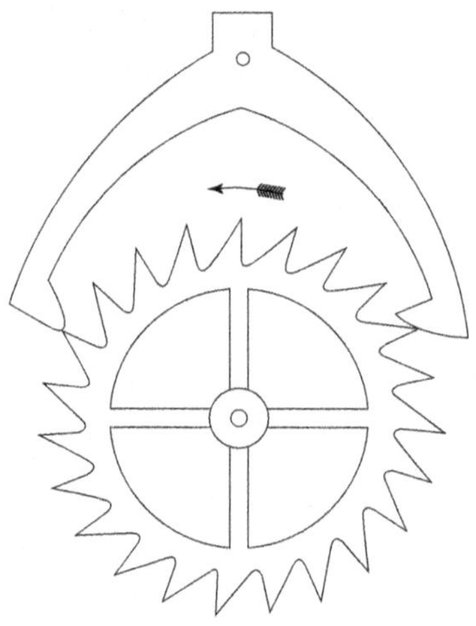

Figure 79-2. Anchor escapement

The anchor is controlled by either a pendulum swinging back and forth, or by a balance wheel. The back-and-forth motion of the anchor allows its teeth to release the escape wheel so that it moves one tooth, and is caught again by the opposite end of the anchor. As it is caught, energy from the escape wheel is transferred to the anchor and on to the pendulum or balance wheel and spring. The anchor both controls the movement of the escape wheel and transfers some of its energy to the controlling mechanism—in this way, it helps keep time and maintain the motion of a pendulum, for example.

Because of Hooke's Law, springs can also be used to create the same uniform motion as a pendulum. This law states that the force required to stretch a spring is proportional to the distance stretched. Formally, this can be written as shown in Equation 79-1.

$$F = -kx$$

Equation 79-1. Hooke's Law

That is, the force, F, is required to stretch a spring distance, x, where k is a constant related to the spring itself. This force F is negative, because it is the force exerted by the spring that would restore the spring to its original pre-stretched position.

If a spring with a weight of mass, m, is stretched and released, it undergoes a harmonic motion. The spring bounces up and down because of the force defined in Hooke's Law, and the frequency, f, of the motion is defined just in terms of the mass and the spring constant (see Equation 79-2).

$$\nu = \frac{1}{2\pi}\sqrt{\frac{k}{m}}$$

Equation 79-2. Frequency of spring oscillation

Hooke used this insight to create a mechanism in which a spring drove a wheel. His design used a straight spring, but in other more common mechanisms the spring is circular and causes a wheel to rotate back and forth (see Figure 79-3). The rotation of this so-called balance wheel controls the movements of the escapement. The frequency of the oscillation of the balance wheel is determined by its mass and the properties of the spring.

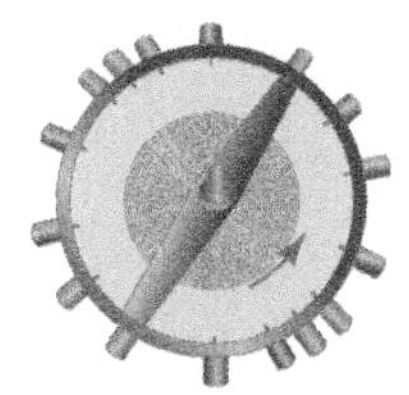

Figure 79-3. A balance wheel and spring

Hooke's anchor escapement had a big disadvantage, however: because of the shape of the teeth and anchor, it had the effect of pushing the escape wheel backward a little each time the wheel moved.

Graham fixed this problem by designing the deadbeat escapement (see Figure 79-4). In the deadbeat escapement there is no backward motion, because the teeth are sharply angled and the anchor is replaced with a pair of curved teeth that smoothly release the escape wheel and capture it again.

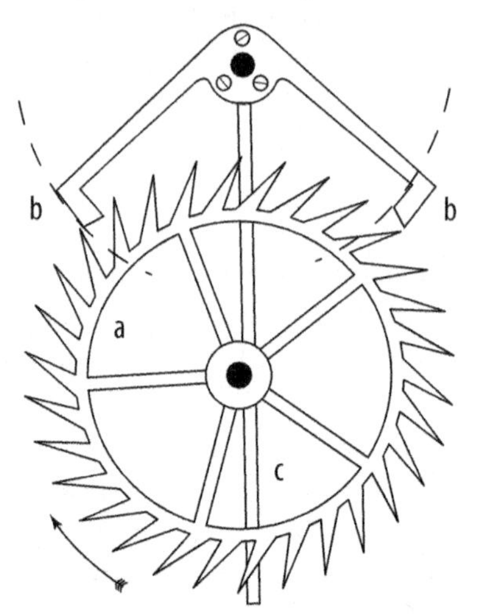

Figure 79-4. Deadbeat escapement

Energy is transferred from the escape wheel after the tooth has been released—it runs along the bottom of the ends of the pallets, which are slanted to receive a push from the escape wheel.

080

Chernobyl Exclusion Zone, Ukraine

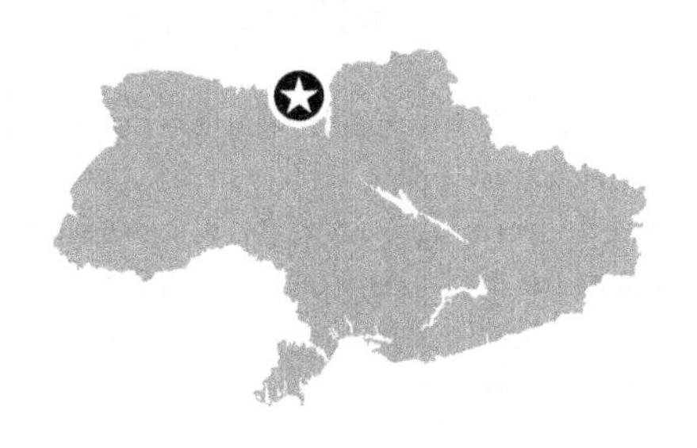

51° 18′ 0″ N, 30° 0′ 18″ E

Aftermath of the Chernobyl Disaster

On April 26, 1986, in the middle of the night, a steam explosion tore the roof off reactor number 4 at the Chernobyl nuclear power plant in Ukraine. The reactor did not have a containment building, so the explosion exposed the reactor core directly to the atmosphere.

The reactor was operating at the time of the explosion, and the graphite blocks that surrounded the reactor fuel were red-hot. With the addition of oxygen from the atmosphere, the graphite began to burn fiercely. To make matters worse, the fuel in the reactor was close to the end of its useful life and was filled with a wide variety of different radioisotopes.

Between the explosion and the fire, the Chernobyl disaster was the worst radiation accident in history. It led to the evacuation of the nearby town of Pripyat, 56 deaths, and a large increase in cancer deaths among the most highly exposed people. Ultimately, a 30-kilometer exclusion zone was created around the reactor and the population within the zone was ordered to leave. More than 350,000 people had to be relocated.

Radioactive fallout from the reactor fire contaminated a wide area. Close to the reactor itself, a large forest of pine trees was killed by fallout and became known as the "Red Forest" because of the color of the dead trees. The forest was bulldozed a year after the disaster and buried.

The radioactive plume spread across Belarus and on to Finland and Sweden, across Northern Europe and into North America. Today, around 5 million people live in parts of Ukraine, Belarus, and Russia affected by radioactive fallout.

The exclusion zone is still in place today and is almost totally uninhabited, except for some elderly people who refused to leave the area or who returned to their homes shortly after the disaster. Still, it is possible to visit the Chernobyl Exclusion Zone on specially organized tours.

Potassium Iodide and the Thyroid

One of the immediate dangers after the explosion was the presence of radioactive iodine in the food system. Radioactive iodine (iodine-131) is produced in nuclear reactors in normal operation as a product of the fission of uranium-235. The uranium-235 breaks apart when its nucleus is hit by a neutron, releasing energy and creating new elements from the split-apart atom. The elements typically created when uranium-235 breaks apart in a nuclear reactor are caesium, iodine, zirconium, technetium, strontium, promethium, and samarium. Typically, about 2.8% of these fission products are iodine-131.

Because the fuel in the Chernobyl reactor was nearing the end of its life at the time of the explosion, its uranium-235 had already been turned into fission products, which led to a large release of iodine-131. When nuclear fuel is extracted from a reactor and cooled for reprocessing or storage, the iodine-131 has enough time to decay, removing any danger of exposure.

Iodine-131 is particularly hazardous to humans because it concentrates in the thyroid gland and is easily passed into human foods. For example, cows that eat grass contaminated with iodine-131 will pass it into their milk, and when the milk is consumed by humans, iodine-131 will pass into the thyroid gland.

The iodine-131 concentrates in the thyroid because the thyroid uses iodine to create two hormones—thyroxine ($C_{15}H_{11}I_4NO_4$) and triiodothyronine ($C_{15}H_{12}I_3NO_4$). These two hormones—typically called T_4 and T_3, respectively—are involved in regulating the body's metabolism.

If radioactive iodine is present, the thyroid will use it to create T_4 and T_3, and the concentration of iodine-131 in the thyroid will likely cause thyroid cancer. The cancer is created because radioactive iodine undergoes rapid decay (its half-life is eight days) and releases beta-radiation (high-energy electrons) inside the thyroid.

To prevent this from happening, workers at Chernobyl were immediately given potassium iodide (KI) tablets. Soon afterward, almost 17 million people in the path of the fallout in Poland were also given potassium iodide as a preventive measure. This was the largest ever "experiment" in the use of potassium iodide to prevent thyroid cancer caused by radioactive fallout, and it was shown to be effective.

If taken before radioactive iodine enters the body, potassium iodide works as a prophylactic, overwhelming the thyroid with a source of iodine so that it will not attempt to use iodine-131 if it is later ingested.

Conversely, radioactive iodine-131 can be used to treat thyroid cancer once it has begun. Low doses of iodine-131 are also used to test thyroid function; high doses are used to destroy the thyroid completely.

Tours start from the Ukrainian capital, Kiev, about 70 kilometers south of Chernobyl, and typically last an entire day. Visits usually take in the town of Chernobyl (which is about 12 kilometers from the reactor itself), where a few people still live and where you'll find a poignant memorial to the firefighters who attempted to extinguish the burning graphite core.

Tours take visitors close to the power plant itself to see the massive sarcophagus built around the exploded reactor, and the partially constructed reactors 5 and 6 (reactors 1, 2, and 3 were kept operating until 2000). The tour also stops in the town of Pripyat, where time stopped in 1986. A ferris wheel sits abandoned and motionless in a disused amusement park. The blocks of flats, the cultural center, and the school sit with their furniture still in place, waiting for people to return. Books rot slowly in the library.

Above all, Pripyat is heartbreaking. A town of almost 50,000 people evacuated suddenly and with no warning, a testament to the long since passed Soviet Union, and by far the most striking reminder of the destruction a nuclear accident can create.

While you're in Kiev, take the time to visit the National Museum of Kiev. The museum helps explain the disaster and has more than 7,000 objects on display, including original photographs, secret documents from the Soviet era, and items from the abandoned homes. A video presentation describes the disaster and its aftermath.

Practical Information

See *http://www.ukraine.com/museums/chernobyl/* for information on the National Museum of Chernobyl.

081

Aurora Borealis, Fairbanks, AK

64° 50′ 16″ N, 147° 42′ 59″ W

The Northern Lights

The Aurora Borealis (also called the Northern Lights, and shown in Figure 81-1) appears in the ionosphere in northern latitudes and is visible on dark nights. The light displays are caused by the interaction between the solar wind and the Earth's magnetosphere (see sidebar). To see the northern aurora, it is necessary to travel far to the north—the aurora is roughly centered on the Magnetic North Pole (see Chapter 128).

In the U.S., a good place to see the aurora is Fairbanks, Alaska; the nights are dark enough, it's far enough north, and the city has relatively clear nights in the spring and autumn. The aurora is also visible from northerly parts of eastern Canada, Iceland, and northern Scandinavia.

In Fairbanks, spring and autumn (around the equinoxes) are the best times for aurora viewing because the sky is dark and clear. You'll have to stay up until local midnight (ignoring daylight savings time, since you want the actual middle of the night as determined by longitude), when the sky is darkest. For best results, escape Fairbanks by driving north until the city lights are no longer affecting visibility.

The most common aurora displays are sheets of green that hang in the air, but many other colors are also possible, from reds to deep violets. Like the weather, the Aurora Borealis is somewhat unpredictable, although aurora forecasts are available based on activity of the Sun and solar wind. To see the aurora, you need a combination of a strong solar wind and a clear sky.

There is an equivalent aurora at the South Pole (the Aurora Australis), but the land it hovers over is mostly uninhabited (apart from hardy souls living in Antarctica), so it is rarely seen.

The planets of our Solar System also have aurora. The Hubble Space Telescope has photographed aurora on both Jupiter (see Figure 81-2) and Saturn, and even moons such as Io, Ganymede, and Europa have auroral displays.

Figure 81-1. Aurora Borealis seen from the International Space Station; courtesy of Image Science and Analysis Laboratory, NASA–Johnson Space Center

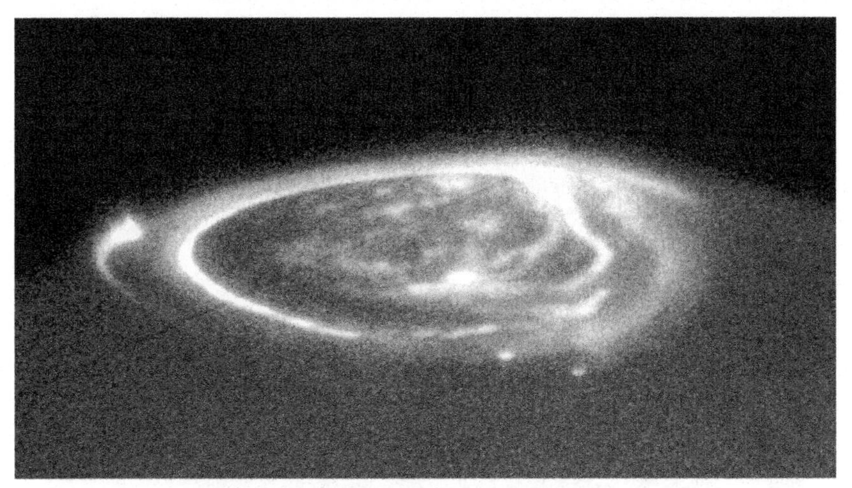

Figure 81-2. Jupiter's aurora; courtesy of NASA/ESA, John Clarke (University of Michigan)

Practical Information

A good starting point for aurora hunting is the University of Alaska at Fairbanks Geophysical Institute at *http://www.gedds.alaska.edu/AuroraForecast/*. The institute has forecasts of auroral activity for Fairbanks and around the world, and practical information about when and how to see an aurora.

The Magnetosphere

The aurora is created when the solar wind interacts with the Earth's magnetosphere. The solar wind consists of charged particles (mostly electrons and protons) that manage to escape the Sun's enormous gravity. These particles start in the Sun's corona (the outer layer of the Sun), which consists of a plasma heated to above 1,000,000 K. Oddly, the middle part of the Sun (the part we usually see) is considerably cooler at around 6,000 K.

Because of the incredible heat in the corona, the particles become separated from the atoms in which they originated and move rapidly away from the Sun (at between 400 kps and 750 kps) in all directions. When the particles approach the Earth, they encounter the Earth's magnetic field.

The portion of space dominated by the strength of the Earth's magnetic field is known as the magnetosphere. The particles in the solar wind are deflected around the Earth, resulting in the magnetosphere being shaped roughly like a bullet pointing toward the Sun (see Figure 81-3). Some of the particles from the solar wind get trapped into the magnetic field and drawn toward Earth, but the Aurora Borealis is mostly formed behind the Earth.

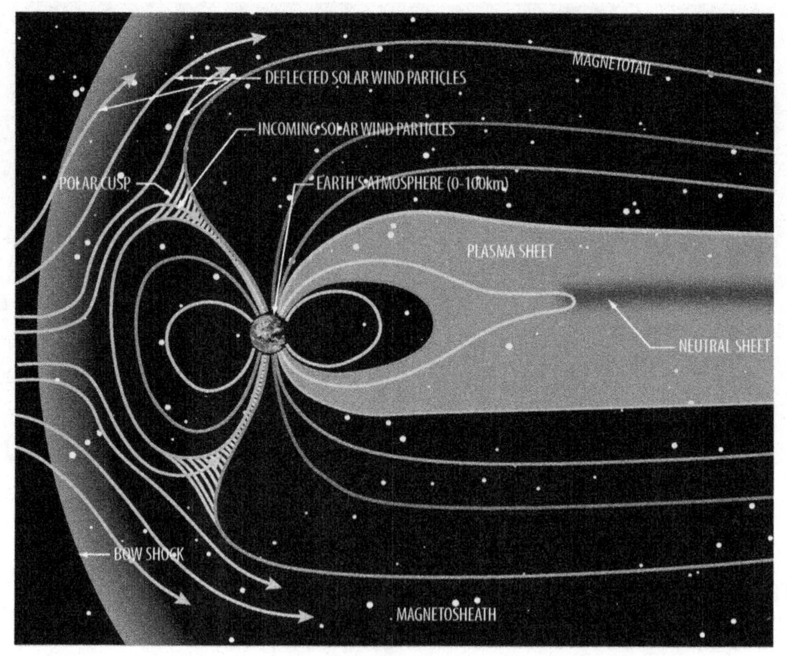

Figure 81-3. The magnetosphere and solar wind; courtesy of NASA/JPL-Caltech

Behind the Earth is the magnetotail, an area where the solar wind does not penetrate and the Earth's magnetic field stretches out into space. The magnetotail contains a plasma, where the flowing solar wind induces a large electrical field. The electric field is able to flow to the Earth along the magnetic field lines in the magnetotail. It's these electrons, from the plasma in the magnetotail, that create the aurora.

The magnetic field lines in the magnetotail flow through the two magnetic poles and essentially create a large electrical circuit in space. When the electrons hit the Earth's atmosphere, they create the aurora display by exciting oxygen and nitrogen.

When oxygen is excited, it eventually returns to its stable state, giving off photons with wavelengths corresponding to red or green visible light. The nitrogen undergoes the same excitation and also returns to a stable state, giving off red or blue light.

082

Trans-Alaska Pipeline Visitor Center, Fox, AK

64° 55′ 45.35″ N, 147° 37′ 47.05″ W

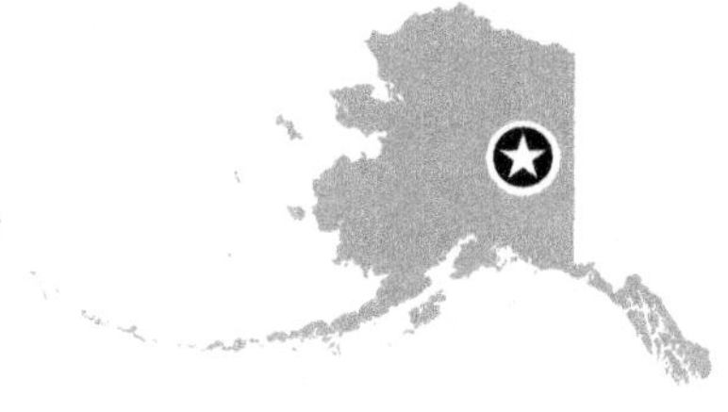

TAPS

The Trans-Alaska Pipeline System (TAPS) runs for almost 1,300 kilometers north-south across Alaska, bringing oil from the Arctic Ocean to the port of Valdez where tankers load up with crude to be shipped to the rest of the U.S. for refining. The pipeline crosses difficult terrain—between the north and south of Alaska, there are mountains, permafrost, hundreds of rivers, and active fault lines.

Most of the pipeline is difficult to access, and the pipeline company unfortunately no longer offers tours of its pump stations. But on the highway outside Fairbanks, you can stop off near the small commune of Fox, where a visitor center tells the story of the pipeline. From there you can walk right up to the pipeline itself before it disappears into the ground.

The pipeline was opened in 1977 and is 1.22 meters in diameter. The combination of Alaska's climate, landscape, and the temperature of the oil flowing through the pipeline creates many challenges.

Oil typically leaves the ground at around 80°C, and as it passes through the pipeline it has a temperature of around 50ºC. The above-ground section of the pipeline (about 675 kilometers) is elevated so that the heat does not melt the permafrost. It's impossible to place the pipeline completely underground because some parts of the permafrost are susceptible to thawing, which would endanger the pipeline as it moved. Even so, heat conduction through the supporting legs could melt the ground, causing the pipeline to become unstable. For this reason, the ground is protected by heat pumps that ensure that the permafrost does not melt. Anhydrous ammonia flows through pipes in the legs; if the ground becomes too hot, the ammonia vaporizes, cooling the ground and rising upward. The ammonia is then cooled again by the cold Alaska air, by fins sticking up above the pipeline.

Another 605 kilometers of pipeline is buried underground; an additional 6 kilometers is buried and surrounded by refrigerated brine to prevent melting.

Galvanic Corrosion and Cathodic Protection

The ability of batteries to make electricity by immersing two different metals in an electrolyte (see page 97) is very useful, but metal objects can accidentally become batteries and end up corroding. This galvanic corrosion is a problem for everything from the Statue of Liberty to a ship's rudder, and pipelines are particularly vulnerable because of their size. The problem can even occur when only one type of metal is present, because of slightly different compositions of the same base metal.

For example, the Statue of Liberty is made from a copper skin with an iron structure. When Gustave Eiffel built the statue, he anticipated that galvanic corrosion would be a problem, so he insulated the copper and iron from each other using the natural plastic shellac (see page 364). Over time the shellac insulation gave way, and between the metals an electrolyte (moist, salty marine air) was able to create a simple battery.

Galvanic corrosion occurs when metallic ions move from the anode to the cathode. This leads to the anode corroding more quickly than it would normally. In the Statue of Liberty, electrons flow from the copper skin to the iron support via the salty air, while iron ions move in the opposite direction. This results in corrosion of the iron.

On ships and on the Alaska Pipeline, galvanic corrosion is turned around to create cathodic protection. The pipeline is deliberately turned into a sort of battery by attaching strips of zinc, which will corrode before the metal to be protected.

In the same way, boat owners often affix zinc blocks (commonly called sacrificial anodes) to propeller shafts, rudders, and hulls to create batteries made from the boat's metal, the zinc block, and seawater. The zinc blocks corrode long before the more valuable parts of the boat (Figure 82-1).

Figure 82-1. Sacrificial anode on a boat's propeller shaft; courtesy of N. H. Anthony

For very large structures, applying zinc (or another metal) to create sacrificial anodes may be insufficient, because not enough current is created to protect the metal. In that case, a current may be created by connecting the structure and a sacrificial anode to a DC electrical supply. Parts of the Alaska Pipeline use such an "impressed current" cathodic protection system to prevent corrosion.

For protection from expansion and contraction damage and to cope with earthquakes, the pipeline zig-zags across the ground, and in places is mounted on sliding rods that allow it to move relative to the ground. Zinc ribbons are used throughout the pipeline to protect it from electrically induced corrosion (see sidebar).

Oil flows through the pipeline at a rate of around 1 million barrels per day (a barrel contains almost 160 liters), and the entire pipeline can contain over 9 million barrels. To inspect the pipeline while it is operating, autonomous robots called pigs (or "pipeline inspection gauges") are inserted at pump stations, and then removed further down the pipeline.

These pigs use both ultrasound and magnetic flux to look for corrosion in the pipe. Other simple pigs clean the interior of the pipeline, removing the waxy buildup of paraffin that can clog it. Still others check for dents or changes in the curvature of the pipeline's walls.

The oil is kept flowing in the pipeline by 11 pump stations (not all of which operate at the same time).

If you are in Fairbanks (perhaps to see the Aurora Borealis; see Chapter 82), it's well worth the short drive out to Fox to see the visitor center and observe the technology behind pumping oil across the state.

Practical Information

To find the visitor center, follow the Steese Highway north from Fairbanks for about 18 kilometers.

083

Titan Missile Museum, Sahuarita, AZ

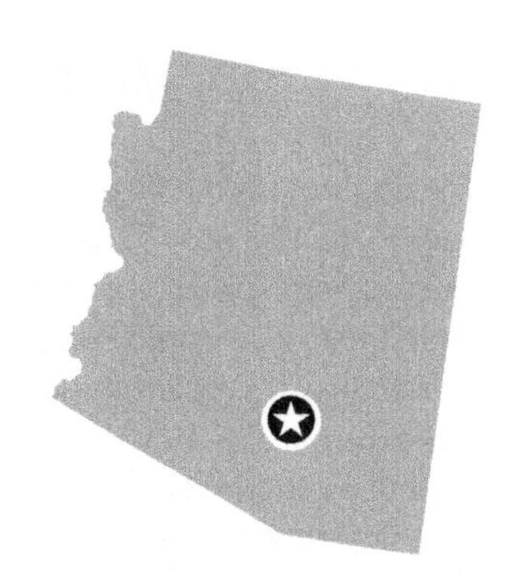

31° 54′ 10.35″ N 110° 59′ 55.53″ W

"Skybird, this is Dropkick with a red dash alpha message"

The Titan Missile Museum is genuinely unique: it's the only remaining Titan missile silo, its missile is still in the silo, the museum is run by a former Missile Combat Crew Commander, and all tours involve visiting the missile and the underground control room for a simulated launch.

The museum is about 30 kilometers south of Tucson in the Arizona desert. Everything at the museum is genuine—the silo is the last of 54 Titan missile sites that were dotted around the U.S. The others were destroyed at the end of the Cold War.

On the surface, there's a small museum of the Cold War, but everything that's really interesting is underground. After an introductory video, visitors enter the complex through a massive blast door more than 30 centimeters thick. When the silo was in use, this door was kept sealed shut. Everything inside the silo is mounted on springs, so if a Soviet missile were to land nearby, its explosion wouldn't damage the missile or its controls.

Down in the control room, the equipment looks dated (the missiles were in place from the 1960s to the 1980s) but the double-padlocked red box stenciled ENTRY RESTRICTED TO MCCC AND DMCCC ON DUTY stands out. That's where the orders and keys were kept for use in a real launch. The silo was manned by two people—each knew the combination of his own padlock, and each had a key inside the red box. The missile could be launched only if they turned their keys at the same time.

Missile and Rocket Engines

The engines on the Titan II missile are an example of a rocket engine. Rocket engines use Newton's Third Law (see Chapter 69) to produce movement—they eject at high speed (and high temperature) the gas resulting from burning fuel. Unlike a jet engine (see page 171), which sucks in air and accelerates it to create thrust, a rocket engine burns fuel to create thrust without needing any external air—because of this, rocket engines are capable of working in a vacuum. In the absence of air as a source of oxygen, they use an oxidizing chemical that breaks down to produce the oxygen necessary for the fuel to burn.

The Titan II rocket used hydrazine (N_2H_4) as fuel and dinitrogen tetroxide (N_2O_4) as the oxidizer. It had a pair of engines with identical parts, each consisting of a combustion chamber (where the fuel and oxidizer were mixed), and a pair of pumps connected to the separate fuel and oxidizer tanks.

When the pumps were turned on, the hydrazine and dinitrogen tetroxide entered the combustion chamber and ignited. Hydrazine breaks down and releases hydrogen, and dinitrogen tetroxide breaks down and releases oxygen. The pressure in the combustion chamber rose enormously, and some of the gases escaped from the chamber into the rocket engine nozzle. As the gases then left the nozzle, they generated thrust. (See Figure 83-1.)

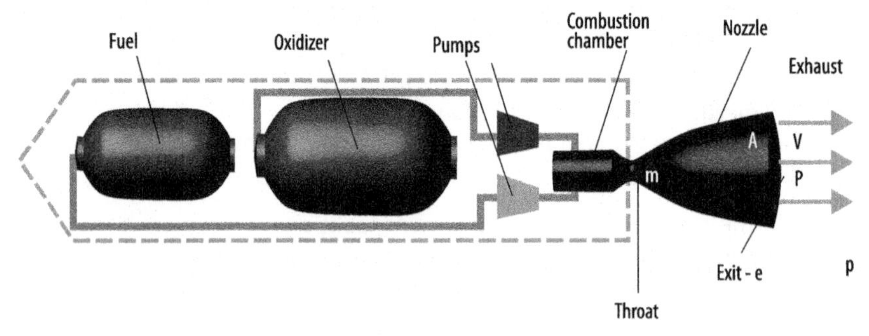

Figure 83-1. Rocket engine thrust

The amount of thrust is determined by the amount of gas leaving the nozzle and the difference in pressure between the exhaust and the atmosphere. The thrust, F, generated by the engine comes from two separate forces. The gas leaving the exhaust has a specific speed, V, and per unit of time a certain mass, m, of gas is ejected. The force generated by the escaping gas is mV.

But there's also a pressure difference between the gas and the surrounding atmosphere. This causes a second force on the missile as the gas expands from high pressure, P, to ambient pressure, p, over the area of the nozzle, A. The second term below means that rocket engines have additional thrust as they rise through the atmosphere because of reduced atmospheric pressure.

The formula for calculating overall thrust is shown in Equation 83-1.

$$F = mV + A(P - p)$$

Equation 83-1. Thrust equation

It's possible to simplify this equation further by defining a velocity, v, that is known as the equivalent exit velocity and takes into account the change in pressure. The equation then becomes F = mv, where v is given by the definition in Equation 83-2.

$$v = V + \frac{A(P - p)}{m}$$

Equation 83-2. Equivalent exit velocity

Newton's Third Law (that every action has an equal and opposite reaction) causes the missile to move because of the force generated by the gases exiting the nozzle.

The six-story Titan II missile was defanged and left outside so that it could be examined by Russia via satellite before being placed back into its original position. The most dangerous part of the missile was not the warhead (unless it was set off), but the propellant used. The propellant was a mixture of A-50 hydrazine (the fuel) and dinitrogen tetroxide (the oxidizer), which was toxic, corrosive, and hypergolic: mixed together, they would spontaneously ignite. The museum shows the procedures and protective clothing needed when refueling the missile.

Outside the silo at ground level, there's a display of the missile's engines and reentry vehicle (which contained the bomb), a missile refueling vehicle that was needed to keep the oxidizer cool to prevent it from boiling, and the entrance to the silo itself, from which the missile would be launched.

The museum offers a variety of tour options, from a one-hour guided visit to an extensive five-hour tour that takes in all parts of the silo, missile, control room, and crew kitchen and bedroom. By special request, it's even possible to sleep in the crew bedroom overnight—four times a year, and with plenty of notice, up to four people can sleep just feet from the missile itself.

Practical Information

The Titan Missile Museum website has full visiting and booking information; see *http://www.titanmissilemuseum.org/*.

084

391 San Antonio Road, Mountain View, CA

37° 24′ 17.68″ N, 122° 6′ 38.62″ W

Shockley Semiconductor

It's not unusual to hear people say of Silicon Valley, "there's no there there" (although that phrase is actually a quotation and refers to Oakland, California, just across the San Francisco Bay). It's quite true—Silicon Valley is one long industrial suburb that has encroached on the towns between San Francisco and San Jose. Finding its iconic places is no mean feat.

Since the founders of Hewlett-Packard were early pioneers of Silicon Valley, the garage where they started the company is a landmark (see Chapter 90), and Apple takes its name from the area's apple orchards (now replaced by featureless buildings). But the place that puts the silicon in Silicon Valley is 391 San Antonio Road in Mountain View. Today, the building houses a grocery store selling the sorts of produce that the area used to grow. To understand its significance to Silicon Valley, we need to go back to the 1950s.

In 1956, William Shockley and a group of engineers opened the Shockley Semiconductor Laboratory at #391. Shockley was one of the co-inventors of the transistor (more on transistors on page 332) and won the Nobel Prize the same year he helped found Silicon Valley. The new company (actually a division of a larger firm) planned to build semiconductors using silicon, instead of the more commonly used germanium.

The company did not last for long. By 1957, a combination of Shockley's personality and his focus on pure research instead of product creation led eight members of his staff to quit and form Fairchild Semiconductor (see Chapter 85). These men, sometimes called the traitorous eight, were Julius Blank, Victor Grinich, Jean Hoerni (who invented the planar process used to make silicon chips), Eugene Kleiner (who went on to found the seminal Silicon Valley venture-capital firm Kleiner Perkins), Jay Last, Gordon Moore and Robert Noyce (who founded Intel), and Sheldon Roberts. Fairchild produced the first integrated circuit in 1959, and many ex-employees of the company went on to found the first Silicon Valley startups.

The First Transistor

When the original inventors of the transistor (Shockley, John Bardeen, and Walter Brattain) tried to patent their invention, they hit a problem—the transistor had been patented in the 1920s by the Austro-Hungarian physicist Julius Lilienfeld. Lilienfeld had described a field-effect transistor in a U.S. patent in 1925, and another in 1928; Shockley and his team didn't create their point-contact transistor until 1947.

The point-contact transistor (created by Bardeen and Brattain) consisted of a block of germanium (which is a semiconductor), topped with a V-shaped piece of plastic. Around the plastic were two sheets of gold, which touched the germanium at the point of the V, very close to each other without being electrically connected (see Figure 84-1).

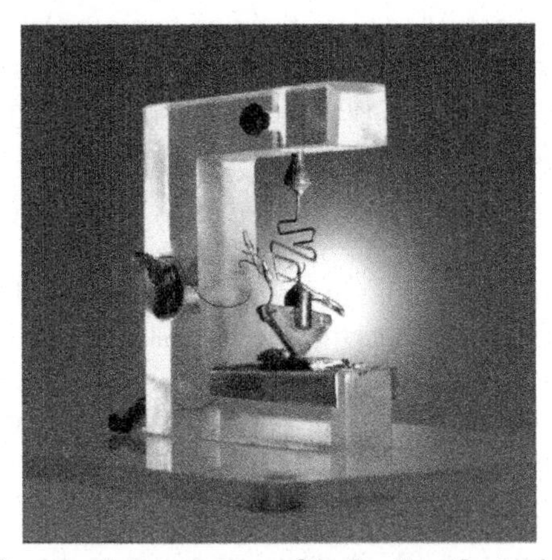

Figure 84-1. The first transistor; courtesy of the Porticus Centre (www.porticus.org)

The germanium sat on a metal plate. The metal plate and the gold foil formed three connections to the germanium. A current applied between one of the gold connections and the metal base controlled the amount of current able to flow between the base and the other piece of gold—the transistor was able to act both as a switch and as an amplifier.

This happened in the point-contact transistor because when a current was applied between a gold contact and the base, the surface of the germanium became a P-type semiconductor (one with too few electrons, hence P for positive) and the rest of the germanium was a natural N-type semiconductor (with an excess of electrons, hence N for negative). So applying a current across one pair of contacts created a PN-junction (an area where P-type and N-type material meet) that controlled the flow of electricity between the second gold contact and the base.

Shockley subsequently invented the bipolar junction transistor, which had three pieces of semiconducting material arranged as a pair of NP and PN junctions (either as NPN or PNP; see page 114). This transistor had the advantage of using only a very small current to control its operation, whereas the point-contact transistor used relatively large currents.

Back in the 1920s Lilienfeld had described yet another form of transistor—the FET, or field-effect transistor (Figure 84-2). Unlike the junction transistor, or the point-contact transistor, where the contact that controls the transistor is electrically connected to the semiconductor, a FET has a "gate" connection that is insulated from the rest of the device.

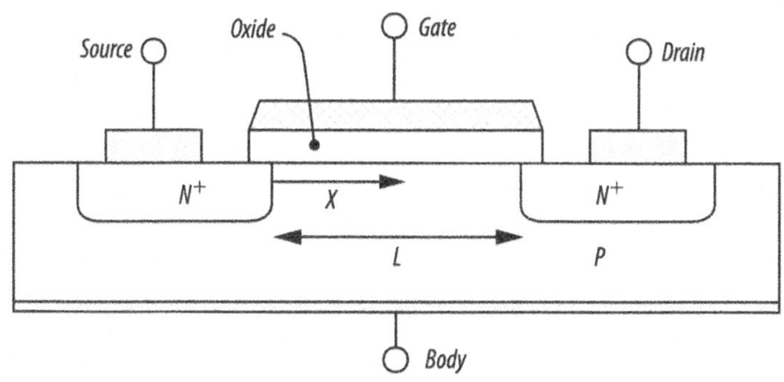

Figure 84-2. Field-effect transistor

In a FET, the gate controls the current flowing between two contacts called the source and drain. The source and drain are connected to an N-type semiconductor (which has free electrons); between the source and drain is a P-type semiconductor that acts as an insulator.

When a voltage is applied between the gate and the source (the source being negative), it creates a "channel" in the P-type semiconductor through which electrons can flow. Since the gate is insulated from the P-type, the electrons will flow from source to drain. The channel is created because the electric field creates a depletion layer (an area of imbalance between positive and negative charge) in the P-type, with the positively charged "holes" being forced away from the gate. Within the depletion layer, electrons are free to flow.

The FET has a big advantage over the other two types of transistor—there's no current flowing between the gate and source, which reduces its power consumption.

The point-contact transistor was the first to be created, and then came the junction transistor. But Lilienfeld's field-effect transistor (which was finally fabricated in the 1960s) gets the last laugh—it's the transistor that's packed into microprocessors and other silicon chips in your computer, cell phone, or MP3 player.

The product that Shockley was most interested in creating was the thyristor. A regular transistor consists of three layers of semiconductor material; the thyristor has four (see Figure 84-3). A transistor will act as a switch only when it is turned on by a current on its base connection—if the current stops, the switch turns off. A thyristor only needs a momentary current on its gate connection to turn the switch on; the switch will remain on until the current flowing through it stops.

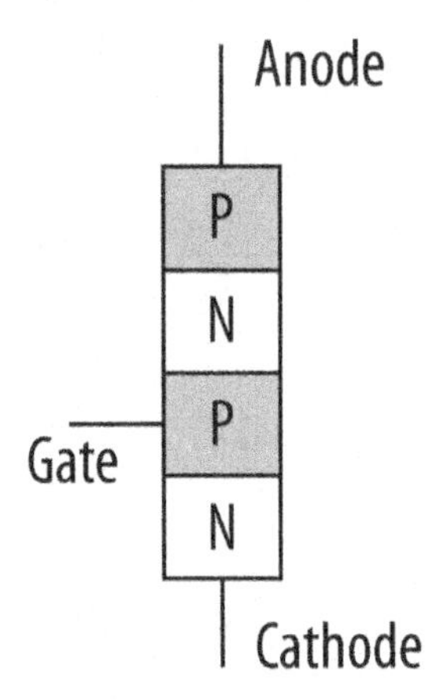

Figure 84-3. A thyristor

Shockley failed to develop a commercial thyristor—although others later did—and he went on to become a professor at Stanford University. Nevertheless, the company he founded is the birthplace of Silicon Valley through its alumni and focus on silicon semiconductors.

Unfortunately, there's not much technology to see at #391 today, but it's a short drive from the HP Garage and the Fairchild building, making it an important stop on a Silicon Valley day out. Don't forget to pick up some apples while you're there!

Practical Information

Punch the address into your GPS (or Google Maps), and you'll find it with ease. There's a small website maintained by former Shockley employees and friends at *http://www.shockleytransistor.com/*.

085

844 E. Charleston Road, Palo Alto, CA

37° 25′ 19.02″ N, 122° 6′ 12.18″ W

Fairchild and the Fairchildren

Shockley Semiconductor (see Chapter 84) spawned the prototypical Silicon Valley startup company, Fairchild Semiconductor, founded in 1957. Innovative yet focused on products, not just research, Fairchild went on to spawn many other Silicon Valley greats (including Intel, LSI Logic, and AMD), and its offspring are frequently referred to as the Fairchildren. And it was at Fairchild that the planar process for silicon chip manufacturing was invented.

Around the same time, in 1958/1959, Jack Kilby (at Texas Instruments) and Robert Noyce (at Fairchild) invented the integrated circuit, where multiple transistors (and other components) were made on a single piece of silicon. Kilby built a circuit based on the mesa process (see sidebar) and connected the components with fine gold wires; Noyce's circuit used the planar process, which had superior wiring. The planar process won the day, but Kilby won the Nobel Prize.

Using the planar process, Fairchild was able to build working integrated circuits and started supplying prototypes in 1960. In 1961 it announced a line of integrated circuits with the claim that they would reduce the size of a computer by up to 70%. The simple circuits included logic and flip flops (which can act as a single bit of memory, being either on or off), and were the forerunners of the microchip revolution that continues to this day. Each chip had just a few transistors on it—a flip flop can be made with just two—but the number of transistors on a chip would rapidly increase.

In 1965 Gordon Moore (one of Fairchild's founders) wrote an article in *Electronics* entitled "Cramming More Components onto Integrated Circuits." In it he noted that the number of components on an integrated circuit would double every two years (he estimated that by 1975 there would be around 65,000 components on a single chip). This prediction is now known as Moore's Law and still

The Planar Process

Making integrated circuits (or silicon chips, as they are commonly called) requires building layers of semiconducting material (P- and N-type; see pages 114 and 324) to make transistors and other components. The components are then linked together with metal wiring, and external wires are connected so that the chip can be wired up to a circuit board.

Chips were initially created by the "mesa process." A wafer of silicon was made with layers of N- and P-type silicon on it; then, powerful chemicals were used to etch valleys into the silicon, leaving flat-topped sandwiches of N- and P-type silicon in the right shape to form transistors. These sandwiches were named mesas, after the flat-topped hills found in some parts of the southwestern U.S.

At Fairchild, Jean Hoerni developed the more flexible planar process. The planar process starts with a wafer of silicon, which is covered with a layer of silicon oxide (SiO_2) by oxidizing the silicon at high temperature in the presence of either steam or pure oxygen. The silicon oxide layer acts as an insulator.

Then, parts of the original silicon are exposed by etching through the silicon oxide. This is done by a photolithographic process in which a layer of a photoresist chemical is applied to the silicon oxide. This chemical is then exposed to ultraviolet light through a mask that defines the pattern of areas on the wafer that should be exposed. Where the light falls, the photoresist chemical changes. The wafer is then washed with a chemical that eats away the photoresist layer where it was exposed to ultraviolet light, and eats away the silicon oxide below, exposing the silicon wafer.

Silicon can be turned into a semiconductor by exposing it to chemicals that create impurities in the wafer. By exposing the silicon to boron, free electrons in the silicon are removed, and the exposed areas become a P-type semiconductor. If arsenic is used instead, an excess of free electrons is created in the silicon, and it becomes an N-type. This process is called *doping*.

Once doped, the silicon is covered over again with silicon oxide, and a new pattern can be etched into it using photolithography. In this way, different parts of the silicon can be doped to produce layers of N- or P-type semiconducting material in patterns that build up transistors and other components.

Finally, the components can be linked together by exposing just the areas where metal should connect with the wafer again using photolithography, spraying on a layer of aluminum, and etching away the aluminum to create wires. In modern integrated circuits, this process is applied repeatedly to build up layers of components and connections.

In Figure 85-1, an NPN transistor is created from a silicon wafer (which has been predoped so that it is N-type). This is done by first covering the wafer with silicon oxide (a), etching a hole in the oxide layer (b), doping to make a P-type area and reoxidizing (c), etching another hole (d), doping and reoxidizing an N-type area (e), etching holes for contacts (f), and finally applying aluminum to make contact points (g).

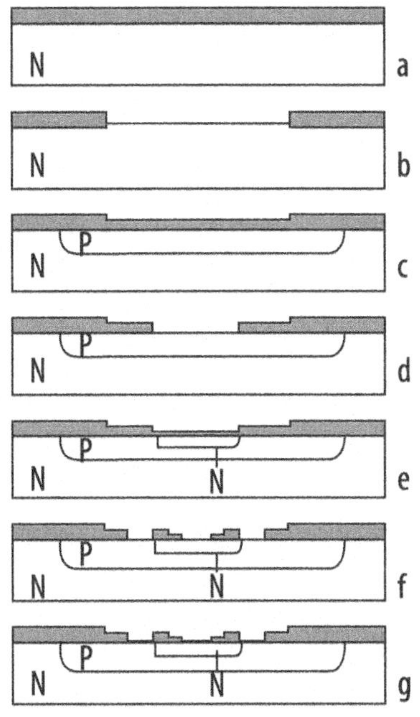

Figure 85-1. Building an NPN transistor by the planar process

holds true today (for example, the Intel Itanium Tukwila processor has 2 billion transistors on a single chip). The law has also been extended into other areas of digital electronics, including hard disk capacity and the number of pixels per dollar on digital cameras.

Fairchild grew and was successful in the new integrated circuit marketplace (playing second fiddle to Texas Instruments), but in 1968 Gordon Moore and Robert Noyce left after the company got into financial trouble. Their next start-up, Intel, is a household name. Fairchild Semiconductor exists to this day.

The actual Fairchild building is still standing at 844 E. Charleston Road in Palo Alto. Unfortunately, like the Shockley Semiconductor Laboratory building, there's no actual technology to be seen (the building is not even occupied by a technology firm), but a photograph of the scientific tourist next to the state historic monument marker is a must for any nerd's photo album.

After you've had your photo taken at the Fairchild building, check out the HP Garage (see Chapter 90), which is just a short drive away, also in Palo Alto.

Practical Information

The Fairchild building is now occupied by a design firm called Collective Creations. Outside the building is a plaque erected by the California Department of Parks and Recreation, commemorating it as the site of the first "commercially practicable integrated circuit."

086

The Computer History Museum, Mountain View, CA

37° 24′ 51.74″ N, 122° 4′ 36.54″ W

Silicon Valley's Finest

Almost all major science museums around the world have sections tracing the history of the computer, but they all pale in comparison to the Computer History Museum in Mountain View, California. The museum is housed in the old Silicon Graphics headquarters (itself an iconic Silicon Valley building), which the museum bought in 2002. Silicon Graphics was once a maker of high-performance computers for computer graphics that made possible 1990s films like *Jurassic Park*; its former headquarters is located on North Shoreline Boulevard next to Charleston Park, and the design and architecture of the building speak of the extreme wealth of Silicon Valley prior to the end of the Internet boom.

The museum is a work in progress, with as much information available online through its excellent website (*http://computerhistory.org/*) as in the building itself. But unlike many computer science museums (including the worth-avoiding Tech Museum of Innovation in San Jose), the Computer History Museum brings computing to life through exhibits that are explained both in general terms for the casual visitor, and in glorious detail for the gearheads. You'll get the most out of your visit by going on a docent-led tour. These tours are not canned recitals from a script, but truly bring to life the machines in the collection. In addition, everything is free—the museum is supported by donations of money, machines, and software. Consider making a donation before you leave.

The Transistor and NAND Logic

Fundamental to the electronic computer revolution is the transistor. A transistor is a small electronic device that can act as an amplifier (they were used to make the "transistor radios" of the 1950s and 1960s, which were the first truly portable music players) or as a switch controlling the flow of electricity in a circuit.

A semiconductor is a material (or more commonly, a combination of materials) that is neither a conductor of electricity nor a complete insulator. The first semiconducting material (silver sulphide) was discovered in 1833 by the great British scientist Michael Faraday (see Chapter 75). Faraday discovered that when heated, the resistance of silver sulphide was lowered and the compound conducted better (and the effect reversed as it cooled).

In a transistor, the amount of electricity flowing through a semiconductor, typically silicon, is controlled electrically. This control means that the transistor can operate as either a switch (turning on or off the flow completely), or an amplifier (since the amount of flow is proportional to the voltage applied).

A simple silicon transistor, called an NPN, is used to make the electronic switches that are the building block of computers. The NPN transistor consists of three pieces of silicon known as the base (often written B), collector (C), and emitter (E), as shown in Figure 86-1.

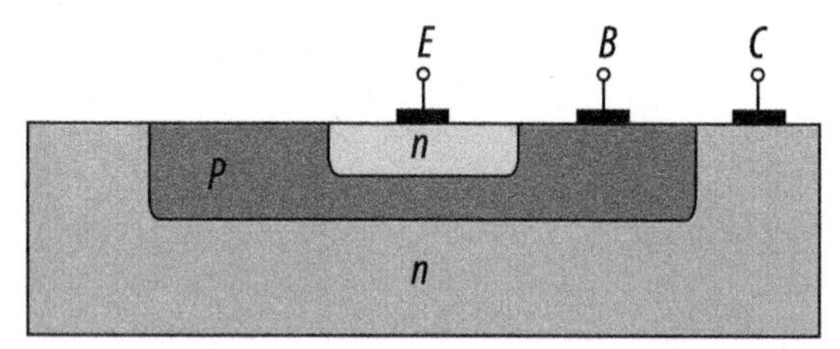

Figure 86-1. Parts of an NPN transistor

The base of the transistor is connected to the P; this area is silicon that has been modified in a process called doping, where the silicon is deliberately exposed to a chemical such as boron, removing free electrons. The P stands for positive, reflecting the dearth of electrons. It is the base B that controls whether the transistor switch is on or off.

The collector and the emitter are made of N-doped silicon; this is silicon that has been exposed to a chemical like arsenic to increase the number of mobile electrons available to pass a current. The N indicates that this region is filled with negatively charged particles: electrons. With no voltage applied between the base and the emitter, the base material acts as an insulator and the transistor switch is off. Once a sufficient voltage is applied between B and

E, electrons are able to move between the collector and the emitter and the switch is on.

NPN transistors are usually drawn using the symbol in Figure 86-2 showing the emitter, base, and collector. The base, on the left, controls the flow of electricity between the collector and emitter.

Figure 86-2. A transistor

Computers use binary logic consisting of just 0s and 1s for all calculations (see section on Binary; page 218). This is partly because binary is easy to work with electrically: no current flowing is 0, current flowing is 1. A transistor is simply a switch that's on when current is flowing through the base (when it's handling a binary 1) and off when there's no current (a binary 0).

Using transistors, it's possible to build up the basic logical functions needed by a computer. Computers combine two or more binary numbers using so-called logic gates to make decisions. For example, an AND gate expresses the concept "If inputs A and B are both 1 then output a 1, otherwise 0"; an OR gate expresses "If inputs A or B or both are 1 then output a 1, otherwise 0"; and a NOT gate expresses "If input A is 0 then output a 1 and vice versa." From these simple, logical decisions, it's possible to build entire chips.

But one such logic gate, called a NAND, trumps all the others because it's the only one necessary. NAND means NOT AND and expresses the concept "If either of input X or Y is 0 then output a 1, otherwise 0." It's the opposite of AND: where AND would say 1, NAND says 0. It turns out that any other logic gates (AND, OR, NOT, and some other more exotic ones) can be made just from NANDs, and a NAND gate can be made from two transistors (Figure 86-3).

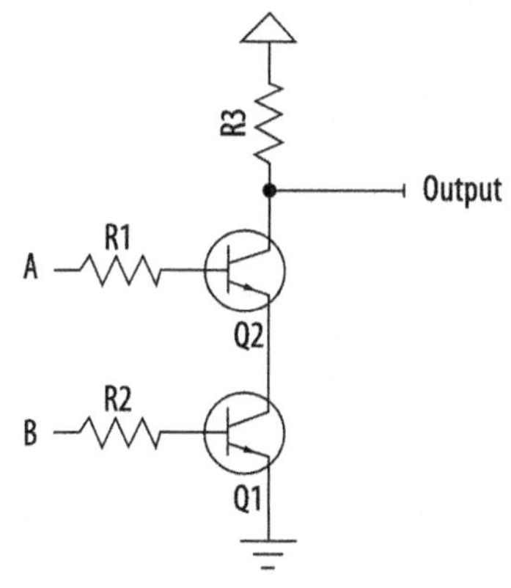

Figure 86-3. A NAND gate

If both of the inputs A and B are turned on (a binary 1), then both transistors turn on, and no current flows from the output because it's flowing through both transistors: the output is 0. If either of the inputs is turned off (a binary 0), then one of the transistors will be off, and the current will flow through the output, giving a binary 1.

NAND gates are usually drawn with the symbol inFigure 86-4.

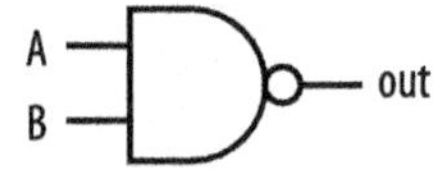

Figure 86-4. Symbol for a NAND gate

Making the other logic gates from NAND gates is fairly simple: to make a NOT gate (which outputs 1 when its input is 0 and vice versa), just join the inputs A and B together (Figure 86-5).

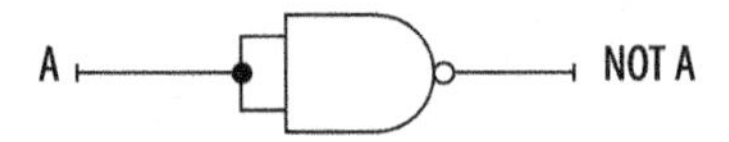

Figure 86-5. A NOT gate from a NAND gate

From that, it's easy to make an AND gate by taking a NAND gate and flipping its output using a NOT (since NAND is NOT AND, and NOT NOT AND is just AND). See Figure 86-6.

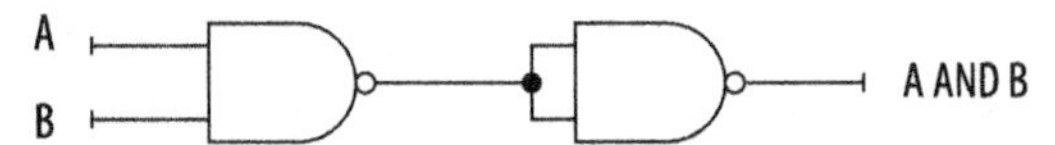

Figure 86-6. An AND gate from two NAND gates

OR is the most complex, requiring three NAND gates (the first two work as NOT gates, flipping the inputs A and B). See Figure 86-7.

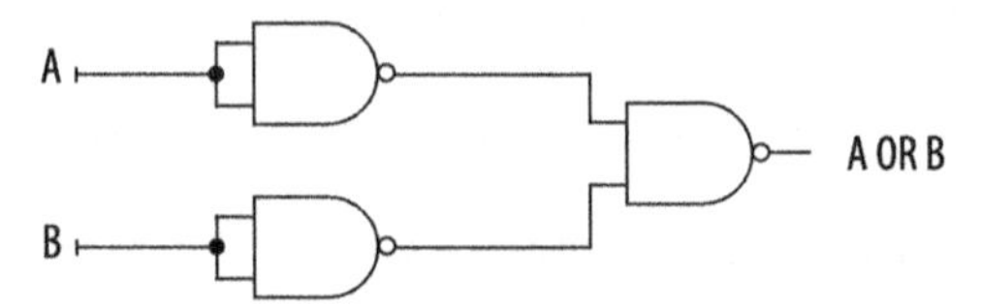

Figure 86-7. An OR gate from three NAND gates

A tour of the 500 pieces of the collection on display is a good way to get oriented. The collection includes bits of 1954's SAGE (Semi-Automatic Ground Environment; see Figure 86-8), which was built as part of the U.S. air defense mechanism during the Cold War. SAGE weighed 300 tons, and incorporated just one comfort item for the operator looking out for Soviet bombers: a cigarette lighter. Some parts of the SAGE system were still in use in 1983, with replacement parts purchased from Soviet-bloc countries because they were unavailable in the U.S.

Figure 86-8. SAGE on display at the museum; courtesy of Flickr user Jurvetson

The 1960s collection includes the industry-dominating IBM 360 mainframe computer with its spinning tape drives. 1960s attitudes are on display with the Honeywell Kitchen Computer sold by upscale department store Neiman Marcus in 1969: the advertising tagline reads, "If she can only cook as well as Honeywell can compute." The computer incorporated a cutting board and was preprogrammed with a collection of recipes; it's not clear if any were sold.

The museum has an excellent collection of supercomputers including a CRAY-1, CRAY-2, and CRAY Y-MP, as well as a Connection Machine CM-1. More recent computer history covering the personal computer revolution includes exhibits of the Altair 8800 (the machine that made Bill Gates leave Harvard and start "Micro-Soft"), an Apple I, the original IBM PC, the almost-but-not-quite portable Osborne 1 Portable Computer (only 24 pounds!), the Commodore 64, the Apple Lisa and Macintosh, and a French Minitel. Along with the machines, the museum has also preserved software including Microsoft Windows 1.0, and peripherals such as the 1978 Dover Laser Printer.

Pre-computer history is covered with a collection of mechanical devices (such as slide rules), a Hollerith Census Machine from 1889 that used punch cards, and a Jacquard Loom (also using punch cards for weaving).

A demonstration well worth seeing is the operation of the only working DEC PDP-1 in the world (this happens twice monthly, on a Saturday afternoon). The PDP-1 was first built in 1959 (the example on display was made in 1963 and restored to full working condition in 2004) and was the first commercial, interactive minicomputer. Data is fed in using a punched or magnetic tape, and the user interacts with the machine using a typewriter/printer combination, or the screen and lightpen. The PDP-1 has a memory of about 9 kilobits, but that was enough for serious scientific applications, playing music, and a game: *Spacewar!*.

Spacewar! is demonstrated along with the PDP-1, with two players flying ships orbiting a star, and trying to shoot each other while avoiding the star's gravity.

Practical Information

The Computer History Museum is located on North Shoreline Boulevard in Mountain View, California, and is accessible directly from US-101 at the Shoreline Boulevard exit. It can be reached from the Mountain View Caltrain Station via the Caltrain Shoreline Shuttle. Full details can be found on the museum's website, *http://computerhistory.org/*.

087

Goldstone Deep Space Communications Complex, Fort Irwin, CA

35° 23′ 22.21″ N, 116° 50′ 57.79″ W

Apollo Valley

When probes sent out into deep space get in contact with Earth, they are usually talking to one of three places—Fort Irwin, California; Madrid, Spain; or Canberra, Australia. These three locations form NASA's Deep Space Network of dishes and antennas communicating with space probes far from home.

The three locations provide 24-hour coverage of space as the Earth rotates, and the entire network is controlled from the Goldstone Deep Space Communications Complex in California. The complex is situated at the Fort Irwin military base, about 60 kilometers north of Barstow. Despite its location on a military base, it is open for public tours.

The first spacecraft tracked by the complex were Pioneer 3 and 4 in 1958/1959. These small probes sent back information about radiation between the Earth and the Moon in the Van Allen radiation belt. Ever since, the Deep Space Network has been involved in receiving pictures and data from spacecraft and sending back commands.

Some of the spacecraft have been flying away from the Earth for decades, making their signals very weak, but the Deep Space Network still stays in communication with them. Pioneer 10 was launched in 1972 and last communicated with the Earth in 2003; it is still flying out in the direction of the star Aldebaran. Earlier Pioneers 6, 7, and 8 are still capable of contacting Earth.

Voyager 1 was launched in 1977 and is still in contact, despite being over 15 billion kilometers from home (which means that a radio transmission from Voyager 1 takes 14 hours to reach Earth). It is now the most distant man-made object in space, and has moved outside the influence of the Sun's gravity. Voyager 2 launched the same year and is also still talking to ground control.

More recent missions, such as those to Mars that left the plucky Spirit and Opportunity rovers on the planet's surface, were all controlled from Goldstone and its satellite stations in Spain and Australia. As well as receiving data from these probes and rovers, the Deep Space Network has on many occasions performed extraterrestrial software upgrades by sending out improved software.

The history and work of the Deep Space Network is covered by a visitor center on site. Just outside the visitor center is a 34-meter dish that was used initially for Project Echo (see page 411). To get a feeling for the scope of the site, it's best to take the tour—the Goldstone complex covers 137 square kilometers and contains many large dishes.

The oldest dish on site (26 meters in diameter) was used for the early Pioneer mission and is now decommissioned. Part of the complex known as Apollo Valley was initially used for training Apollo astronauts in the bleak desert landscape. Six dishes (two of which are no longer used) stand in Apollo Valley. The largest dish on site (70 meters in diameter) is called the "Mars antenna" (see Figure 87-1) because of its regular communication with multiple Mars-bound spacecraft. It's also the most sensitive dish on site, and is used to contact the Voyagers.

Figure 87-1. The Mars antenna; courtesy of NASA/JPL-Caltech

Practical Information

When you are planning a visit to the Goldstone Deep Space Communications Complex, it's absolutely essential to book a tour in advance. Details are at *http://deepspace.jpl.nasa.gov/dsn/features/goldstonetours.html*.

Error Detecting and Correcting Codes

Noise is a constant problem when communicating with distant spacecraft, and can result in the corruption of the binary signal being received—some of the 0s become 1s and some of the 1s become 0s. Without ways to detect such corruption and fix the resulting errors, communication with deep space would be impractical.

To solve the noise problem, two sorts of codes can be used: error detecting (which indicates when noise has corrupted a stream of 1s and 0s) and error correcting (which allows the receiver to fix the corruption).

Error detection is common in everyday life. A normal 16-digit credit card number, such as 4417 1234 5678 9113, actually consists of 15 digits, plus one digit that is used for error detection. The final digit, 3 in the example, is used to verify that all the other digits are correct. Specifically, this digit can spot any incorrect digit in the number, as well as swapping two numbers by accident.

To check a credit card number for validity, the following steps are applied:

- Reading from right to left, double every other digit, leaving intermediate digits alone. So 4417 1234 5678 9113 becomes 8,4,2,7,2,2,6,4,10,6, 14,8,18,1,2,3.
- If any of the digits has become a number greater than or equal to 10, add the digits together to get a single-digit number. So now the number becomes 8,4,2,7,2,2,6,4,1,6,5,8,9,1,2,3.
- Add all the digits together and see if the answer is divisible by 10. In this example, the sum is 70 and therefore the number is valid.

If any of the digits are incorrect or swapped, the sum is not divisible by 10 and the card number is invalid. When creating a new card, the first 15 digits are the actual account number, and the last digit is calculated so that the sum will be divisible by 10.

A similar technique is used in ISBNs. This book's ISBN is 978-0-596-52320-6. Once again, the final digit is there to check for errors in the other digits. This algorithm is slightly different:

- Take the first 12 digits of the ISBN (i.e., excluding the check digit). Multiply every other digit by 3. For this book, the numbers 9,7,8,0,5,9,6,5,2,3,2,0 become 9,21,8,0,5,27,6,15,2,9,2,0.
- Add the digits together and work out the remainder when dividing by 10. The sum of this book's digits is 104, which gives a remainder of 4.
- If the remainder is 0, then the check digit is 0. Otherwise, subtract the remainder from 10. For this book, 10 – 4 = 6 is the check digit.

These error detecting codes work well when it's possible to ask again for the information (such as repeating a credit card number over the phone, or rescanning a book's barcode), but they are not very useful when a distant spacecraft's transmission is received with garbled digits. The spacecraft is so far away that it's impractical to ask for retransmission. That's where error correcting codes come into play.

In an error correcting code, the idea is to transmit a piece of data followed by a small amount of error correction data that can be used to both detect errors and correct them, even though both the data and the error correction data could be corrupted. One such error correcting code is the Hamming Code.

The Hamming Code builds on the simplest error detecting code of all—the parity bit. Imagine a simple stream of bits, say 1001101, being transmitted by a spacecraft. The simplest code of all would be to count the number of 1s in the stream (in this case there are 4) and add a single bit indicating whether there was an even or odd number of 1s. In this example, a 0 would be added to indicate that an even number of bits are 1 (if there were an odd number, then a 1 would be added). The spacecraft would transmit 10011010.

On the ground, the parity bit can be easily checked to see if there's any corruption. But parity bits aren't very useful when used alone, because they can only reliably detect when a single bit has been corrupted (flipped from 1 to 0, or 0 to 1). If more than one bit has been corrupted, the parity bit could be incorrect. And the parity bit doesn't show which bit has been flipped.

The Hamming Code uses multiple parity bits to not only detect single bits but also double-bit errors, and correct any flipped single bit. It works by transmitting seven bits of information for every four bits of data; this is called a Hamming (7,4) code.

Suppose that a spacecraft wants to transmit four binary bits: d1, d2, d3, and d4. It calculates a parity bit for groups of three bits selected from the original four. Parity bit p1 is the parity of d1, d2, and d4; parity bit p2 is the parity of d1, d3, and d4; and parity bit p3 is the parity of d2, d3, and d4. This is illustrated in Figure 87-2.

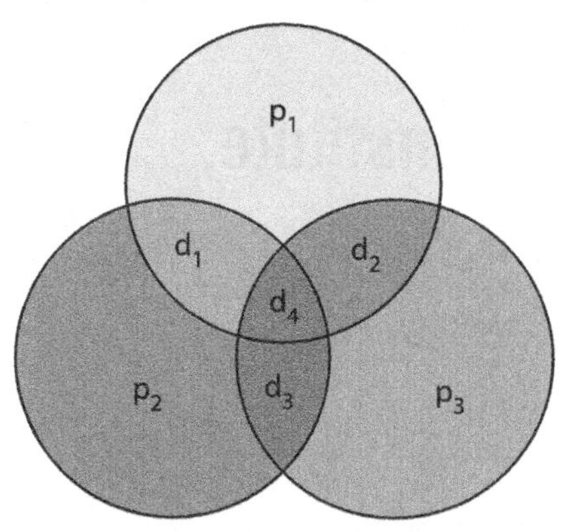

Figure 87-2. Hamming (7,4) code

Now suppose that one of the bits, say d3, has been corrupted and flipped. d3 is used to calculate the parity bits p2 and p3, so they will be incorrect, but p1 will still be correct. Looking at the overlapping circles in the diagram, it's easy to see that if p2 and p3 are incorrect, then one of the bits d1, d2, d3, or d4 has been corrupted, but since p1 is correct, then d1, d2, and d4 can be eliminated, showing that the error is in d3.

So not only does the receiver know that d3 is corrupt, but it can correct it. The same logic applies to any other single bit—they are each uniquely tracked by some combination of p1, p2, and p3.

Now suppose a single parity bit, say p2, has been flipped. This will indicate that one of d1, d3, and d4 is corrupt. But since the other two parity bits are correct, the receiver can conclude that there's nothing wrong with the transmitted data, and that it's the parity bit that's been corrupted.

Today the Hamming Code has been displaced by much more sophisticated error correcting schemes, which are able to correct more than the single-bit corruption fixable by a Hamming Code. Error correcting codes have become a hidden part of daily life—they are used in CDs and DVDs, satellite TV broadcasts, hard disks, and cellular telephones.

088

Joint Genome Institute, Walnut Creek, CA

37° 55′ 55.37″ N, 122° 1′ 20.1″ W

Gattaca

The Joint Genome Institute's Production Genomics Facility in Walnut Creek, California, doesn't look like ground zero of a biological revolution, but set back from leafy Mitchell Drive is a 5,600-square-meter industrial building where a large section of the human genome was sequenced.

In 1997, the U.S. Department of Energy set up the Joint Genome Institute to bring together work on DNA sequencing performed by Department of Energy laboratories around the country. The space leased in Walnut Creek was used to build a new laboratory that took part in the Human Genome Project.

In 1990, under the guidance of James Watson (co-discoverer of the structure of DNA; see Chapter 71), the Human Genome Project was created with the goal of sequencing the human genome by 2005. Sequencing a piece of DNA involves breaking it down into its most basic elements, the four nucleotides thymine (T), adenine (A), guanine (G), and cytosine (C), and finding the order in which they appear.

The nucleotides join together in pairs (A joins with T and C joins with G) called base pairs. The pairs can join in either direction (i.e., it's possible to have an AT pair or a TA pair), and in reporting the sequence for the DNA, only one half of the pair is mentioned.

The sequence of base pairs, and hence the sequence of the letters A, T, C, and G, is the genetic code of a living organism. The pairs are chained together in the DNA and are used to produce amino acids. In turn, the amino acids are used to create proteins.

For example, the Bovine circovirus (a simple virus) has a DNA sequence with 1,768 entries:

ACCAGCGCACTTCGGCAGCGGCAGCACCTCGGCAGCACCTCAGCAGCAACATGCCCAGCAA
GAAGAATGGAAGAAGCGGACCCCAACCACATAAAAGGTGGGTGTTCACGCTGAATAATCCT
TCCGAAGACGAGCGCAAGAAAATACGGGAGCTCCCAATCTCCCTATTTGATTATTTTATTGT
TGGCGAGGAGGGTAATGAGGAAGGACGAACACCTCACCTCCAGGGGTTCGCTAATTTTGTGAA
GAAGCAAACTTTTAATAAAGTGAAGTGGTATTTGGGTGCCCGCTGCCACATCGAGAAAGCCA
AAGGAACTGATCAGCAGAATAAAGAATATTGCAGTAAAGAAGGCAACTTACTTATTGAATGTG
GAGCTCCTCGATCTCAAGGACAACGGAGTGACCTGTCTACTGCTGTGAGTACCTTGTTGGA
GAGCGGGAGTCTGGTGACCGTTGCAGAGCAGCACCCTGTAACGTTTGTCAGAAATTTCCGCG
GGCTGGCTGAACTTTTGAAAGTGAGCGGGAAAATGCAGAAGCGTGATTGGAAGACCAATGTA
CACGTCATTGTGGGGCCACCTGGGTGTGGTAAAAGCAAATGGGCTGCTAATTTTGCAGACCCG
GAAACCACATACTGGAAACCACCTAGAAACAAGTGGTGGGATGGTTACCATGGTGAAGAAGTG
GTTGTTATTGATGACTTTTATGGCTGGCTGCCGTGGGATGATCTACTGAGACTGTGTGATCGA
TATCCATTGACTGTAGAGACTAAAGGTGGAACTGTACCTTTTTTGGCCCGCAGT

ATTCTGATTACCAGCAATCAGACCCCGTTGGAATGGTACTCCTCAACTGCTGTCCCAGCTGTA
GAAGCTCTCTATCGGAGGATTACTTCCTTGGTATTTTGGAAGAATGCTACAGAACAATCCACG
GAGGAAGGGGGCCAGTTCGTCACCCTTTCCCCCCCATGCCCTGAATTTCCATATGAAATAAAT
TACTGAGTCTTTTTTATCACTTCGTAATGGTTTTTATTATTCATTTAGGGTTTAAGTG
GGGGGTCTTTAAGATTAAATTCTCTGAATTGTACATACATGGTTACACGGATATTGTAGTC
CTGGTCGTATTTACTGTTTTCGAACGCAGTGCCGAGGCCTACGTGGTCCACATTTCTAGAG
GTTTGTAGCCTCAGCCAAAGCTGATTCCTTTTGTTATTTGGTTGGAAGTAATCAATAGTG
GAGTCAAGAACAGGTTTGGGTGTGAATTAACGGGAGTGGTAGGAGAAGGGTTGGGGGATTG
TATGGCGGGAGGAGTAGTTTACATATGGGTCATAGGTTAGGGCTGTGGCCTTTGTTACAAAGT
TATCATCTAGAATAACAGCAGTGGAGCCCACTCCCCTATCACCCTGGGTGATGGGGGAGCAG
GGCCAGAATTCAACCTTAACCTTTCTTATTCTGTAGTATTCAAAGGGTATAGAGATTTTGTTG
GTCCCCCCTCCCGGGGGAACAAAGTCGTCAATATTAAATCTCATCATGTCCACCGCCCAGGAG
GGCGTTCTGACTGTGGTAGCCTTGACAGTATATCCGAAGGTGCGGGAGAGGCGGGTGTTGAA
GATGCCATTTTTCCTTCTCCAACGGTAGCGGTGGCGGGGGTGGACGAGCCAGGGGCGGCGGCG
GAGGATCTGGCCAAGATGGCTGCGGGGGCGGGTCCTTCTTCTGCGGTAACGCCTCCTTGGA
TACGTCATAGCTGAAAACGAAAGAAGTGCGCTGTAAGTATT.

The Human Genome Project set out to determine the sequence of nucleotides in humans. This was an enormous task because of the number of base pairs present—around 3 billion. Using a technique called shotgun sequencing (see sidebar), the task was completed in 2003, two years ahead of the original schedule. The entire human genome can be browsed on the Web at *http://www.ensembl.org/Homo_sapiens/*.

Happily, the Joint Genome Institute offers free tours of its facility. Tours take about an hour, and visitors get to see close up the machinery that performs DNA sequencing. Tours begin with an explanation of the institute, the history of DNA sequencing, and the Human Genome Project. Visitors then get to see the robotic machinery that automatically dissects DNA to determine its sequence.

Practical Information

Go to *http://www.jgi.doe.gov/education/tours.html* for information on the Joint Genome Institute. It is imperative to book tours in advance.

Shotgun Sequencing

Dealing with a genome as long as the human one required devising a method that could cope with billions of base pairs; methods that worked for organisms with short genomes (like the Bovine virus described previously) were not up to the task. By the end of the 1970s, the idea of splitting a long chain of DNA into small fragments and sequencing them was in use. This technique made use of computer power to reconstruct the original DNA from the fragments, and became known as shotgun sequencing.

Shotgun sequencing starts by taking large strands of DNA to be sequenced and splitting them in a HydroShear machine. The DNA is diluted and forced through tiny holes in a piece of ruby. As the DNA passes through the holes, the liquid pressure causes it to stretch and snap in random places. Unfortunately, this process also mixes up the fragments of DNA, so reassembly isn't as simple as putting the pieces back together in the order they snapped.

After the fragments of DNA are collected, they are inserted into a plasmid (a piece of DNA that a bacterium is willing to receive and incorporate into its own DNA). The specially prepared plasmid pUC18 has two extra pieces of useful DNA—one that provides antibiotic resistance, and another that causes bacteria that incorporate it to turn blue.

The specially prepared plasmids are then mixed in a solution with E. coli bacteria. The bacteria are persuaded (by electric shock) to accept the plasmid and its three pieces of DNA. The E. coli are then treated with an antibiotic to kill off any bacteria that didn't accept the DNA plasmid. This leaves a population of specially prepared E. coli that can reproduce to replicate the original fragments of DNA.

Next, the E. coli are allowed to grow and form colonies. Some colonies will be blue (because they accepted the plasmid and incorporated it), and others will be white (because they survived the antibiotic despite failing to incorporate the plasmid correctly). A camera-toting robot picks up the blue bacteria and lets them grow further until billions of E. coli are present, containing the many copies of the fragments of DNA to be sequenced.

Another robot extracts the bacteria and places them in a machine that heats them to 95°C, causing them to burst open and free the plasmids containing copies of the DNA to be sequenced. Next, further chemical reactions (mostly handled by robots) cause the DNA base pairs to be colored with fluorescent dyes. Each base pair is dyed a different color, so that the base pair sequence can be seen by observing the colors.

The DNA is then passed through a capillary into a machine that "reads" the colored base pairs using a laser, thus obtaining the DNA sequence for that fragment of DNA. But the process is only capable of reading the first 1,000 or so base pairs in any fragment. Therefore, the entire process is designed to split many copies of the DNA into many different fragments that are then partially sequenced. Because the fragments are random chunks of DNA, the same piece of DNA will appear in many different fragments. Using these overlapping fragments, a computer can then be used to piece together the matching sequences to produce a complete picture of the original DNA.

Specially developed software (see Figure 88-1) examines the sequences determined from the fragments and assigns a quality measure to each base pair (to determine the reliability of the identification of that base pair). Other software, all developed by the University of Washington, then pieces together the matching fragments.

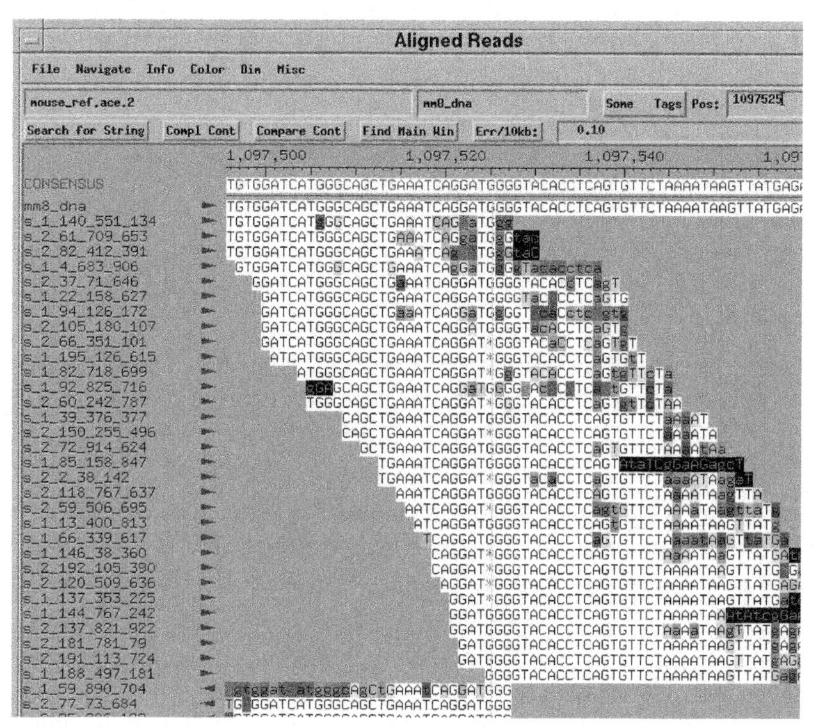

Figure 88-1. Aligning fragments of DNA by computer; courtesy of David Gordon, Howard Hughes Medical Institute at University of Washington

089

1 Infinite Loop, Cupertino, CA

37° 19′ 54.16″ N, 122° 1′ 50.46″ W

Visiting the Mothership

For the true Apple fan, visiting an Apple store isn't enough: a pilgrimage to Apple headquarters in Cupertino, California, is a must. Apple's campus is located just off Interstate 280 south of the De Anza Boulevard exit. Starting from San Francisco International Airport, the building is about a 45-minute drive down the self-proclaimed "World's Most Beautiful Freeway," which avoids the grubby industrial parks of Silicon Valley and takes the scenic route down the San Francisco Peninsula.

Arriving at Apple's campus off De Anza Boulevard, you'll come first to 1 Infinite Loop, the main office of Apple's six HQ buildings. The road circles around, surrounded by parking lots; between the buildings are landscaped gardens for Apple employees.

In front of number 1 is a sign written in two Apple fonts: the number 1 is in Apple's old Chicago font (still seen on older iPods) and "Infinite Loop" is in Apple's version of Garamond. But that's not the only secret message for the initiated: the very name of the street is joke for a computer programmers.

Fundamental to the operation of computers is the ability to repeat operations. And the most common way to repeat is to loop: to go back to the beginning of some procedure and do it again.

For example, a computer might sort a list of names by comparing the first two names and swapping them if they were out of order; it would then compare the second and third names, and so on, repeating the process until the end of the list was reached. The computer would then start again from the beginning, stopping only when the list was ordered (see Figure 89-1).

Loops are fundamental to all computers, including the first theoretical computers known as Turing Machines. A theoretical question posed by the machine's inventor, Alan Turing, was whether it was possible to find out if a loop ever stopped: this was known as the Halting Problem (see Chapter 66). If a loop never stops, it's an infinite loop, and usually causes a problem. (Next time your computer stops responding, be it an Apple or a PC, it's probably stuck in an infinite loop.)

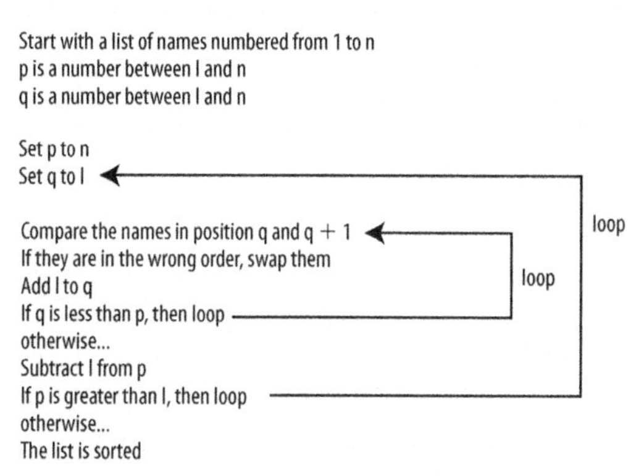

Figure 89-1. A computer program using two loops

The only thing an outsider can visit at 1 Infinite Loop is the company store, but it's unlike any other Apple store. The store sells the usual range of Apple products (although not including computers), books, and accessories, plus what can only be described as Apple swag: miscellaneous products adorned with an Apple logo. You can come away with an Apple notebook (filled with sheets of paper!), mousepads, T-shirts, and coffee mugs.

Pride of place goes to the black Apple shirt bearing the slogan: "I visited the Mothership" (Figure 89-2).

Figure 89-2. The Mothership shirt; courtesy of Kenneth Yan

Practical Information

There's more information on the Apple Company Store at 1 Infinite Loop, including directions, at *http://www.apple.com/companystore/*.

Deadlock

Computers sometimes get into infinite loops because of programmer error, but a common and more subtle problem occurs when two programs try to operate at the same time. In the right (or wrong) circumstances, two programs can get into a deadlock where each program is waiting for the other to do something. While waiting, they both sit in loops, constantly asking, "Has the other program finished yet?"

Imagine two programs that both need to print a document at the same time. Since a printer can only print one document at a time, one of the programs will need to have priority over the other. The programs might also need access to the hard disk to read the documents they are printing.

The following sequence of events can lead to a deadlock:

1. Program A asks the computer for exclusive access to the printer and is granted it.
2. Program B asks the computer for exclusive access to the hard disk and is granted it.
3. Program A now starts waiting for the hard disk to become available so it can read the file it wants to print.
4. Program B starts waiting for the printer to become available so that it can print.

Programs A and B now both sit in loops, constantly checking to see if the printer or hard disk is available. Since both programs need both devices, and neither is willing to give up a device once it has been granted access, the computer is stuck in infinite loops.

This particular problem doesn't happen in today's machines, because the hard disk can be accessed by multiple programs at once, and because the printer has a queue. The print queue means that exclusive access to the printer is not needed; programs can just add the document to be printed to the queue. But deadlock problems still exist, because computers are doing more than one thing at once.

Many current PCs have multiple processors (sometimes called cores) that allow them to work on more than one thing at the same time. And even if they don't, operating systems like Microsoft Windows and Mac OS X enable multiple programs, or parts of programs called threads, to operate at the same time.

This is not a problem, unless those threads need to share access to some resource.

For example, a user might ask a photo program to change the size of a digital picture. For speed, the computer breaks the photograph into four parts and starts four threads (Figure 89-3). The four threads all do the same thing: they change the size of one-quarter of the picture.

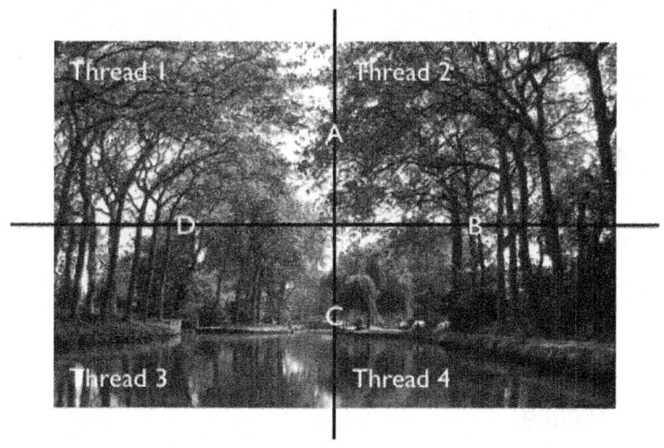

Figure 89-3. Four boundaries in a photograph

At the boundaries of the four parts, the threads need to cooperate so that the picture lines up correctly. That means the threads will need to cooperate on which one is working on which boundary.

To cooperate, the threads use a mechanism called a lock. For each boundary there's a corresponding lock, and the thread can't work on a boundary until it has "grabbed the lock." A thread grabs a lock by asking for exclusive access: it either gets access or sits in a loop waiting for access to be granted.

Here's how four threads can get themselves into a deadlock:

1. Thread 1 wants to work on boundaries D and A, so it first grabs the lock for A.
2. Thread 2 wants to work on boundaries A and B, so it first grabs the lock for B.
3. Thread 3 wants to work on boundaries B and C, so it first grabs the lock for C.
4. Thread 4 wants to work on boundaries C and D, so it first grabs the lock for D.
5. All the threads then try to grab the other lock they need, only to find that some other thread has already taken it. They are now all stuck waiting in a circle for another thread to "release" a lock.

A deadlock won't necessarily occur in this situation; for example, if thread 1 had managed to grab the locks for A and D, then it could get its work done, release them, and the other threads could operate.

The main difficulty in fixing deadlock problems is that they are affected by timing. If programs or threads operate at slightly different speeds, the deadlocks may not occur, or may occur very infrequently. That makes tracking them down time consuming and frustrating.

090

The HP Garage, Palo Alto, CA

37° 26′ 35.05″ N, 122° 9′ 17.32″ W

367 Addison Avenue

A common piece of Silicon Valley mythology is the company that started in a garage and went on to become the Next Big Thing. But some companies really did start in garages: Apple was started in Steve Jobs's parents' garage in Los Altos, California, and Google's founders rented a garage in Menlo Park, California, and worked for five months in it (and in the owner's hot tub).

But almost 40 years before either of Google's founders was born, Bill Hewlett and Dave Packard decided to "make a run for it" and try to build their own business. They found a house for rent, complete with a garage. Dave Packard and his wife, Lucile, lived in the ground-floor apartment, and Bill Hewlett got to tinker in the garage and sleep in a shed.

That was 1938, and Hewlett and Packard's first product was a simple audio oscillator dubbed the Model 200A, to make it sound like the newly formed Hewlett-Packard (or HP) had been in business for a while. The Model 200A undercut and outperformed other oscillators on the market, and Walt Disney became an early, happy customer. The cheap audio oscillator was put together in the garage with the paintwork on its case baked on in the oven tended by Lucile Packard. Lucile also worked at Stanford University to help support the new startup, and in the evenings did all the domestic work and handled Hewlett-Packard's correspondence and bookkeeping.

In 1989, the house at 367 Addison Avenue was dedicated as the Birthplace of Silicon Valley. In 2000, the Hewlett-Packard company bought it and restored the house, the garage, and the shed.

The HP Garage is not open to the public, although HP has opened the doors on some occasions to reveal that the interior contains period books, diagrams, and HP products. The garage can be easily photographed from the street, though, and visitors are welcome to do so, but try to contain your excitement: Addison Avenue is in a quiet residential area of Palo Alto known as Professorville.

The Wien-Bridge Oscillator

The Model 200A oscillator was based on a circuit called a Wien-bridge oscillator, with a significant modification by Bill Hewlett that made its name: it was free of distortion, and the amplitude of the wave form it generated was constant. Bill Hewlett patented the circuit in 1939.

A Wien-bridge oscillator is a simple oscillator circuit capable of producing a sine wave output with a specific frequency. In the Model 200A, the frequency could be adjusted by changing the capacitance of a pair of capacitors via a dial on the front, and it could produce sine waves between 35 Hz and 35 kHz.

The Wien-bridge oscillator is based on a measurement device called, unsurprisingly, a Wien bridge (see Figure 90-1). The Wien bridge is designed to measure an unknown resistance and capacitance (in series) in the same way that the Wheatstone bridge is used to measure an unknown resistance. Where the Wheatstone bridge uses DC, the Wien bridge uses an AC power supply.

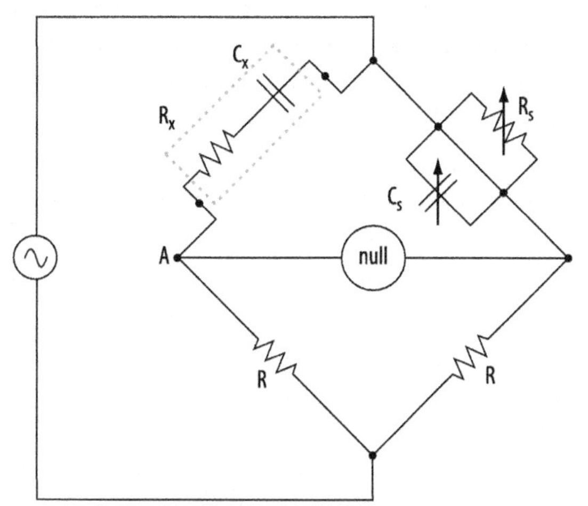

Figure 90-1. A Wien bridge

In the Wien bridge, the variable resistance R_s and capacitance C_s are adjusted until no AC voltage is present across the bridge (measured between A and B). From these values, the unknown resistance R_x and capacitance C_x can be calculated.

In a Wien-bridge oscillator (Figure 90-2), the same circuit is used to generate an oscillating signal into an amplifier.

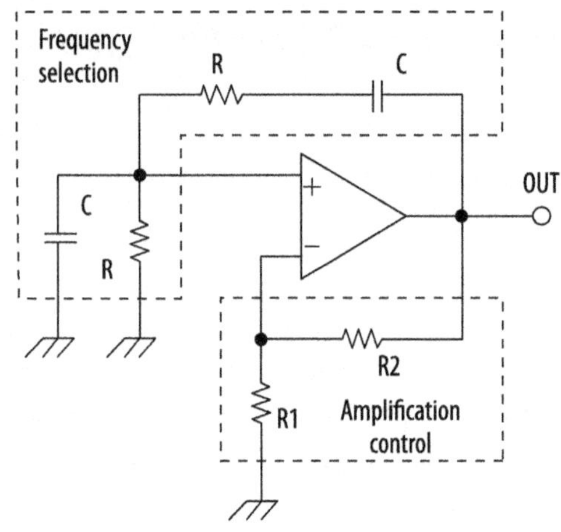

Figure 90-2. Wien-bridge oscillator

The resistors, R, and capacitors, C, determine the frequency at which the oscillator operates, and the voltage divider circuit made from R1 and R2 determines the amount of gain from the amplifier. By altering either R or C, the circuit can be made to oscillate at different frequencies determined by the formula: $f = 1 / (2 \pi R C)$. In the Model 200A, the capacitors C were variable and the other resistances fixed.

The two pairs of resistors R and capacitors C are acting as filters: one pair cuts off oscillations below frequency f and the other above frequency f. Together they act to fix the oscillation frequency of the circuit at $1 / (2 \pi R C)$. This network feeds back the output signal into the amplifier, creating and maintaining the desired frequency. This positive feedback effect is most commonly seen when a microphone is brought too close to a loudspeaker, generating a distinctive howling sound.

Bill Hewlett's clever modification to the circuit was the replacement of the resistor R1 with a light bulb. A light bulb is a form of resistance, but its resistance varies with the current flowing through it. As the current increases, the light bulb shines brighter and heats up, and its resistance increases.

A common problem with oscillators (including the Wien-bridge oscillator) is a tendency for the amplitude of the oscillation to increase until the signal becomes clipped because it is too large. Hewlett's light bulb solved this problem. The voltage divider gets its power from the output of the amplifier; if the amplitude of the output increases, the current flowing across the light bulb increases and its resistance increases.

The resistance changes faster than the amplitude and changes the voltage divider, bringing the amplitude back under control. This simple change made the Model 200A extremely stable.

Practical Information

HP's web page on the garage has more historical information; go to *http://www.hp.com/hpinfo/abouthp/histnfacts/garage/*.

091

U.S. Navy Submarine Force Museum, Groton, CT

41° 23′ 15.22″ N, 72° 5′ 12.91″ W

"Underway on Nuclear Power"

The USS *Nautilus* was the first nuclear submarine, and the first submarine to pass submerged over the Geographic North Pole. When the submarine launched, the captain signalled "Underway on nuclear power," and then proceeded to break the underwater sustained speed record and the longest submerged cruise covering 2,100 kilometers.

Launched in 1955, *Nautilus* remained active until 1980, and was then decommissioned. The submarine is now the prize exhibit at the U.S. Navy Submarine Force Museum, where visitors can climb aboard the historic vessel and take an audio tour.

When entering the museum, visitors pass through a pair of metal circles representing the circumference of the hull of the first U.S. submarine (the USS *Holland*, built in 1897) and the currently active USS *Ohio*, launched in 2007. The USS *Holland* had a beam (the width of the widest part of the craft) of just over 3 meters and could carry a total of 6 people; the USS *Ohio* is over 12 meters across with a crew of 150.

Inside the museum are scale models of every type of U.S. submarine. There's a full-size replica of Bushnell's *Turtle*, a failed one-man submarine designed to attack British ships in 1776. Two films document the history of submarines and the USS *Nautilus* itself. And there are exhibits explaining the operation of submarines, submarine power, warfare, and life aboard.

The *Nautilus* tour begins by crossing a small bridge onto the forward end of the floating submarine. A staircase takes visitors down into the torpedo room where Mk 14 torpedoes are stowed. Climbing through a watertight door, visitors enter the wardroom where the ship's officers worked and socialized. Close to the wardroom are small staterooms, and the only private room on the boat—the Commanding Officer's.

Behind the wardroom is the Attack Center, from which torpedoes could be fired and the periscopes raised. Next stop is the Control Room, with controls for diving, surfacing, and steering, and which also contains the submarine's radio equipment.

Finally, visitors enter the Crew's Mess, the kitchen and sleeping areas. The visit does not cover the nuclear reactor (which has been removed) or the engines. Much of that equipment is still a secret.

Outside the museum are numerous interesting exhibits, including the launch tube for a Trident missile, a Japanese two-man submarine from the Second World War, and a Swimmer Delivery Vehicle used by Navy SEALs to come and go from a submerged submarine.

Practical Information

Full details on the U.S. Navy Submarine Force Museum are available from *http://www.ussnautilus.org/*.

The Gyrocompass

Ships and submarines commonly use gyrocompasses for navigation instead of magnetic compasses. Magnetic compasses are affected by ferrous materials used to construct ships and by electromagnetic fields created by electrical equipment. And magnetic compasses point to Magnetic North (see Chapter 128), not the Geographic North Pole.

Gyrocompasses work by exploiting the behavior of a gyroscope. If a gyroscope is set spinning with its rotational axis pointing in a specific direction (for example, toward the Pole Star) and is mounted inside a gimbal (so it is free to move in any direction; see Figure 91-1), then the gyroscope will keep pointing in the same direction regardless of external movement, such as the rotation of the Earth. The gyroscope remains pointing in the same direction because of the conservation of angular momentum.

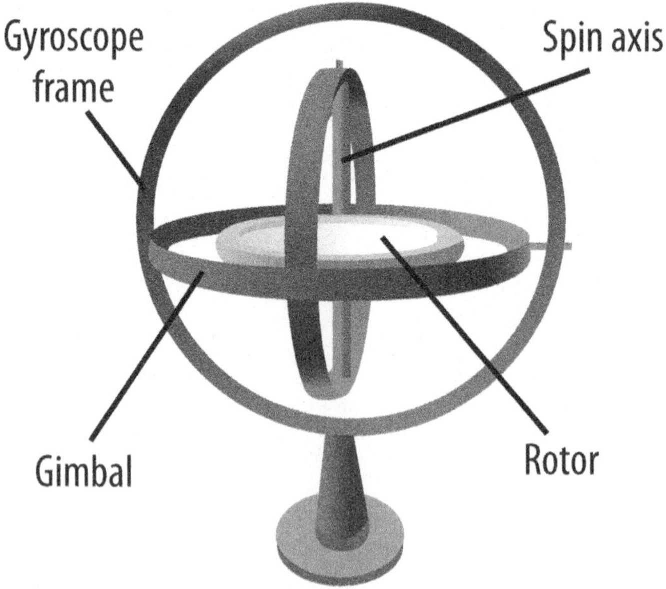

Figure 91-1. Gimbal-mounted gyroscope

The gyrocompass (Figure 91-2) exploits this behavior to create a navigational aid by aligning its axis with the Earth's axis of rotation. As a ship moves, the axis remains pointing toward the Geographic North Pole.

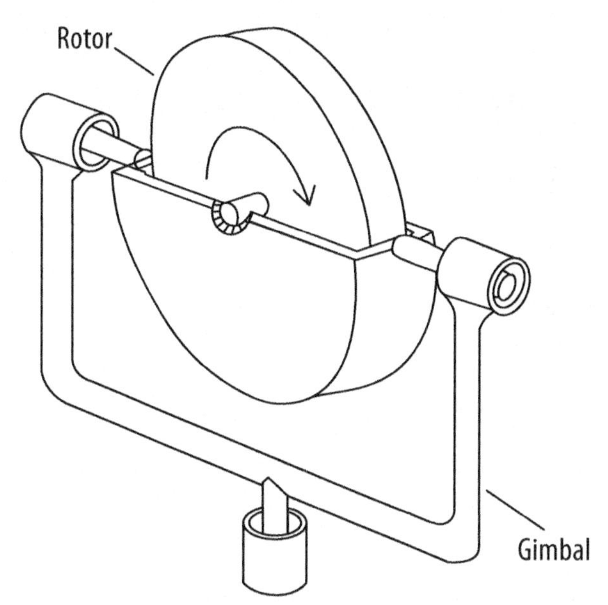

Figure 91-2. Gyrocompass

The rotor spins at high speed, and a pair of gimbals allow the rotor to move in any direction to compensate for the rotation of the Earth. To keep the axle pointing at the Geographic North Pole (i.e., aligned with the Earth's axis), its casing is weighted. The weight causes the Earth's gravity to keep the axle horizontal with respect to the Earth's surface.

Imagine a gyrocompass aligned so that its axle is pointing along a local meridian (i.e., it is aligned with the north-south axis). If the axle is horizontal, then the weighted case has no effect. If the ship then moves longitudinally (to a different meridian), two separate forces are at work on the compass.

Firstly, the gyrocompass moves relative to the Earth to keep its axle aligned with the original local meridian. Relative to the Earth's surface at the new location, the axle will no longer be level and gravity will come into play, pulling on the weighted case.

Gravity causes the spinning rotor to *precess* (its axis of rotation changes direction); the outer gimbal rotates until the axle once against points along the local meridian and is horizontal.

092

National Air and Space Museum, Washington, DC

38° 53′ 18″ N, 77° 1′ 12″ W

The Best Air and Space Museum in the World

Where can you see the Wright Brothers' original 1903 Flyer, the Apollo 11 Command Module that took Neil Armstrong and Buzz Aldrin to the Moon (and left Michael Collins in orbit; see Figure 92-1), a spare mirror for the Hubble Space Telescope, a Lockheed U-2C spy plane, the "Spirit of St. Louis" aircraft that made the first non-stop, transatlantic flight, and unused parts of the original Skylab space station? Nowhere but the National Air and Space Museum in Washington, DC. It's one of the treasures of the capital and is filled with objects that mark pioneering moments in aviation and space travel.

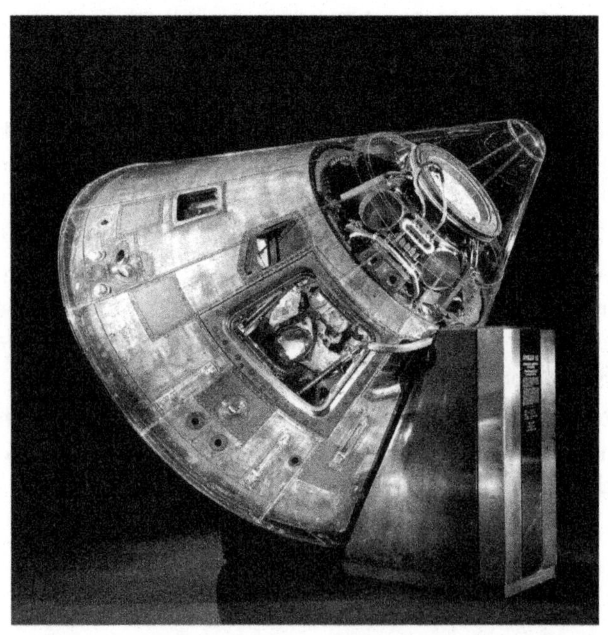

Figure 92-1. The Apollo 11 Command Module; photo by Eric Long/NASM, National Air and Space Museum, Smithsonian Institution

In addition to original aircraft and spacecraft, there are full-size models (often built for testing) of spacecraft such as Voyager 2, the Mars Exploration Rover, an Iridium satellite, Pioneer 10 (the space probe that for 30 years sent pictures back to Earth, and now continues its voyage with a plaque describing where it came from and who built it), Robert Goddard's 1926 liquid-propelled rocket, and one of the Viking landers that went to Mars in 1976. There's also a chunk of rock brought back from the Moon by the astronauts on Apollo 11, which visitors are invited to touch.

Other phenomenal objects include the Bell X-1 in which Chuck Yeager broke the speed of sound, a model of Sputnik 1, and an X-15 aircraft, which was the first aircraft to reach Mach 4, 5, and 6. There's the first American spacecraft to take a man, John Glenn, into orbit around the Earth. And there's Gemini IV, the capsule from which the first American spacewalk was performed.

A recent addition is SpaceShipOne, the small rocket that won the Ansari X Prize by becoming the first private spaceship to take a human into space and back twice within one week.

The museum has an outpost in Virginia called the Udvar-Hazy Center, close to Washington Dulles airport, where a large hangar once used by Boeing is now used to display even more aircraft and spacecraft. The centerpiece of the center is the space shuttle Enterprise; there's also a Concorde, the Boeing B-29 Superfortress Enola Gay (which dropped the atomic bomb on Hiroshima), a Lockheed SR-71A Blackbird spy plane, and a Boeing 367-80 "Dash 80" (the prototype for the Boeing 707 and precursor of all commercial aircraft).

The Udvar-Hazy Center also has a collection of rockets and missiles (including the Bell No. 2 Rocket Belt made in the 1960s as a "jet pack" for soldiers), a collection of VTOL (vertical take-off and landing) aircraft, a backup for the first communications satellite Echo-1 (which was nothing more than a metal balloon off which radio signals could be bounced), and, curiously, the model for the mothership from the film *Close Encounters of the Third Kind*.

Even if you visit only one of the museum's two sites, it's an exhausting and overwhelming experience. So many significant objects are packed into such a small space.

Practical Information

Exploring the National Air and Space Museum's website is almost as good as an actual visit. It has extensive information and photographs of almost everything that's on display. The website is at *http://www.nasm.si.edu/*. Best of all, entry to the museum is free.

The Temperature of Space

The museum has a large collection of pressure suits worn by astronauts including John Glenn, Buzz Aldrin, and Sally Ride. Pressure suits are obviously necessary because in space (and on the Moon) there's no oxygen to breathe. But the pressure suits also protect the wearer from the cold of space.

Space is cold because it's a vacuum—there are no (or very, very few) particles around to transfer heat. Since temperature is usually defined as the average kinetic energy of moving particles (such as in a gas), space appears to have no temperature at all.

But, even though space is cold, it's not at Absolute Zero because the remnants of the Big Bang continue to provide a small amount of warmth throughout the universe. To understand this requires a detour via physics as understood before Einstein.

Sir Isaac Newton wrote in 1704 in his book *Optiks*, "Do not all fixt Bodies when heated beyond a certain degree, emit Light and shine; and is not this emission performed by the vibrating motions of their parts?" Once again he was well ahead of his time, because it was not until the early 20th century that the mechanism by which heated bodies radiate heat was understood. Nevertheless, the concept of emitted radiation and temperature is familiar to anyone who refers to something being red- or white-hot.

In 1860, the German physicist Gustav Kirchoff showed that when heated, an ideal black body (i.e., anything that absorbs all light hitting it; the closest real material to this ideal black body is graphite powder, which absorbs 97% of light) will be an ideal source of thermal radiation (electromagnetic radiation, such as infrared rays from an electric fire, or light from an incandescent bulb).

An ideal black body emits thermal radiation across the electromagnetic spectrum (since it absorbs electromagnetic radiation at any wavelength, it also emits at any wavelength), and for any particular temperature there's a characteristic peak intensity at a specific wavelength. Roughly speaking, the hotter the black body, the shorter the wavelength at which the peak is seen.

The Sun can be considered to be approximately a black body, with a temperature of 5,780 kelvin at the surface. That temperature corresponds to a range of colors (which we see as white) with a peak wavelength of 502 nanometers; 502 nanometers is in the green portion of the visible spectrum, which is also the part of the spectrum where human eyes are most sensitive to differences in hue.

Returning to the void of space, there's radiation left over from the Big Bang in the form of cosmic microwave background radiation (see page 413). This radiation peaks at a wavelength of 1.9 millimeters and a curve drawn of its intensity by wavelength shows an almost perfect black body, radiating with a temperature of 2.7 kelvin. Figure 92-2 actually has two curves on it—one shows the theoretical black body radiation in the universe, the other the observed curve taken using the Far Infrared Absolute Spectrophotometer (FIRAS) instrument aboard NASA's Cosmic Background Explorer (COBE). The two curves are so close as to be indistinguishable.

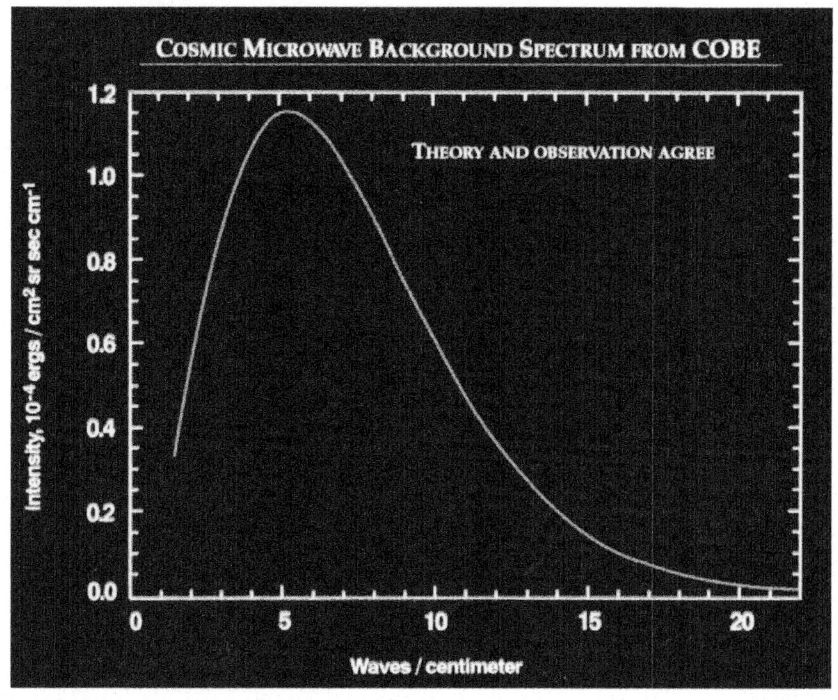

Figure 92-2. The observed and predicted spectrum of cosmic microwave background radiation

Thus the temperature of space is 2.7 kelvin. Pretty chilly, but not Absolute Zero.

093

National Museum of American History, Washington, DC

38° 53′ 28.68″ N, 77° 1′ 48″ W

The Collections

It's easy to overlook the National Museum of American History because its name conjures up mental images of rows of dusty paintings of American pioneers. But, in reality, the museum is full of scientific, mathematical, and industrial artifacts that make it one of the three great science museums in the world alongside the Science Museum in London (Chapter 77) and the Deutsches Museum in Munich (Chapter 19).

The major collections of interest cover Communications, Computers and Business Machines, Energy and Power, Health and Medicine, Industry and Manufacturing, Measuring and Mapping, Science and Mathematics, and Transportation.

The Communications collection has one of Alexander Graham Bell's Large Box telephones, which was used for an 1876 demonstration of the telephone between Boston and Salem. (For more on Alexander Graham Bell, see Chapter 4.) There's also Robert Metcalf's prototype Ethernet board, which was made almost 100 years after the telephone. Ethernet is the standard for connecting computers into networks, and is used in homes and offices worldwide.

Much of the Computers and Business Machines collection is duplicated in other computer museums (such as the Computer History Museum; Chapter 86), including the Altair 8800 Microcomputer (which helped launch the personal computer revolution), the Apple II, part of the ENIAC machine, and a Hollerith Tabulating Machine (see page 41).

But there's also the first computer bug—inside a 1947 notebook used by computer scientist Grace Hopper when working on the Mark II computer at Harvard University. The Mark II was behaving unpredictably, and a moth was found inside one of its relays. The moth is pasted into the notebook with the label "first actual case of bug being found."

The Energy and Power exhibit contains an 1805 Voltaic Pile (see page 96), an early GE fluorescent lamp, and a Carrier centrifugal refrigeration compressor from 1922 that was used for practical and efficient air conditioning.

The Health and Medicine collection has the first electro-hydraulic heart to be implanted in a human, an early CT scanner, a 1931 iron lung, and one of Roentgen's early X-ray tubes. In the Industry and Manufacturing collection there's a Bakelizer (the machine used to make Bakelite; see sidebar) and a Ford Model T.

Measuring and Mapping contains a 1940 mass spectrograph and a number of navigational devices including compasses, octants, and marine chronometers. The Science and Mathematics collection has thousands of well-explained devices from slide rules to lasers. The Transportation collection has some of the biggest items—there are cars, trucks, trams, and locomotives.

And of course, the museum covers much more than science. If you are not already exhausted, the non-science parts of American history are covered in the museum's historical collections.

Practical Information

Information about the National Museum of American History is at *http://americanhistory.si.edu/*. Admission to the museum is free.

Bakelite

Bakelite was the world's first synthetic plastic. It was created in 1907 by Dr. Leo Baekeland by reacting phenol and formaldehyde under heat and pressure. At the museum, his Bakelizer is still in working order—it consists of a heavy iron pressure chamber and boiler needed to control the reaction between phenol and formaldehyde.

Bakelite is a polymer: it is made up of large molecules that are themselves made up of repeating groups of elements called structural units. Prior to the creation of Bakelite, naturally occurring polymers were used. Early gramophone records were made from shellac, which is a polymer made from the excretions of the insect *Kerria lacca*. The raw excretions are called lac (from which the word "lacquer" is derived).

DNA (see Chapter 71) is also a polymer, as are plastics. The common synthetic plastic polypropylene (used for everything from packaging to bank notes) is a polymer consisting of repeating blocks of C_3H_6 in a long chain joined by carbon bonds (Figure 93-1).

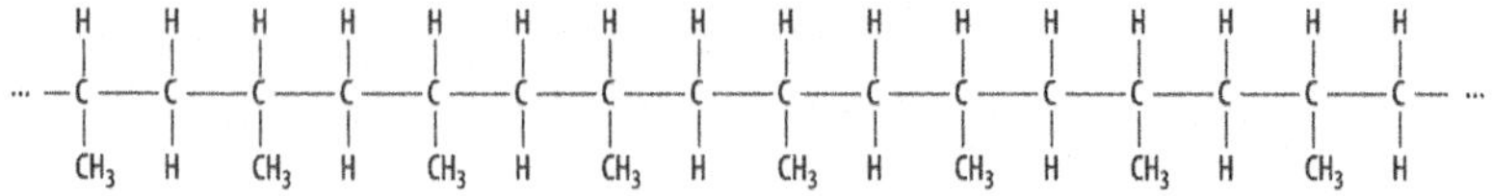

Figure 93-1. Polypropylene

In contrast, Bakelite is a cross-linked polymer instead of a chain—its structural units are linked together in a complex and irregular pattern. The Bakelite reaction starts with a mixture of phenol and formaldehyde. Phenol (Figure 93-2) is a crystalline solid with chemical formula C_6H_5OH, which was used as an early antiseptic by the English surgeon Sir Joseph Lister.

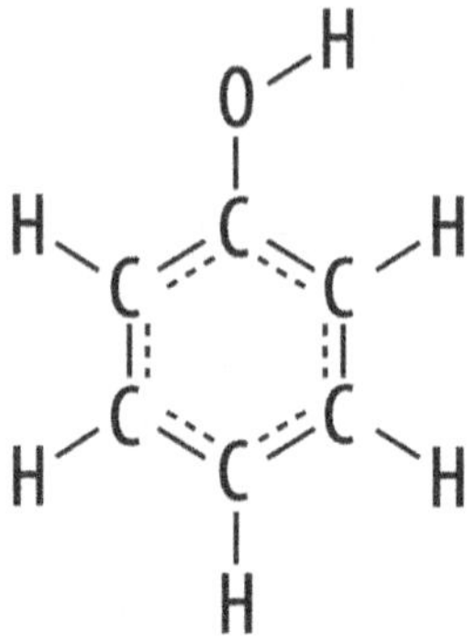

Figure 93-2. Phenol molecular structure

Formaldehyde (Figure 93-3) is colorless gas with chemical formula H_2CO, which, when dissolved, is used in embalming.

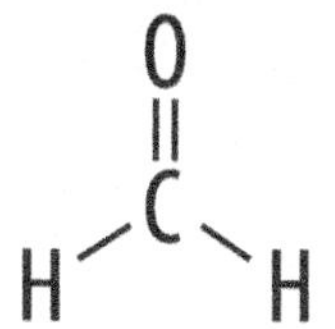

Figure 93-3. Formaldehyde molecular structure

Formaldehyde and phenol react together to form an intermediate molecule that, when heated, will react with phenol a second time to produce the Bakelite structural unit and water. The structural units then link together to form the polymer (Figure 93-4).

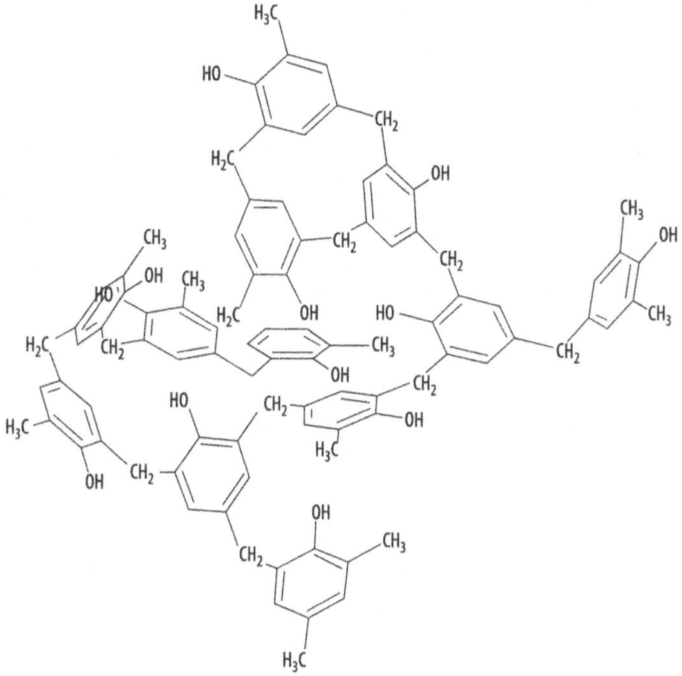

Figure 93-4. Bakelite molecular structure

Bakelite is a thermo-setting polymer. Once cooled, its structure remains fixed and hard; heating Bakelite will not cause it to melt again, but instead promotes further strengthening of the polymer structure.

094

Kennedy Space Center, Merritt Island, FL

28° 35′ 6″ N, 80° 39′ 3.6″ W

Not the Magic Kingdom

If you were to sit down and make a list of great places for the scientific tourist to visit, the Kennedy Space Center would likely be near the top of the list. Not only is it the site of the launch of the Apollo 11 mission to the Moon, the Space Shuttle, and countless other spacecraft, but it is also a major tourist attraction.

The specially built Kennedy Space Center Visitor Complex is the starting point for any visit. But taking a tour of the Kennedy Space Center is essential—it's an enormous site, and just staying at the visitor complex turns a visit to the center into just another tourist day out. The standard tour included in the ticket price takes in the International Space Station Center (where components for the station go through final checks before being launched and where visitors can enter a mockup of the actual station), the Launch Complex 39 Observation Gantry (from which the two space shuttle launch pads can be seen, as well as the special road along which the Shuttle crawls to its launch pad from the enormous Vehicle Assembly Building), and the Apollo/Saturn V Center (with its complete Saturn V rocket to walk around and under).

There are two other tours, which must be booked in advance: the NASA Up-Close Tour and the Cape Canaveral: Then & Now Tour.

The Up-Close Tour takes visitors as close as possible to the shuttle launch pads, around the Vehicle Assembly Building, and on to view the gigantic Crawler Transporters used to move the Space Shuttle.

The Cape Canaveral: Then & Now Tour takes visitors back to the launch pads used for the Mercury, Gemini, and Apollo missions. This includes Launch Pad 34, where the Apollo 1 fire killed astronauts Gus Grissom, Ed White, and Roger Chaffe (it is now a memorial to that tragedy).

Back at the Visitor Center, there's another memorial to astronauts who have died during the course of their missions called the Space Mirror. The names of the crews of Apollo 1, the space shuttle Challenger, and the space shuttle Columbia are engraved on slabs of black granite that reflects the Florida sky, giving the impression that the names are projected into the heavens.

Also at the center is a Rocket Garden filled with unused rockets from the Mercury, Gemini, and Apollo era. The garden is complemented by the Early Space Exploration exhibit inside that covers the same era.

The Kennedy Space Center provides the opportunity to meet a real astronaut at the Astronaut Encounter (the schedule of astronaut appearances is on the website). There's also the Space Shuttle Plaza, where there's a full-size mock-up of the Space Shuttle, its two rocket boosters, and the main fuel tank.

If you plan your trip right (or if you're just lucky), you'll get to see a launch. NASA publishes information about launch dates and times on its website.

There are also a few pure tourist attractions, including the Shuttle Launch Experience, which is intended to give visitors the sensation of a shuttle launch (including being rotated to a horizontal position and strapped into the seat) with a few moments of (phony) weightlessness. There are also two IMAX theaters showing space-related films, often with actual footage taken by cameras on the Space Shuttle or International Space Station.

Practical Information

You can find information about visiting the Kennedy Space Center at *http://www.kennedyspacecenter.com/*. Book ahead to avoid disappointment, as it is very popular.

Escape Velocity

To escape from the Earth's gravity, you need to be moving at 11.2 kps, no matter how large or small your spacecraft. This is the speed at which the kinetic energy of the spacecraft is equal to the potential energy caused by the Earth's gravity at the surface. Even though it's a speed, it's generally called the escape velocity.

Because the Moon is so much smaller than the Earth, it has a weaker gravitational field, so the potential energy of a spacecraft trying to escape from the Moon is lower. It only requires a speed of 2.6 kps to have sufficient kinetic energy to escape the pull of the Moon's gravity. And because the Sun is so much larger, a much higher speed (617.5 kps) is required to escape its gravity.

Of course, the gravitational pull weakens as a spacecraft moves away from the body it is trying to escape. The Sun's enormous escape velocity only applies to spacecraft at the surface of the Sun (in which case they've got bigger problems than trying to achieve 617.5 kps). The Voyager 1 space probe that was launched in 1977 has traveled about 16 billion kilometers from the Sun and is currently travelling at 17.1 kps.

To calculate the escape velocity for a body (such as the Sun), it's only necessary to go back to Newton's Law of Universal Gravitation. This law states that the force due to gravity between two bodies (such as a spacecraft with mass m_1 and the Sun with mass m_2) is proportional to the mass of each body and inversely proportional to the square of the distance, r, between them. See Equation 94-1.

$$F = G\frac{m_1 m_2}{r^2}$$

Equation 94-1. Newton's Law of Universal Gravitation

G is the gravitational constant. To calculate the potential energy, P, from the force, simply integrate the force of gravity to discover that the potential energy due to gravity (i.e., the amount of work you'd have to do to move the body the required distance against the force) is inversely proportional to the distance from the body. This calculation is shown in Equation 94-2.

$$P = \int \frac{Gm_1m_2}{r^2} = -\frac{Gm_1m_2}{r}$$

Equation 94-2. Gravitational potential energy

For a spacecraft of mass, m, moving with velocity, v, and with no other force acting on it (other than gravity), its kinetic energy will be given by the standard formula 1/2 mv^2. So the escape velocity can be calculated by finding the velocity, v, at which the kinetic energy is equal to the gravitational potential energy (see Equation 94-3).

$$\frac{1}{2}m_1v^2 = \frac{Gm_1m_2}{r} \text{ or } v = \sqrt{\frac{2Gm_2}{r}}$$

Equation 94-3. Escape velocity

Thus the escape velocity only depends on the distance to and the mass of the body exerting a gravitational pull. The mass of the spacecraft is irrelevant.

So how fast does Voyager 1 need to be moving to escape the Sun's gravitational pull? The distance is 16,000 billion meters, the Sun's mass is 1.9891×10^{30} kilograms, and the gravitational constant is $6.674\,28 \times 10^{-11}$. A quick calculation shows that at Voyager 1's current position, it needs to be traveling at 2.88 kps.

The escape velocity isn't only important for spacecraft—if a planet is too small, its atmosphere will literally escape from it because the molecules of gas in the atmosphere are moving too fast. For this reason, the Moon has no atmosphere. The speed with which gas molecules move is proportional to the gas temperature—the hotter the gas, the faster the molecules move. So the presence or absence of an atmosphere depends on the escape velocity of the planet and its temperature.

095

Kalaupapa National Historic Park, Molokai, HI

21° 11′ 19.01″ N, 156° 58′ 53.62″ W

The Leper Colony

Between 1866 and 1969, people suffering from leprosy in Hawaii were forcibly removed to the leper colony on the island of Molokai. Today the leper colony is a national park and home to leprosy survivors. The colony is isolated below the highest sea cliffs in the world, with a drop of 1,010 meters. Reaching the colony entails a mule ride or hike down the cliffs following a 4.5-kilometer trail.

The trail provides a stunning view across the sea, and prior to the arrival of aircraft was the only way to reach the colony below. All supplies had to be brought in down the cliff, or on the occasional boat: to this day a large boat docks only once per year, carrying heavy items like televisions, refrigerators, or cars.

If you want to visit Kalaupapa today, taking a tour is mandatory. The small town at Kalaupapa is home to an aging population of leprosy survivors, many of whom have lived there for more than 50 years. The tour is run by a resident of Kalaupapa who survived leprosy and decided to remain.

Just seven years after the colony opened, the Norwegian scientist Gerhard Armauer Hansen isolated the bacterium responsible for leprosy: *Mycobacterium leprae*. It was the first bacterium shown to be responsible for a human disease. Shortly afterward, the German physician Robert Koch isolated the bacteria responsible for anthrax, tuberculosis, and cholera. The isolation of these bacteria put an end to the miasma theory of disease, which held that diseases came from breathing in air from decomposing matter.

To this day, leprosy's exact mode of transmission is unclear, although it is suspected that close contact between infected people is the primary mechanism. It is thought that leprosy is airborne through droplets from the nose and mouth.

In the late 1800s and early 1900s, governments around the world concluded that the safest way to deal with lepers was to isolate them in colonies. Isolation was partly driven by the physical disfigurement that sufferers endured, and partly because leprosy was believed to be highly contagious.

The first treatment for leprosy didn't appear until the 1930s. The antibiotic dapsone (sometimes referred to as sulfone treatment) cured leprosy after years of treatment, but the disfiguring effects were irreversible. Many leprosy survivors decided to remain in Kalaupapa rather than attempt to reintegrate with Hawaiian society.

Subsequently, *Mycobacterium leprae* developed resistance to dapsone, and only in 1985 did effective treatment become available. To treat dapsone-resistant leprosy, three drugs are used: dapsone, rifampicin, and clofazimine. This multidrug therapy has been provided worldwide and free of charge by the World Health Organization since 1995.

In 2000, the WHO achieved its goal of bringing the prevalence of leprosy down to fewer than 1 case per 10,000 people. Nowadays, leprosy affects fewer than 1 in 50,000 people and is completely curable.

The remaining residents of Kalaupapa are elderly and the population is dwindling. Eight thousand people died on this triangle of land, and just a handful of survivors remain. The history of Kalaupapa spans a period of enlightenment in medicine, from the discovery of the bacterial cause of disease, to cures by antibiotics, to the realization that diseases could mutate and resist the cure.

The long hike down the cliff gives you plenty of time to reflect on medical scientific progress and the treatment of patients.

Practical Information

Information about visiting Kalaupapa is available from the U.S. National Park Service at *http://www.nps.gov/kala/*. There's also a small museum dedicated to leprosy in Carville, Louisiana; details of the National Hansen's Disease Museum can be found at *http://www.hrsa.gov/hansens/museum/*.

Bacterium, Cure, and Drug Resistance

Bacteria are single-cell organisms and are the most abundant group of organisms on the planet. They are found almost everywhere on Earth (including in human intestines, in the sea, air, and soil, and even in rocks). The bacteria responsible for leprosy, *Mycobacterium leprae* (Figure 95-1), is rod-shaped and covered with a thick, waxy coating that cannot survive without a host cell.

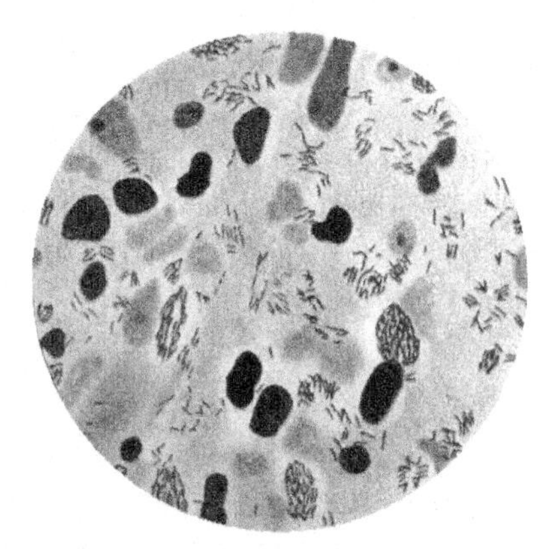

Figure 95-1. Mycobacterium leprae

Once in a host cell, *Mycobacterium leprae* reproduces in the same manner as almost all bacteria: the bacterium splits in two, with each half holding a copy of the bacteria's DNA. In humans, *Mycobacterium leprae* attacks nerves, mucus membranes, and the skin. One form of the leprosy bacterium causes large lumps and bumps on the skin called nodules.

Antibiotics fight bacteria by affecting their life cycle: they may prevent the bacteria from making proteins, interfere with the cell wall of the bacteria so it disintegrates, or inhibit their ability to reproduce. The best-known antibiotic, penicillin, works by weakening the cell wall of bacteria; the bacteria then literally burst open because the weakened cell wall allows water inside.

The first antibiotic used against leprosy, dapsone, interferes with the production of folic acid, which bacteria need to be able to make DNA. Rifampicin prevents bacteria from reproducing by preventing DNA from being able to make proteins. The exact operation of clofazimine is unknown, but together all three drugs cure leprosy.

Before rifampicin and clofazimine became available, *Mycobacterium leprae* developed resistance to the dapsone antibiotic. Bacteria develop drug resistance through two main mechanisms: mutation and gene transfer.

Since bacteria reproduce rapidly, they can evolve through natural selection very quickly. If a bacterial infection is treated with an antibiotic, some mutations may resist the antibiotic. Since the rest of the population is killed by the drug, these resistant bacteria are able to survive and thrive (see Figure 95-2).

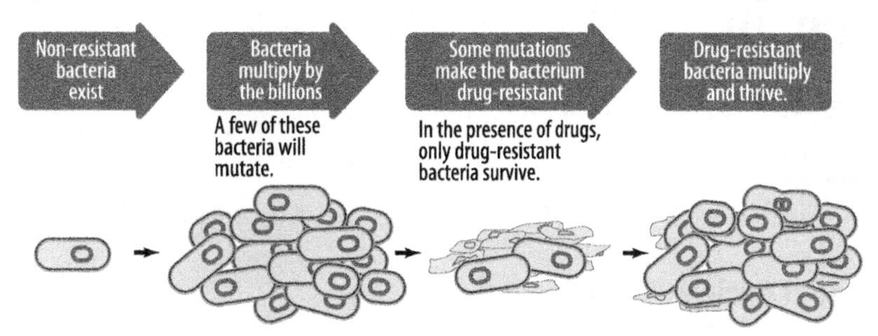

Figure 95-2. Resistance through mutation

Mutation occurs when a bacteria divide. In contrast, gene transfer happens between bacteria without division. Bacteria contain a circular DNA molecule called a plasmid, which is independent of the DNA controlling the bacteria. The plasmid can be duplicated and then transferred to another bacterium, taking with it antibiotic properties (Figure 95-3).

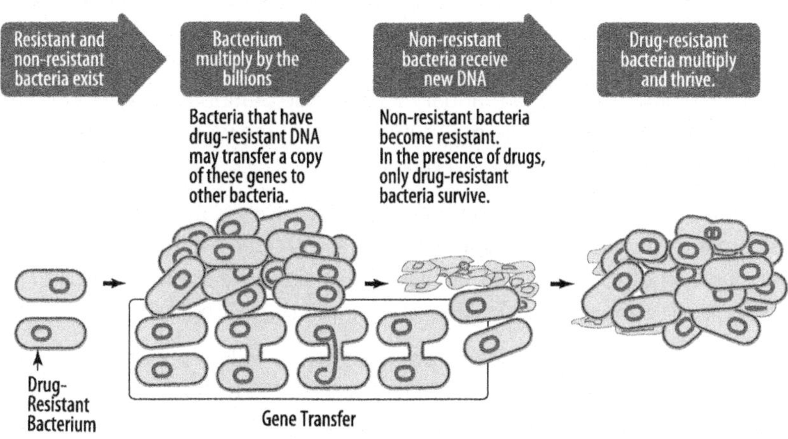

Figure 95-3. Resistance through gene transfer

Plasmids are also important in genetic engineering: they can be extracted, modified, and reinserted to change a cell's DNA.

096

Experimental Breeder Reactor No. 1, Arco, ID

43° 30′ 41.12″ N, 113° 0′ 20.29″ W

The First Nuclear Power

Think of Idaho, and potatoes are more likely to come to mind than nuclear reactors. However, the Idaho National Laboratory (INL), created in 1949, covers 2,300 square kilometers and has the largest concentration of nuclear reactors in the world: more than 50 reactors have been built on the site.

One of those reactors is the Experimental Breeder Reactor No. 1 (or EBR-1), the first nuclear reactor to produce electricity. Close to EBR-1 is the town of Arco, which boasts of being the first town powered by nuclear energy (for a few hours on July 17, 1955, Arco received its electricity from a reactor belonging to the INL). Arco doesn't boast about being close to the site of the first major nuclear accident, though: on January 3, 1961, the Stationary Low-Power Plant Number 1 experienced a core meltdown, killing three workers who were so irradiated they were buried in lead coffins.

The most striking sight in the little town of Arco (population around 1,000) is the conning tower of the nuclear submarine USS *Hawkbill*, known as the Devil Boat because of the large 666 painted on its side. The submarine tower celebrates Arco's connection to nuclear power and the U.S. military.

But the real interest lies 18 miles outside Arco, where EBR-1 is open to the public. EBR-1 first produced electricity on December 20, 1951, when it made enough electricity to power four 200-watt light bulbs (replicas and one of the original light bulbs are on display). The next day, enough electricity was generated to power the entire building. But EBR-1 wasn't built to generate electricity—it was created to test the idea of a breeder reactor, and to use plutonium instead of uranium as fuel.

A breeder reactor is capable of producing more fuel than it consumes, generating heat (to be used to make electricity) and fuel (for later use). EBR-1 demonstrated that the breeder technique worked, and in early 1953 analysis of the reactor showed that it was creating one new atom of nuclear fuel for each atom used. In other reactors, the breeding ratio (the amount of fuel produced per atom used) was later improved to 1.27 new atoms per atom consumed, proving that a breeder reactor could indeed create more fuel than it used.

Breeder reactors were attractive in the 1950s because of their fuel economy, and because they could use common radioactive elements such as thorium instead of the difficult-to-obtain uranium U-235 used in non-breeder reactors.

Nuclear Aircraft Engines

Smack dab in the parking lot of EBR-1 are two nuclear aircraft engines, looking like something a mad scientist dreamed up. The U.S. government originally built the two reactors as an energy source for modified jet engines. The plan was to build an aircraft that could fly longer and farther without needing to refuel. Such an aircraft would be able to quickly fly anywhere in the world from a safe base inside the U.S. At least, that was the plan.

In 1955, two J-47 General Electric jet engines (which normally burned jet fuel to create heat and were in use by the USAF at the time) were modified to take hot (radioactive) air from the Heat Transfer Reactor Experiment 1 (HTRE-1; see Figure 96-1) and were tested on the ground at the EBR-1 site. The compressed air from the engine passed through the reactor core and back into the J-47 engines (renamed to X-39 to indicate their experimental nature).

Figure 96-1. Heat transfer reactors; courtesy of Paul Mitchum (Mile23)

Breeder Reactor

A non-breeder nuclear reactor works by nuclear fission: the nucleus of the fuel used (generally uranium or plutonium) splits, releasing neutrons. These neutrons impact other atoms, causing them to split, releasing more neutrons and continuing the reaction. This is the famous chain reaction, which, in addition to releasing neutrons, creates an enormous amount of heat that can be used to turn turbines and generate electricity (and if left uncontrolled will cause a nuclear explosion).

Simple nuclear reactors use uranium (in the form U-235). When the nucleus of U-235 is hit by a neutron, the U-235 nucleus splits apart, releasing three neutrons (which can go on to hit other U-235 atoms and continue the chain reaction). The neutron is initially absorbed by U-235, creating the unstable element U-236, which breaks apart. The three neutrons are released, and the rest turns into a barium atom and a krypton atom, plus a burst of energy in the form of gamma radiation (see Figure 96-2).

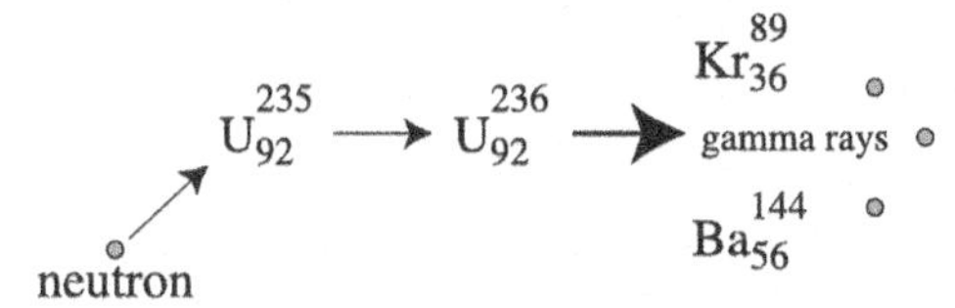

Figure 96-2. U-235 fission

Because of this property, U-235 is said to be "fissile." The other commonly used fissile atom is plutonium (in the form Pu-239). Both U-235 and Pu-239 can sustain a chain reaction, and since both will by themselves break down and release neutrons, a sufficient quantity of either U-235 or Pu-239 (the so-called "critical mass") will be enough to get the reaction going.

Unfortunately, both U-235 and Pu-239 are very rare. Less than 1% of naturally occurring uranium is U-235; the rest is almost entirely U-238, which cannot by itself sustain the chain reaction needed. Plutonium is extremely rare and needs to be made inside nuclear reactors.

However, Pu-239 can be made (or bred) from easily available U-238. A breeder reactor works by mixing U-235 and U-238. The U-235 creates a chain reaction: some of the neutrons it gives off go on to hit other U-235 atoms and continue the reaction. But some of the neutrons hit U-238, which then undergoes a transformation into Pu-239 (see Figure 96-3).

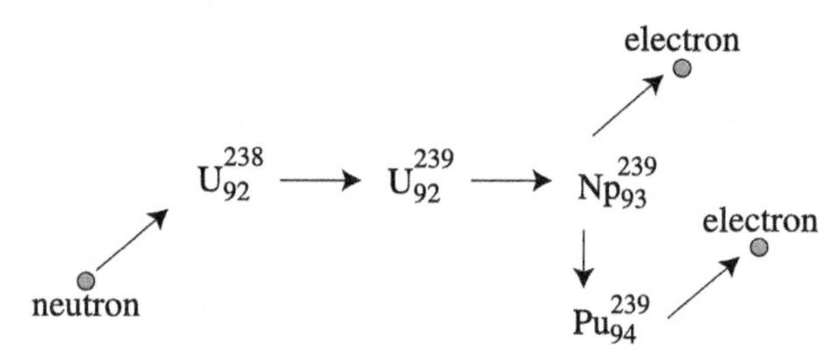

Figure 96-3. U-238 becomes Pu-239

First, U-238 absorbs the neutron and becomes U-239. U-239 then decays by emitting an electron, turning into neptunium (Np-239). The neptunium then decays (again with the release of an electron) to produce plutonium. This plutonium is capable of sustaining a nuclear chain reaction. In this way, a nuclear reactor using U-235 turns U-238 into Pu-239.

HTRE-1 proved that the engines could work, but showed that there were many problems: the reactors were cumbersome, heavy shielding was required to protect the crew, and the engines were letting out radioactive air. Nevertheless, experiments called HTRE-2 and HTRE-3 were carried out. These versions were lighter, and produced enough thrust that calculations showed they could power an aircraft for 30,000 miles at 460 miles per hour.

While work on the reactors and engines was going on, the USAF was also testing an aircraft capable of flying with a nuclear engine. A modified Convair B-36 bomber (renamed the NB-36H) flew 47 times, carrying a small 1-megawatt nuclear reactor and four tons of lead shielding to protect the crew. The reactor was not used to power the plane—it was there to test the amount of radiation produced during flying—but it was operational.

Had the aircraft project not been canceled, the bomber would have required a miles-long runway because of its weight. The runway was never built, but a huge hangar with radiation shielding was prepared on land belonging to the Idaho National Laboratory.

The entire nuclear aircraft project was overshadowed by the creation of the Intercontinental Ballistic Missile (ICBM). The use of missiles to deliver nuclear bombs eliminated much of the need for long-range bombers, and the project was abandoned by President Kennedy in 1961.

EBR-1 is entirely preserved in its 1950s state; a tour includes the pristine control room, the reactor chamber, turbines, and equipment for manipulating radioactive material. A clear explanation of the operation of the reactor is provided on panels throughout. From the outside, EBR-1 looks like a simple brick building, but inside it's a world of fascination for the nuclear tourist.

Practical Information

EBR-1 is 18 miles southeast of Arco on Highway 26. It is open from Memorial Day weekend to Labor Day weekend, from 9 a.m. to 5 p.m. every day. Guided tours are available; for more information, visit *http://www.inl.gov/factsheets/ebr-1.pdf.*

097

Fermilab, Batavia, IL

41° 49′ 55″ N, 88° 15′ 26″ W

The Fermi National Accelerator Laboratory

After CERN (see Chapter 32), Fermilab is the most famous particle accelerator and high-energy physics laboratory in the world. It has a significant outreach program that includes educational events, exhibits aimed at the under-12 set, Sunday talks by physicists, and the ability to take a guided tour.

Fermilab was created in 1967 and named in honor of the Italian physicist Enrico Fermi (who won the Nobel Prize in Physics in 1938). The laboratory operates particle accelerators (Figure 97-1) to do research into subatomic particles. Fermilab was the site of the discovery of the bottom quark in 1977 and the top quark in 1995, and in 2008 the lab announced the discovery of the Omega-sub-b particle made from two strange quarks and a bottom quark.

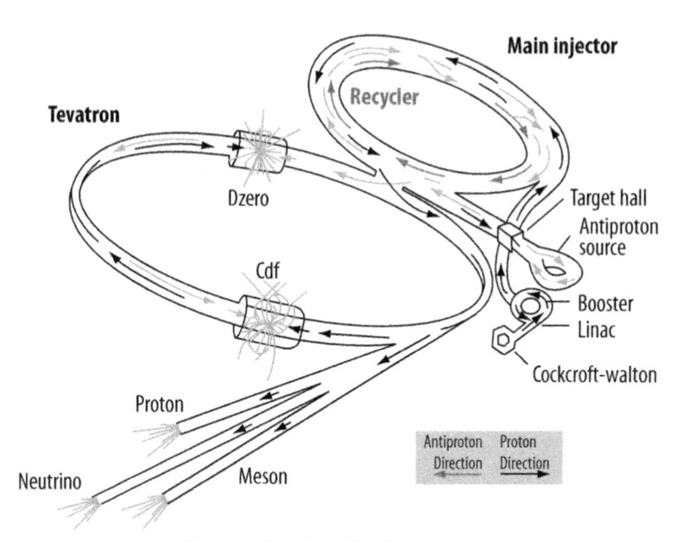

Figure 97-1. Fermilab's accelerators

Cockcroft-Walton Generator

Fermilab's Tevatron accelerator sends protons and antiprotons around a 6.3-kilometer ring and causes them to collide with a combined energy of almost 2 TeV. (This sounds like a lot, but is in fact close to the energy involved in two mosquitoes colliding.)

To create the protons necessary, hydrogen gas is ionized into H^- ions. Negatively charged hydrogen consists of two electrons orbiting a single proton. Next, the ions are accelerated, which can be achieved by simply attracting them toward a large positive voltage through a thin carbon foil, which strips off the two electrons from each H^-, leaving H^+ ions. H^+ ions are simply protons.

To create the H^- ions from hydrogen gas, a very large voltage is needed; at Fermilab, a Cockcroft-Walton generator is used to produce hundreds of thousands of volts. The British scientist John Cockcroft and Irish scientist Ernest Walton won the 1951 Nobel Prize in Physics for their work in creating early particle accelerators. As part of that work, they had to come up with a method of generating huge voltages. The circuit they created is now called the Cockcroft-Walton generator.

The Cockcroft-Walton generator is a form of voltage multiplier. It takes in an alternating current and outputs a high-voltage direct current, which can be used in a particle accelerator. In the generator, the alternating current charges up the capacitors in parallel, but when they discharge, the current flows in series through the diodes. This has the effect of taking a low voltage and multiplying it by the number of stages in the circuit.

In Figure 97-2, each of the bottom two capacitors, C1 and C2, is charged to twice the input AC voltage in the following manner. On the first half-cycle of AC current, the leftmost diode/capacitor pair, D1 and C4, results in C4 charging to the input voltage; on the next half-cycle, C1 charges via D2 from C4 and the incoming voltage, resulting in it charging to twice the input voltage. The remaining capacitor, C3, acts to even out the voltage across the bottom two capacitors, resulting in four times the input voltage across the bottom capacitor pair.

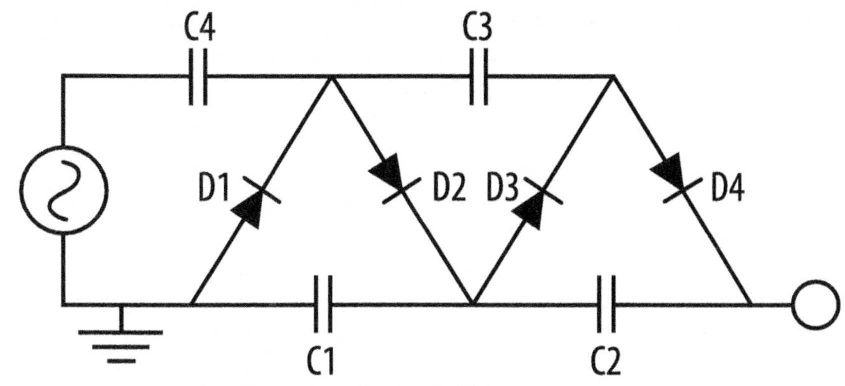

Figure 97-2. Cockcroft-Walton generator

Adding more stages increases the output voltage. With this generator, Cockcroft and Walton produced voltages of hundreds of kilovolts and were able to accelerate protons (H^+ ions) into lithium and observe the creation of alpha particles (which are just helium ions: He^{2+}).

Creating antiprotons is another matter altogether. Very high-energy protons are accelerated and hit a nickel target; for every million protons hitting the target, around 20 to 30 antiprotons are created. The nickel collision creates many other particles, and these are separated from the antiprotons by bending them away using powerful pulsed magnets.

With protons and antiprotons injected in opposite directions into the Tevatron, they are accelerated to close to the speed of light, and a small adjustment in the magnetic field holding the particles in place causes them to collide with one another.

The lab also does extensive work on neutrinos, including sending a beam of neutrinos through the Earth to a mine in Minnesota 735 kilometers away. The Tevatron, Fermilab's main particle accelerator, is the second largest accelerator in the world (after the LHC; see Chapter 32) and performs proton/antiproton collisions.

Around the Tevatron are two detectors called DZERO and CDF. The DZERO detector was responsible for the detection of the Omega-sub-b particle. CDF (which simply means the Collider Detector at Fermilab) is used to study the particles that are released when the protons and antiprotons collide by recording their trajectories and energy, and is also used to look for evidence of phenomena that do not fit the Standard Model of Physics (see page 118). The trajectory of a particle depends on its momentum; as the particles shoot out from the collision they are deflected by magnets—the greater the deflection, the lower the momentum. From the momentum, the charge or the mass of the particle can be inferred.

Casual visitors to Fermilab are welcome to wander the extensive prairie grounds (where a herd of bison live) and can follow a nature trail. The Lederman Science Center is open to the public and explains the work of Fermilab. The center is mainly aimed at schoolchildren, but it provides a good understanding of the basics of particle physics for those unfamiliar with protons, neutrons, electrons, and quarks.

The best way to visit Fermilab, however, is by attending an Ask-a-Scientist talk. These talks take place on the first Sunday of the month; as the name suggests, a Fermilab scientist gives a talk about the lab's work and answers questions from the public. The talk is followed by a tour of the lab. Since Fermilab does no secret work, the tour is a fascinating behind-the-scenes look at what particle physicists get up to.

If you can't attend a talk, there are weekly tours of the facility on Wednesday mornings. All tours and talks are free, but be sure to book in advance because space is very limited.

Fermilab also frequently hosts cultural events such as movies, art exhibits, and concerts.

Practical Information

Fermilab's education website can be found at *http://ed.fnal.gov/*.

098

MIT Museum, Cambridge, MA

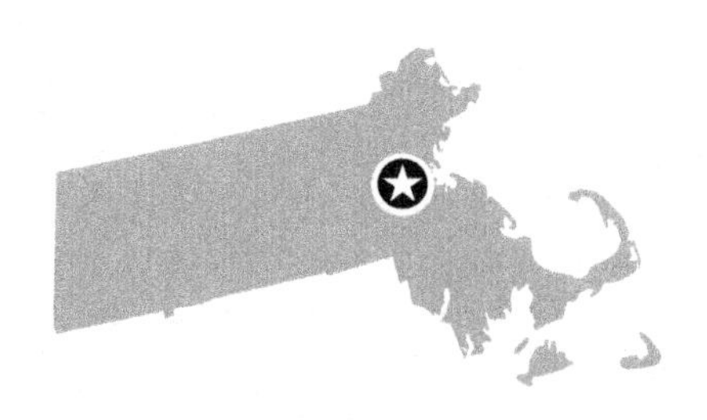

42° 21′ 42.50″ N, 71° 5′ 51.64″ W

Ideas in the Making

It will come as no surprise that the Massachusetts Institute of Technology has a museum, a good one at that. Instead of being filled with ancient scientific instruments (there are plenty of those to be found elsewhere—see Chapters 17, 19, and 77) the MIT Museum deals with recent scientific activity associated with the institute.

The major exhibitions at the MIT Museum cover robotics and artificial intelligence, holography, the work of Harold Edgerton, and the education students receive at MIT. The museum also contains an exhibition of kinetic sculptures, a hands-on lab centered on DNA, and a collection of model ships.

MIT is well known as a center for robotics and artificial intelligence, and the museum doesn't disappoint. Many robots are on display, including Kismet (a robot with realistic facial expressions) and Cog (a humanoid robot). The entire robotics collection and exhibition is cutting edge, and well worth lingering over.

The holography collection is simply the best in the world. There's an entire gallery of holograms to look at, including some (such as a woman blowing a kiss) that change as you move around them, and one featuring the artist Keith Haring. The complete hologram exhibit is also available online at the museum's website, but to see them properly you have to be there.

The intersection of art and science is represented by the fantastic moving sculptures of Arthur Ganson. These include a machine that oils itself, flapping bird-like bits of paper, and a violin that plays itself using a feather.

Harold Edgerton's photographs and films, made using a stroboscope for very high-speed photography, are the most important part of the MIT Museum collection. The strobe lights were able to stop the wings of a hummingbird in flight, study the motion of a golfer's swing, and capture a single drop of milk creating a splash. Even the pattern of smoke around a turbine blade is revealed.

Holograms

Holograms are familiar to everyone—most credit cards now have holograms on them as an anti-forgery device, and holographic images showing 3D scenes have been commonplace since the 1970s. But holography actually dates back to 1947, when Dennis Gabor invented them while working on improving the images created by electron microscopes. For his invention he received the Nobel Prize in 1971.

A hologram uses the basic processes of diffraction and interference of light to reproduce a three-dimensional image. Interference is a fundamental property of waves and turns up in radio astronomy, electronic music, and even supersonic booms. When two waves are superimposed, their relative phase (when the peaks and troughs appear) causes an interference pattern. That is, the exact way in which the peaks and troughs of each wave coincide leads to varying peaks and troughs in the resulting combined wave.

In 1803, the British scientist Thomas Young showed that light was made up of waves by splitting a beam of light with a very thin card, resulting in an interference pattern (of dark and light bands) between the two parts of the light beam. In another experiment, a beam of light was sent through a pair of tiny slits to produce a similar interference pattern. In holography, interference between light beams is at the heart of recording the image.

A hologram is recorded on a photographic medium by illuminating the subject to be photographed with a source of coherent light (i.e., light that is in phase, such as a laser). The light is first split into two (using a half-silvered mirror)—one half of the light falls directly on the photographic medium; the other half illuminates the object, bounces off, and hits the photographic medium, where it interferes with the light that came straight from the laser and creates an interference pattern (see Figure 98-1).

Each dot of reflected light interferes with the light coming straight from the laser, and dots of reflected light from all over the object come together all over the photographic medium. Each part of the hologram contains light from all over the object joined together by interference.

Viewing the hologram (Figure 98-2) means using it as a diffraction pattern to reproduce the original object. When a wave such as light encounters a very small object (such as the thin paper in Young's experiment), it appears to bend around the object and spreads out in slightly different directions. This is called diffraction. In Young's experiment the light is diffracted at the sheet of paper, and then recombines and interferes to produce the interference pattern.

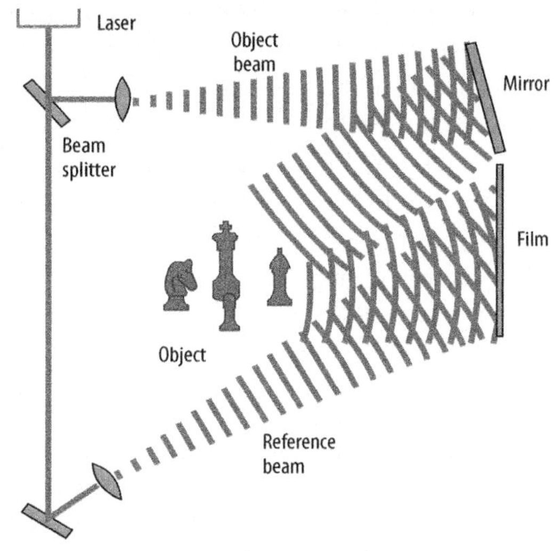

Figure 98-1. Creating a hologram

Figure 98-2. Viewing a hologram

Young's interference pattern can be thought of as being a "picture" of the paper around which the light was diffracted. In a hologram, the interference pattern is illuminated with the same wavelength of light that was used to make it. The pattern acts to diffract the light and reproduce an image of the original object.

This happens because the interference pattern recorded in the hologram captured not just the intensity of light reflected from the object (as would happen with a black-and-white photograph) but also the phase of the light that gives the depth of each part of the object. By interfering, the brightness and depth are recorded from all over the object.

When illuminated with the right light, the hologram appears to be floating in mid-air.

Serious computer scientists will be blown away by CADR (also called the Knight Machine), a computer entirely based around the Lisp language. Lisp is unusual among computer languages in that its notation is essentially mathematical (as opposed to many languages that consist of telling the computer "do this; if something happens, then do that"), and this purity leads many computer scientists to think of it as the greatest computer language of all.

The Hart Nautical Gallery features a large number of model boats, and also covers technologies for ocean exploration. These include autonomous underwater vehicles as well as side-scan SONAR systems that give incredibly detailed pictures of the sea bed and anything lying on it.

The museum also has a large architectural collection.

The MIT Museum bills itself as being about "ideas in the making"—it's a museum of past triumphs, but also focuses on current questions and problems. Visitors are invited not only to view its collection, but also to take part in MIT's present-day work.

Practical Information

Full details of the MIT Museum are online at *http://web.mit.edu/museum/*. In addition, Gabor's Nobel Prize lecture on holography is a joy to read; you can find it at *http://nobelprize.org/nobel_prizes/physics/laureates/1971/gabor-lecture.html*.

099

Gaithersburg International Latitude Observatory, Gaithersburg, MD

39° 8′ 12.00″ N, 77° 11′ 57.00″ W

An Innocuous White Building

Latitude—the number of degrees from the Equator toward either pole—seems like something that should be fairly fixed. But when measuring latitude by observations of the stars, the apparent latitude varies from day to day. This happens because the Earth does not spin uniformly on its axis—it wobbles.

The International Latitude Observatories were set up in 1899 to determine how much the Earth wobbles by observing a group of 12 stars. The observatories were located in Mizusawa, Japan; Tschardjui, Central Asia; Carloforte, Italy; and in the U.S. at Gaithersburg (Figure 99-1), Cincinnati, and Ukiah. All six observatories lay along the same line of latitude at 39° 8′ N.

Figure 99-1. The Gaitherbsurg Observatory; courtesy of Amy Fredericks (etacar11)

Precession, Nutation, and the Chandler Wobble

The Earth orbits the Sun once a year and rotates on its axis once a day, but the daily rotation isn't simply along the Geographic North to South Pole axis. In fact, the Earth's axis of rotation is wobbling because of three effects: precession, nutation, and polar motion (see Figure 99-2).

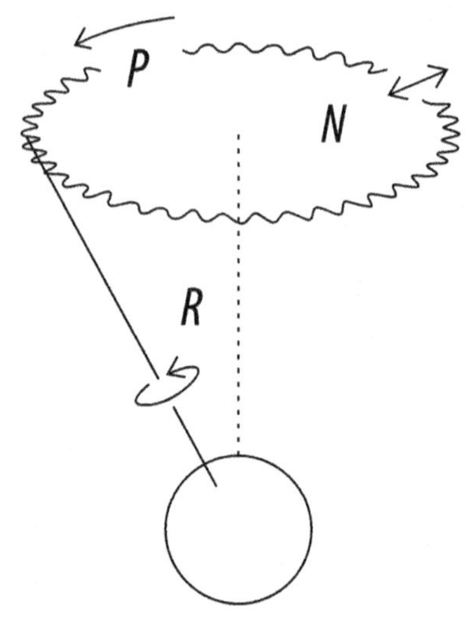

Figure 99-2. Precession and nutation

An imaginary line drawn out into space through the Earth's axis traces out a wobbly circle with a period of about 26,000 years. The circle is caused by precession, a slow wobble similar to the motion of a spinning top. The precession is caused by the gravitational pull of the Sun and Moon on the Earth's Equator. The Earth is not a sphere; it has a bulge at the Equator of about 42 kilometers that is caused by its rotation. Because the Earth's axis of rotation is at an angle relative to the orbital plane, the bulge is tilted either closer or further away from the Sun. The gravitational pull on this off-center bulge causes the precession.

But the precession circle isn't perfect—the path of the imaginary line oscillates with a period of almost 19 years. This nutation is caused by the gravitational pull of the Moon on the Earth.

Both the precession and the nutation are relatively slow phenomena and can be predicted. The Chandler Wobble is a different matter.

The Gaithersburg Observatory helped to calculate the Chandler Wobble, which was described in 1891 by American astronomer Seth Carlo Chandler. The wobble is one component of polar motion—a small motion of the pole of a few meters per year. Polar motion can be forecast only with accuracy for a few months in the future.

The Chandler Wobble has a period of 14 months and combines with a second wobble with a period of one year. The overall polar motion is a spiraling movement that starts and ends at roughly the same position for a period of almost seven years. The Chandler Wobble is believed to be caused by changing pressure at the bottom of the oceans that keeps the Earth vibrating.

On top of these three motions affecting the location of the pole, there's a westward drift of 20 meters per year because of the motion of the Earth's core and mantle. Figure 99-3 shows the average position of the North Pole since 1900, and the spiraling polar motion between 2001 and 2006.

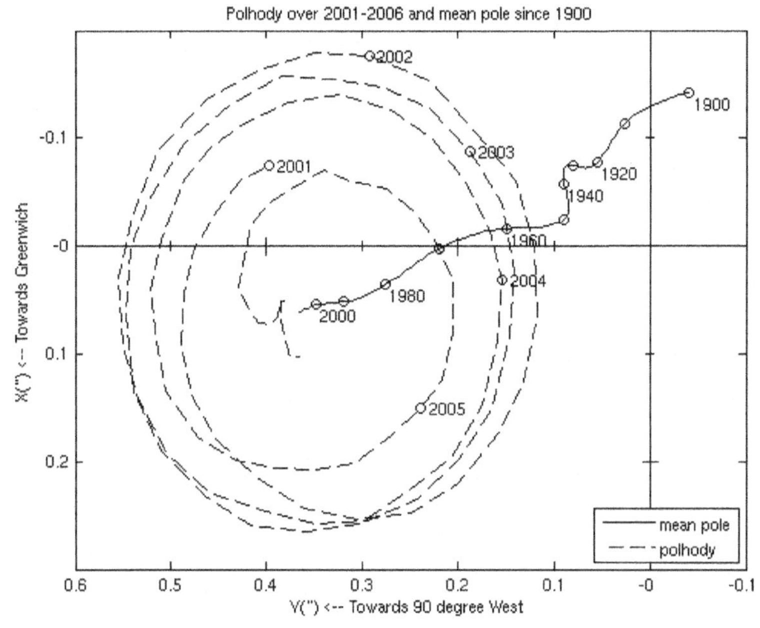

Figure 99-3. Polar motion; courtesy of IERS

Today, the Earth's rotation is determined accurately by radio telescopes using interferometry (see page 424).

The apparent latitude varies because of the Earth's wobble (see sidebar) and because of the refraction of light in the atmosphere. The wobble causes the pole to move as much as six meters from its average position on a daily basis. Determining the exact nature of the wobble was done by observing a pair of stars from the group of 12 each night to calculate the apparent latitude of each observatory.

Even though the wobble is very small, it is important to astronomers observing very distant objects: a small error in pointing a telescope caused by the irregular movement of the Earth can cause the observer to completely miss the object.

The Gaithersburg Observatory was used until 1982 (with a brief pause during the First World War) and consists of a small building that contained a telescope. The observatory building is white (to keep it cool) and has slatted walls to reduce any air disturbances inside. The roof can slide open to allow the telescope to make an observation.

Close to the observatory building is a one-meter-high marker used to align the telescope with the local meridian. There are five circular markers set into the ground that enable the precise latitude, longitude, and elevation of the observatory to be determined.

The telescope originally used at the observatory is on display in the city of Gaithersburg at the Community Museum.

Practical Information

The observatory is only open during special events organized by the city of Gaithersburg, but it can still be viewed from the road when closed. The city is planning to open a small park around the observatory to make it more accessible.

Gaithersburg is about a one-hour drive from the National Electronics Museum (Chapter 100) and the National Cryptologic Museum (Chapter 101). A full day's worth of scientific sightseeing can take in all three.

100

National Electronics Museum, Linthicum, MD

39° 11′ 44.01″ N, 76° 41′ 4.69″ W

Electronic Defense and Countermeasures

The National Electronics Museum focuses on the application of electronics to defense, and has the most important collection of radar equipment in the world.

The museum begins with an introduction to electronics and magnetism designed for beginners. The Fundamentals Gallery explains the relationship between electricity, magnetism, and radio waves with hands-on exhibits. The Communications Gallery introduces radio communication, including Marconi's experiments (more on Marconi is covered in Chapter 62), Morse code, and the development of the telephone (for more on Alexander Graham Bell, see Chapter 4).

Three galleries cover the history of radar, starting with British work in the 1930s through the attack on Pearl Harbor and the Second World War. The second radar gallery covers advances in radar to cope with the Cold War—Doppler radar was developed so that surface-to-air missiles could attack nuclear bombers, and eventually became the technology used to trap speeding motorists. The final radar gallery covers modern radar systems including the AWACS rotating dome, airport radar, and a demonstration of phased array radar. The museum also has a gallery dedicated to radar countermeasures, including the use of radio jamming, decoys, and chaff.

The Under Seas gallery shows how SONAR works, and there's a demonstration of passive and active tracking of underwater sounds (a great spot for fans of *The Hunt for Red October*).

Outside the museum are a number of a large exhibits: a TPS-43 transportable U.S. radar used for ground sensing of aircraft, a TPS-70 that replaced the TPS-43, and a Nike AJAX anti-aircraft radar used to steer the Nike AJAX missile to its target. And there's an SCR-270, which was the type of radar used to detect the Japanese aircraft that attacked Pearl Harbor.

Pulsed Doppler Radar

Common radar systems are pulsed: a pulse of radio waves is sent out, the radar system stops transmitting for a time and listens for a reflected radio wave, then another pulse is sent (see Figure 100-1). The characteristics of a pulsed radar are determined by the pulse frequency, the pulse width (how long the pulse is transmitted for), and the pulse repetition period (the time from the start of one pulse to the start of the next).

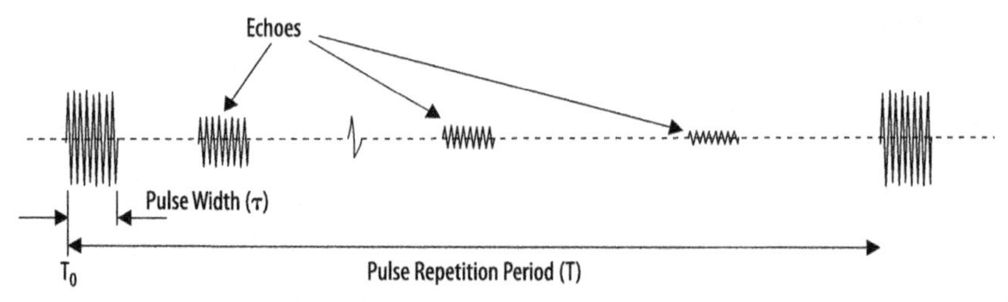

Figure 100-1. Major features of pulsed radar

The pulse width determines the minimum distance at which an object can be detected—while the pulse is being transmitted, an echo cannot be received. The pulse repetition period defines how long the radar will be listening for echoes. For example, a radio wave sent to a target one kilometer away takes 6.7 microseconds to return an echo. If the pulse width was 1 microsecond, then the pulse repetition period must be greater than 7.7 microseconds to be able to detect the echo.

In real radar, the pulse repetition period will be much longer. The TPS-43 radar on display at the museum had a pulse width of 6.5 microseconds, but a pulse repetition period of 4 milliseconds, giving it a range of almost 450 kilometers.

Pulsed radar can detect the distance to an object, and by moving the radar, an object's direction and elevation can be found. Pulsed Doppler radar can, in addition, measure the speed at which an object is moving.

Pulsed Doppler radar uses the same technique of short pulses and long quiet periods as found in pulsed radar, but it also measures the change in frequency of the received echoes. When the radar pulse hits a moving object (such as an aircraft), the frequency of the reflected wave is altered because the object is moving. If the object is approaching the radar, the frequency will have increased slightly, and if it is receding, the frequency will have decreased.

This effect can be heard when a police siren is approaching quickly and its tone alters. Figure 100-2 shows an object producing a continuous wave in all directions on one frequency while moving to the left. To the left, the frequency has increased and the wavelength shortened; to the right, the opposite has occurred.

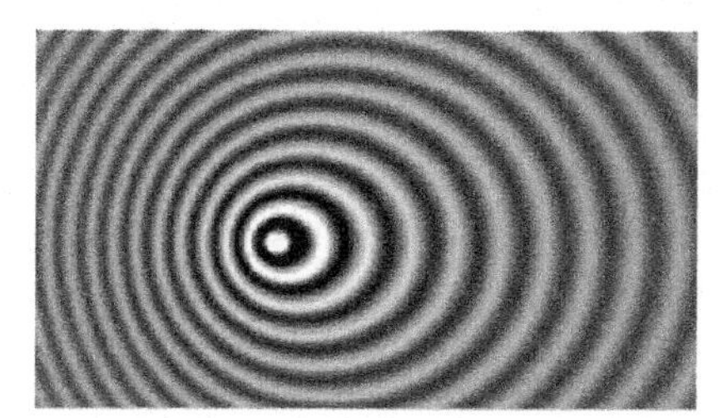

Figure 100-2. The Doppler effect

By measuring the shift in frequency, the radar system can calculate the speed at which the object is approaching or receding from the radar. This is only one component of the velocity, since the object is unlikely to be coming straight at the radar. If other radar stations calculate the velocity for the same object from other locations, the object's true velocity can be found by combining the different radar observations.

An important consequence of Pulsed Doppler radar's ability to detect the velocity of objects is that it can ignore objects that are not moving. This makes it effective in searching for movement and eliminating noise created by radio waves bouncing off fixed objects. For example, Pulsed Doppler radar can be used to find a moving vehicle on a (non-moving) road.

The Doppler Effect is also used in police radar guns to measure the speed of an approaching or receding car.

The National Electronics Museum is about a 30-minute drive from the National Cryptologic Museum (see Chapter 101), and both can be covered in a day. Admission is free, and free guided tours are also available. Tours take about 90 minutes and must be booked in advance through the museum's website.

The museum also has an amateur radio station, W3HEM—if you bring your ham license, they'll let you use their equipment; book ahead to be sure of a slot.

Practical Information

Complete information about the National Electronics Museum, including directions, is available at *http://www.nationalelectronicsmuseum.org/*.

101

National Cryptologic Museum, Fort Meade, MD

39° 6′ 53.28″ N, 76° 46′ 29.28″ W

Looking Backward at the NSA

The U.S. National Security Agency, or NSA, was once so secret that insiders joked that its initials stood for No Such Agency. But Hollywood filmmakers searching for a malevolent force more secretive and more dangerous than the CIA have brought the NSA to public attention through films such as 1998's *Mercury Rising* and *Enemy of the State*. And the NSA's geeky mission of code making and breaking, electronic surveillance, wiretapping, Internet monitoring, and database mining seems increasingly relevant in this age of all-pervasive electronic communication.

Nevertheless, the day-to-day work of the NSA remains shrouded in mystery. That makes the NSA's public museum, the National Cryptologic Museum, a surprising discovery. The museum opened in 1993 and is housed in an old motel just outside the NSA's headquarters at Fort Meade, Maryland.

The small museum was originally open to NSA employees only, and houses a superb collection of artifacts covering the history of codes and codebreaking. Entry to the museum is free of charge.

A great starting point for a visit is the working Enigma machine used by Nazi Germany in the Second World War to encrypt messages. At the time, the Enigma machine's codes were considered to be unbreakable, but they were in fact broken, in great secrecy, in the UK during the war (see Chapter 40). The Enigma machine on display at the NSA museum offers a rare treat, as visitors are invited to actually touch and use it to create encrypted messages.

The SIGSALY exhibit shows part of the equipment used for secure telephone communication between the Pentagon and the British War Office during the Second World War. SIGSALY weighed 55 tons and was too large to install in a government office in London; instead, it ended up installed in Selfridges department store on Oxford Street. It worked by inserting random noise into the actual conversation to mask the words spoken. By adding random noise at one end, and subtracting the same random noise at the other, it was possible to have a secure conversation: any eavesdropper would hear something akin to static. The random noise was stored on a phonograph record; each end would have a copy of the same record (which was essentially the encryption key), and the two records had to be played back at precisely the same speed so that each end stayed synchronized with the other. A single record gave just 12 minutes of secure speech.

Also on display at the museum are many examples of cryptography from well before the Second World War. A somewhat speculative exhibit on possible communication by U.S. slaves, using patterns sewn into quilts, is an early example of steganography (hiding a message so that an eavesdropper doesn't even realize a message is being sent). The U.S. Civil War exhibit has code books and a fascinating cipher cylinder that encrypts using the Vigenère system (see Chapter 121 for details of the cipher).

The Cold War section highlights the importance of aerial reconnaissance with a complete Air Force C-130 aircraft outside the museum. This exhibit is dedicated to the memory of 17 U.S. servicemen killed when the Soviet Union shot down their C-130, which had strayed into Armenia while eavesdropping from Turkish airspace.

The Cold War also drove innovation in satellite-based reconnaissance, and the museum has an exhibit concerning the first U.S. intelligence satellites, the GRAB and GRAB II. Launched in 1960 and 1961, the Galactic Radiation Background Experiment had the overt mission of measuring solar radiation, but covertly the satellites were fitted with equipment capable of picking up and measuring Soviet radar signals. These radar signals were processed and then relayed to ground stations, enabling the NSA to gather data on Soviet radar systems from 500 miles up. Since radar (and other radio signals) travel well beyond the horizon, GRAB simply had to be positioned with a clear view of the radar site to be able to intercept the signal as it continued off into space.

More recently, computer-based encryption has become the norm, and with it computer-based attacks on encryption systems. The museum's supercomputing exhibit includes a Cray YMP supercomputer. Built in 1993, the year the first web browser became available, the computer had 32 Gb of memory. At the time, a standard PC using the newly released Intel Pentium processor had around 8 Mb of RAM. As well as having 8,000 times the memory of a PC, the Cray YMP operated at 2.67 billion operations per second, around 3,000 times faster than the Intel Pentium appearing on desktops.

According to the museum, the NSA has the world's largest supercomputing facility located nearby: however, it isn't open to visitors! Despite giving no insight into the capabilities of today's NSA (its work is top secret, after all), the museum provides a fascinating starting point for speculation about the NSA's power.

FROSTBURG: The Connection Machine CM-5

Part of the supercomputing exhibit is a computer that looks like something straight out of a movie. The black monolith of the Connection Machine CM-5 stands taller than a man and sports a vertical array of constantly flickering red LEDs that were used to check the operation of the machine.

The NSA's CM-5, code-named FROSTBURG after a town near NSA headquarters, was used for unspecified mathematical code breaking and had 512 processors running in parallel, giving a total speed of 65.5 billion operations per second. It also had 500 Gb of RAM and ran a special version of the UNIX operating system named CMost. It was installed at NSA in 1991, at a time when a PC with 4 Mb of RAM was considered "high end."

FROSTBURG was the first massively parallel processing system bought by NSA. A massively parallel computer essentially consists of a large number of independent computers, each with its own CPU and memory, with a means for all the computers to communicate with one another. The individual computers, usually called nodes, communicate over an internal network by passing messages. The FROSTBURG computer had 512 such nodes.

A standard personal computer typically has a single CPU, capable of working on a single task. Even with multitasking operating systems like Microsoft Windows or Mac OS X, the CPU only works on one task at a time with the operating system switching between programs as necessary. In contrast, a massively parallel computer can be used to break a single task (such as code breaking) into many parts, assigning each part to a single CPU (or node).

For code breaking, each CPU could attack a small part of an encrypted message, or try out a small number of keys to see if any of them decrypt the message. Working independently and in parallel, the nodes in a massively parallel machine can quickly work through a task that would take a standard PC months or even years to complete.

Massively parallel machines do have a significant design challenge: although the nodes operate independently, they need to communicate with one another to exchange information. Each node only has access to its own memory, and cannot access the memory associated with any other node. If a node needs information from another node, it sends a message across an internal network requesting the information and receives a data-bearing message in reply. Hence, the design and speed of the communication network are critical to getting maximum performance from the computer.

The alternative to this internal network is to have all the nodes share memory: in a shared-memory design, the nodes don't need to communicate with each other, but they are forced to coordinate which node has access to the memory.

This creates problems of contention (when more than one node tries to access memory at the same time), and these contention problems increase as the number of nodes increases. Typically, shared-memory machines can manage no more than 32 nodes.

With 512 nodes, the FROSTBURG CM-5 is not a shared-memory machine: its nodes are connected to a fast internal network. The simplistic network design would be to connect every node to every other node with a dedicated link forming a mesh of connections. Unfortunately, that kind of architecture runs into trouble as the number of nodes increases.

For example, a 4-node computer needs a total of 6 connections. If the nodes are labeled A, B, C, and D, then you need connections AB, AC, AD, BC, BD, and CD so that any node can communicate directly with any other node. A 5-node computer needs 10 interconnections, a 6-node computer needs 15, and so on (see Figure 101-1).

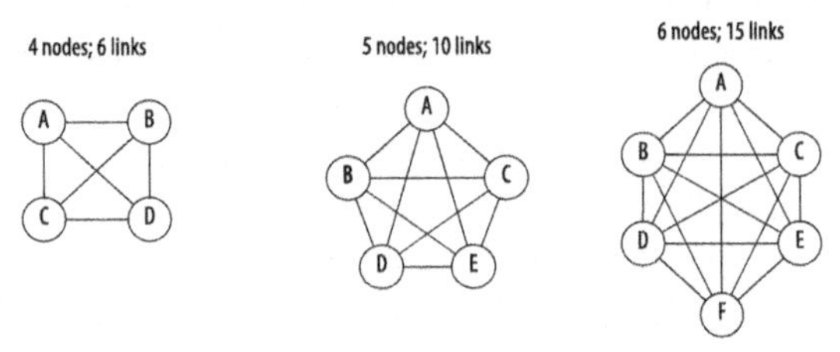

Figure 101-1. Mesh connectivity

The total number of connections for a computer with an even number of n nodes is governed by the formula n × (n – 1) / 2. The first of n nodes needs to be connected to (n – 1) other nodes, the next node to (n – 2) nodes (one less because it's already connected to the first node, and that connection is counted in the (n – 1)), the node after that is connected to (n – 3) nodes, and so on. So the total number of links is (n – 1) + (n – 2) + … + 1.

The German mathematician Johann Gauss is reputed to have astonished his primary-school teacher by instantly adding the first 100 numbers (1 + 2 + … + 100), giving the correct result (5,050), by observing that the sum can be written (100 + 1) + (99 + 2) + … + (51 + 50). There are 50 parts to that sum, and each part sums to 101, so the result is 50 × 101 or 5,050.

The same reasoning applies to summing (n – 1) + (n – 2) + … + 1. This can be written (n – 1 + 1) + (n – 2 + 2) + … and has (n – 1) / 2 parts, yielding the formula n × (n – 1) / 2.

With this design, FROSTBURG would have needed a total of 130,816 connections. And the CM-5 was designed to scale up to thousands of processors, making a mesh of connectivity impossibly expensive. Another problem with large numbers of connections relates to the speed of light. As the physical size of a supercomputer increases, its speed will decrease; the speed with which information can move within the computer is limited by the speed of light, making compact designs a necessity.

The CM-5 solved the connection problem by arranging its nodes into a "fat tree." In computing terms, a tree is a set of points (also typically called nodes) linked together in a structure that resembles an upside-down tree (see Figure 101-2). The root node is at the top, with connections to other nodes branching out below it. There may be several levels of nodes ending with the leaf nodes, which don't have any branches below them.

In a "fat tree," the links (branches) between nodes get thicker the closer the branch is to the root. In the CM-5, the fat tree defines the layout (or topology) of the internal network. The thickness of the link represents the speed of the network link: thicker means faster. The computing nodes were arranged at the leaves of the tree; the other parts of the tree were specialized routing chips capable of routing messages to and from the computing nodes (Figure 101-3).

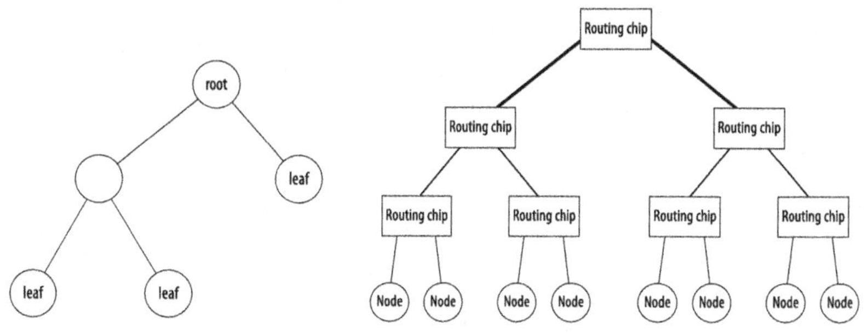

Figure 101-2. A simple tree

Figure 101-3. Simplified CM-5 fat tree

When a node wishes to communicate with another node, it passes a message to the routing chip it is connected to. That routing chip passes the message up the tree until it reaches a point where it can descend to its intended destination node. For adjacent nodes, the message passes through a single routing chip; for more distant nodes it may pass through many routing chips. To ensure that messages move quickly and without contention, the speed of links (and routing chips) is doubled for each step up the tree, with the root routing chip being the fastest and having the fastest connections.

The simplified diagram in Figure 101-3 shows a binary tree: each routing chip is attached to at most two other chips or nodes. The CM-5 actually used a quaternary tree (each chip was connected to four other chips or nodes) and the speeds did not double with each layer of the tree, but the system enabled nodes to communicate with a delay of 3 to 7 microseconds, and specially coated cables transmitted signals at 90% of the speed of light.

It's sobering to think that this massively powerful machine is, to the NSA, a museum piece.

Practical Information

Details of the National Cryptologic Museum are available from the NSA website at *http://www.nsa.gov/about/cryptologic_heritage/museum/*. The museum is open Monday to Friday and the first and third Saturdays of the month. You can contact the museum by phone at 301-688-5849.

102

The Henry Ford, Dearborn, MI

42° 18′ 12.9″ N, 83° 14′ 2.68″ W

"America's Greatest History Attraction"

The name Henry Ford immediately evokes images of cars and the Model T, but Ford built more than cars. In Dearborn he constructed an enormous history museum with a fascinating collection of scientific and technological exhibits alongside American cultural artifacts.

The Henry Ford consists of a museum and an entire village of houses; because of the size of the site, it's worth carefully planning your visit. The Henry Ford's website has interactive tools for deciding what to visit and where to find objects of interest.

The museum contains entire homes and laboratories that were moved to The Henry Ford and put on display. The most interesting of these are the bicycle shop and home of the Wright Brothers and the reconstruction of Thomas Edison's Menlo Park, New Jersey laboratory. There's also Buckminster Fuller's Dymaxion House prototype, a circular home made entirely of aluminum that could be delivered on a truck and was designed to be energy- and space-efficient.

Of course, it's no surprise that the museum also has an extensive collection of cars and car-related memorabilia (including the neon sign used to illuminate the first McDonald's fast food restaurant). And there's a model of the (happily) never-created Ford Nucleon nuclear-powered car, plus the chance to take a ride in a restored Model T.

The transportation exhibit has a large collection of long-distance vehicles, including stage coaches, buses, and steam locomotives. At least one of the steam trains is in operation, pulling passengers around the site.

Two-Stroke and Four-Stroke Engines

Naturally, The Henry Ford also has a variety of gasoline engines on display. Gasoline engines can be largely divided into two types, two stroke and four stroke, depending on how the cycle of drawing gasoline into the cylinder (intake), compressing it, exploding it (ignition), and expelling the exhaust fumes is arranged.

In cars, four-stroke engines are used. Each stroke is a movement of the piston in the cylinder and has a single purpose: there's one stroke each for intake, compression, ignition, and exhaust. Taken together, these four separate actions are known as the Otto Cycle after the German engineer Nikolaus Otto, who invented the internal combustion engine as we know it today.

In the Otto Cycle, a single cylinder in the engine starts with the piston at the top. The top of the cylinder has a pair of valves, one for the intake of fuel and the other for the exhaust fumes. The first step in the cycle is the intake of fuel. A mixture of gasoline and air is forced into the cylinder through the intake valve as the piston descends (Figure 102-1).

In the second part of the cycle, the piston raises again and compresses the air/fuel mixture with both valves closed (Figure 102-2). At this point the fuel is ready to be ignited.

Next, a spark is generated by a spark plug in the cylinder and the fuel explodes, forcing the piston down and generating power (this movement is called the power stroke; see Figure 102-3). (In contrast, a diesel engine [see page 66] has no spark plugs, but the engine goes through a similar cycle, with the fuel injected near the end of the compression stroke, and just air entering during the intake.)

After the explosion, the exhaust valve opens, and the piston comes up again and expels the fumes (see Figure 102-4). With the piston back at the top of the cylinder, the cycle can begin again.

In a car engine, the four pistons are typically linked together by a common rod called the crank shaft. As the engine operates, one piston performs the power stroke, moving the other pistons through different parts of the cycle. Each piston is at a different point in the four-part Otto Cycle, ensuring smooth operation.

In a two-stroke engine, the four parts of the cycle are performed by two piston movements. The cylinder in a two-stroke engine does not have valves; instead, it has a pair of holes (called ports) that allow fuel to enter (the intake) and fumes to exit (the exhaust). The ports are covered and uncovered by the piston during its movement.

With the piston at the top of the cylinder and a fuel/air mixture inside the cylinder, a spark plug causes an explosion that forces the piston down. When the piston is at the top, the two ports are covered by it and the explosion occurs in a confined space.

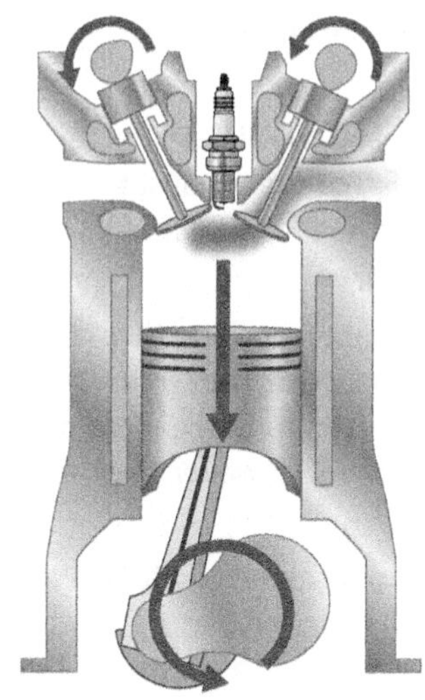

Figure 102-1. Intake stroke

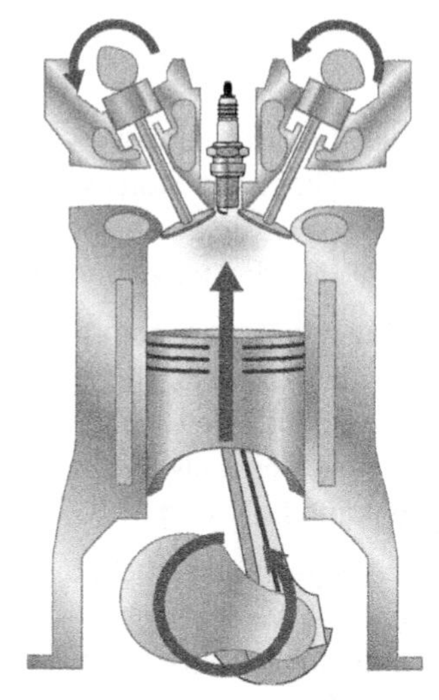

Figure 102-2. Compression stroke

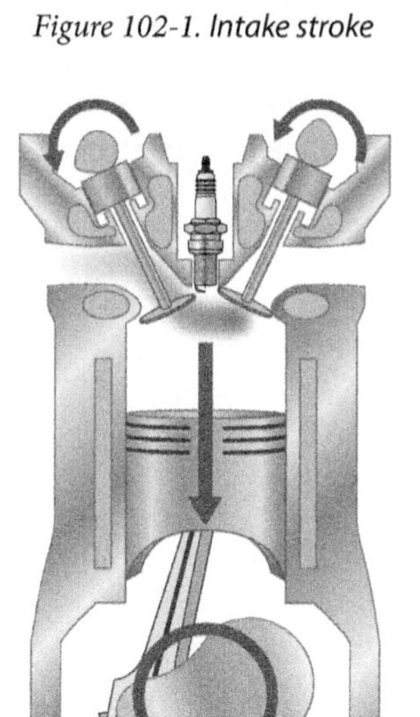

Figure 102-3. Power stroke

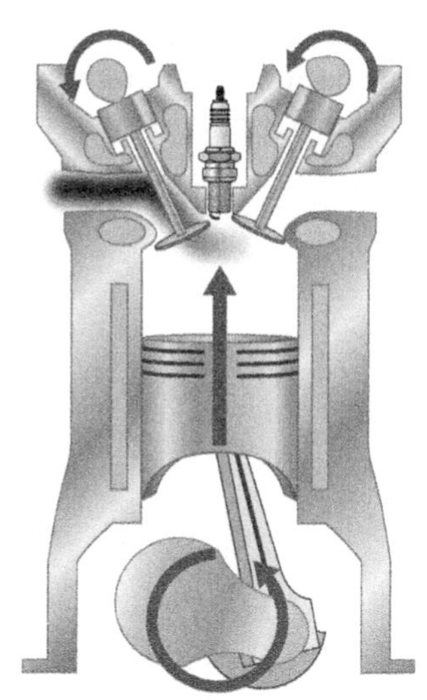

Figure 102-4. Exhaust stroke

As the piston descends, it first uncovers the exhaust port, allowing the gases in the cylinder (which are under pressure from the compression and explosion) to escape. Unlike a four-stroke engine, the space under the cylinder is important; it is filled with the air/fuel mixture, and as the piston descends this mixture is somewhat compressed, readying it to enter the cylinder. This one movement of the cylinder combines the power stroke and the exhaust stroke.

Eventually the piston uncovers the intake port, and the air/fuel mixture enters the cylinder and forces out whatever is left of the exhaust fumes. Then the piston moves up again and further compresses the mixture. This movement combines the intake and compression strokes.

Two-stroke engines are commonly used for portable machines like chain saws and lawn mowers, and for small vehicles like boats with outboard motors and mopeds. They are light, simple to manufacture, and easy to maintain because they have only two cylinders and no valves. However, they are also inefficient and polluting.

In a four-stroke engine, the intake and exhaust phases are carefully timed so that there's no waste; in a two-stroke engine, the fuel/air mixture entering the cylinder inevitably leaks out of the exhaust since both are open at the same time. Also, in a four-stroke engine the area under the piston can be used for lubrication because it is not used for the cycle. In contrast, a two-stroke engine has to be lubricated by adding oil to the fuel/air mixture. Some of this oil burns at the same time as the gasoline, making two-stroke engines more polluting than four-stroke.

Steam power is well represented, with many working steam engines on display including a Newcomen engine taken from Cobb's Engine House in England; it is the oldest surviving steam engine in the world. The steam power collection is the most fascinating part of The Henry Ford, but is sadly downplayed in favor of marketing attractions such as the IMAX theatre. Avoid the crowds and see Ford's wonderful collection of working engines, which includes the enormous Highland Park Engine used to power one of his factories.

Some of the non-science artifacts on display include the car JFK was riding in when he was assassinated, the bus in which Rosa Parks was arrested for refusing to give up her seat, and the seat Abraham Lincoln was sitting in when he was shot.

A complete audio tour of The Henry Ford is available for download from the museum's website free of charge. While at the museum, it's also possible to tour the present-day Ford Rouge factory where the Ford F-150 truck is assembled.

Practical Information

Details of the museum (plus a lot of fancy marketing) are available at *http://www.thehenryford.org/*.

103

Gateway Arch, St. Louis, MO

38° 37′ 28.81″ N, 90° 11′ 5.82″ W

Gateway to the West

The Gateway Arch in St. Louis, Missouri, stands at 192 meters, and is the tallest monument in the U.S. (Figure 103-1). It was opened in 1967, and the view from the top is spectacular. The top is reached by riding in small five-person trams that travel up the legs of the arch to the observation deck at the top. While you're up there, you can contemplate the perfect shape of the arch—it's a catenary.

Figure 103-1. The Gateway Arch; courtesy of Pat Dye (gobucks2)

Even though the arch is man-made, there's something natural-looking about it, because a catenary is a shape made simply by the force of gravity. Take a short piece of rope (or string, or a necklace of uniform thickness) and hold one end in each hand. With your hands level, the rope falls into a graceful curve under its own weight pulled down by gravity. That shape is a catenary (a term coined by Thomas Jefferson based on the Latin word for chain), and the weight of the rope is transmitted by tension through the rope itself.

Invert a catenary and build it out of something rigid, and you've got a catenary arch. The Gateway Arch is one example, but catenary arches are very common. They've been built since antiquity because they are stable—the weight of the arch's components acts downward through the arch's legs.

The arch follows the shape of a catenary curve and is constructed from stacked metal equilateral triangles. The centroid of each of the triangles is aligned with the curve being followed; the triangles start out at the bottom with 16-meter sides that narrow to 5 meters at the top. This narrowing explains why the trams are so small—you should be grateful that they're there at all, though, because the original plan was to have visitors walk 1,076 steps to the top.

The exterior of the Gateway Arch is made of plates of stainless steel that cover 90 meters of reinforced concrete, topped by a steel structure that completes the shape. Because the arch curves, the trams are mounted on gimbals that allow them to stay level as they climb the curved interior of the legs.

The arch is as high as it is wide, and it weighs about 15,600 tonnes. It sways in the wind up to a maximum of close to 50 centimeters in each direction; on very windy days, the observation area is closed.

At the base of the arch is a visitor center that explains its construction and a little of the mathematics behind its shape. There's also the Museum of Westward Expansion, which explains the Lewis and Clark expedition that helped open up the western U.S., and also covers the Louisiana Purchase and the debate over slavery. The arch is sometimes called the Gateway to the West.

The arch is part of the Jefferson National Expansion Memorial, a 37-hectare park along the Mississippi River. Once you've spent a couple of hours seeing the arch inside, and the associated museum, the park makes an ideal spot for a walk, lunch, or bike ride. It's also possible to take helicopter tours of St. Louis for an aerial view of the arch from all angles.

Practical Information

Visiting information (and, I promise, no calculus—unlike the upcoming sidebar) is available from *http://www.gatewayarch.com/*.

The Catenary

A telephone cable strung between a pair of telephone poles hangs down in a curve called a catenary. The curve looks a bit like a parabola, but it is not—it's actually described by a hyperbolic cosine function. What follows is probably the hardest bit of mathematics in this book; it involves a bit of calculus, but try to stick with it if you can!

Start by imagining a cable hanging under its own gravity, with a y-axis passing vertically through the cable's lowest point; call that point A. Pick another point on the cable to the right and call it B. The line from the origin, O, to the point B makes an angle of a with the x-axis. (See Figure 103-2.)

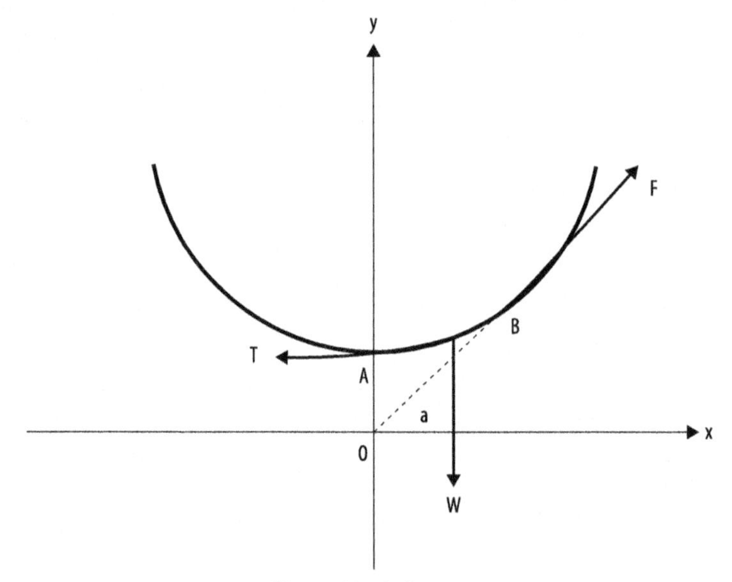

Figure 103-2. A catenary

Three forces act on the segment AB—there's the horizontal tension, T, at point A; the tension along the cable, F, at point B; and the weight of the segment AB acting downward, W. Since the cable is not moving, these three forces are in equilibrium and can be redrawn into a triangle of forces (Figure 103-3).

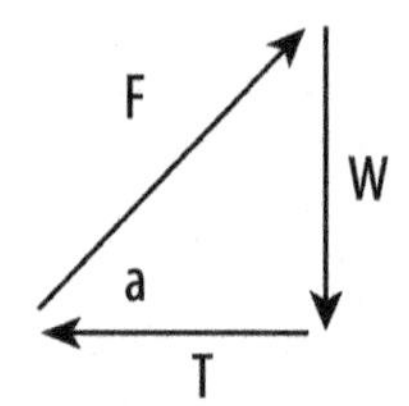

Figure 103-3. Triangle of forces

Basic trigonometry shows that F sin a = W and F cos a = T. The weight of the segment can be expressed by taking the linear density of the cable (call that D) and multiplying it by the segment's length (call that l) to obtain W = Dl. Dividing the two trigonometric equations eliminates the force F to express the angle (which gives the slope of the cable) in terms of the cable weight and horizontal tension (Equation 103-1).

$$\tan a = \frac{Dl}{T} \text{ or } l = \frac{T}{D} \tan a$$

Equation 103-1. Basic catenary equation

The coordinates of B are (x, y), and some basic trigonometry gives that dx/dl = cos a and dy/dl = sin a. Differentiating the equation above gives dl/da = T/D $\sec^2$ a. So it's possible to express everything in terms of differentiating by a (see Equation 103-2).

$$\frac{dx}{da} = \frac{dx}{dl}\frac{dl}{da} = \frac{T}{D} \cos a \sec^2 a = \frac{T}{D} \sec a$$

$$\frac{dy}{da} = \frac{dy}{dl}\frac{dl}{da} = \frac{T}{D} \sin a \sec^2 a = \frac{T}{D} \tan a \sec a$$

Equation 103-2. Catenary differential equations

Now integrate those two equations to get equations for x and y (Equation 103-3).

$$x = \frac{T}{D} \log(\sec a + \tan a) + c_0, y = \frac{T}{D} \sec a + c_1$$

Equation 103-3. x and y expressed in terms of a

Both the constants are in fact 0. c_0 is 0 because the angle a is 0 when x is 0 (see the diagram back in Figure 103-2). c_1 can be made 0 by placing the x-axis at position T/D.

Now flip the equation for x around and raise it to the power of e. Inverting the resulting equation, plus a little trigonometry (recalling that $\tan^2$ a + 1 = $\sec^2$ a), gives the two exponential equations shown in Equation 103-4.

$$\sec a + \tan a = e^{\frac{Dx}{T}}, \sec a - \tan a = e^{-\frac{Dx}{T}}$$

Equation 103-4. Catenary exponential equations

Add those together, and you've got the equation for y based on x (which gives the shape of the curve) shown in Equation 103-5.

$$y = \frac{T}{D}\frac{e^{\frac{Dx}{T}} + e^{-\frac{Dx}{T}}}{2} = \frac{T}{D} \cosh \frac{Dx}{T}$$

Equation 103-5. The catenary equation

104

Horn Antenna, Holmdel, NJ

40° 23′ 26.61″ N, 74° 11′ 5.57″ W

Evidence for the Big Bang

Down a private road on top of Crawford Hill in Holmdel, New Jersey, is a 15-meter horn-shaped antenna built for Bell Labs and designed to pick up faint signals from an early NASA communications satellite. It achieved its purpose, and much more: the antenna ended up detecting signals that confirmed the Big Bang.

The Horn Antenna (Figure 104-1) was built in 1959 as part of NASA's Project Echo. Rather than using powered satellites for communication, as is done today, Project Echo involved placing 30- to 40-meter spherical balloons in orbit and bouncing microwave signals off them. In 1960 the Echo 1A satellite was successfully launched and placed in low Earth orbit. Echo 1A, weighing 76 kilograms and made of a mylar polyester film, showed that satellite communication would work: TV and radio signals were transmitted to it, bounced off it, and received.

Figure 104-1. The Horn Antenna; courtesy of NASA

On August 12, 1960, the Horn Antenna at Holmdel received a message transmitted from California and bounced off Echo 1A: "This is President Eisenhower speaking. This is one more significant step in the United States' program of space research and exploration being carried forward for peaceful purposes. The satellite balloon, which has reflected these words, may be used freely by any nation for similar experiments in its own interest."

In 1964, two scientists working on calibrating the Horn Antenna to remove interference were unable to identify a source of random noise. The antenna's design was intended to eliminate noise from the surroundings; it was cooled to just above Absolute Zero to eliminate noise from the heat of the receiver; and the scientists chased off pigeons that contaminated the antenna interior with their droppings. The scientists had even placed metal tape over protruding rivets to try to achieve silence.

But a microwave noise remained. The scientists, Arno Penzias and Robert Wilson, determined that the noise was present day and night. By moving the antenna, which was built to examine any part of the sky, they found that the noise was present no matter where they looked.

In fact, they had discovered a trace of the Big Bang: cosmic background microwave radiation. The noise being received on a microwave wavelength of 7.35 centimeters (the wavelength at which the antenna's receiver was built to operate) had been predicted since the 1940s. With its detection, the Big Bang theory prevailed over other rival explanations for the origin of the universe, and Penzias and Wilson were awarded the Nobel Prize.

Today the antenna is a National Historic Monument and is easily seen from the road. It sits on a rotating base and is attached directly to a small hut, where the receiving equipment was located. The entire antenna (but not the hut) can rotate on its axis to point at different parts of the sky. It is no longer in use.

Practical Information

At the entrance to Bell Labs' old Holmdel site at 101 Crawfords Corner Road, there's a water tower billed as the World's Largest Transistor (Bell Labs was the home of the first transistor, although it looked nothing like the water tower). Continuing northwest on Crawfords Corner Road, you'll reach a junction where you turn right onto Holmdel Road. The antenna is on a private road off of Holmdel Road; turn right after you've passed Longview Drive.

Cosmic Background Radiation

The Big Bang theory says that about 13 billion years ago, the universe we live in was created from a primordial state. Little is known about this primordial state, but it was highly compressed and extremely hot. When the Big Bang occurred, there was a sudden expansion and cooling that entirely and uniformly filled space. Within a tiny fraction of a second, space was filled with a plasma of quarks and gluons.

Within a second, protons and neutrons were created from the plasma. Within three minutes electrons had appeared, and thereafter simple nuclei were created (such as hydrogen and helium). From then until about 380,000 years later, the universe was filled with simple nuclei and photons. It is the photons that are responsible for the cosmic background radiation.

Initially, the photons (in the form of gamma rays) bounced off the nuclei. After about 240,000 years of expansion, the hydrogen and helium nuclei formed atoms and the photons, whose wavelengths had increased as the universe expanded and cooled, were free to travel throughout the universe. They continue to travel to this day, and can be picked up using a microwave receiver pointed in any direction.

When the photons began to be free to move around the universe, the universe's temperature was around 3,000 kelvin. The universe has continued to cool, and today the photons have a temperature of less than 3 kelvin. The noise that Penzias and Wilson were struggling to eliminate corresponded to a temperature of 3 kelvin.

Since 1964, additional experiments have been performed to verify the presence of the cosmic background radiation. These have confirmed its existence, and have also shown that it is not completely uniform. The radiation is anisotropic—it varies depending on where you look. From this non-uniformity, scientists are able to determine features of the early universe and estimate its age.

A 2001 satellite called the Wilkinson Microwave Anisotropy Probe has surveyed the cosmic background radiation and produced detailed information about its intensity and non-uniformity (see Figure 104-2).

Figure 104-2. Wilkinson Microwave Anisotropy Probe map of the cosmic background radiation; courtesy of NASA/WMAP Science Team

105

Institute for Advanced Study, Princeton, NJ

40° 19′ 54.51″ N, 74° 40′ 4.80″ W

"Small and Plastic"

Since 1930, the Institute for Advanced Study in Princeton, New Jersey, has welcomed some of the greatest theoretical thinkers from around the world. The Institute was created by a brother/sister pair of philanthropists as an independent body, where great minds would be free from academic pressures of "publish or perish" to work in a peaceful and welcoming environment.

Albert Einstein came here in 1933 and remained until his death in 1955. He lived nearby (at 112 Mercer Street—the house is not open to the public) and walked home each day across the Institute's beautiful grounds. During Einstein's first year at the Institute, the Austrian mathematician Kurt Gödel visited and later joined the Institute to do research. He too stayed, until his death in 1978.

Einstein evidently considered Gödel a peer; the two walked to work together and home again, and Einstein told friends that he would go into work "just to have the privilege of walking home with Kurt Gödel." Einstein had revolutionized physics during his Annus Mirabilis (see Chapter 33), and Gödel had revolutionized mathematics with his Incompleteness Theorem (see sidebar). This theorem showed that any system of mathematics had limits—some things would not be provable without inventing new mathematics (in other words, it would be necessary to "think outside the box" of any mathematical system to prove some theorems).

The Institute was home to many other stars, too: J. Robert Oppenheimer ("father" of the atomic bomb) was director of the Institute from 1947 to 1966, and John von Neumann (a prolific mathematician who is well known for helping to found game theory and computer architecture) came to the Institute in 1930 after fleeing from Europe. The list of alumni is long and illustrious and counts many Nobel Laureates, in particular winners of the Fields Medal (the closest thing mathematics has to a Nobel prize).

The Institute is not reserved for just physicists and mathematicians: it has four schools covering Historical Studies, Mathematics, Natural Sciences, and Social Science. It is run by a small group of academics who oversee around 200 people, called members, studying freely toward whatever long-term goals they set for themselves. The first director of the Institute wished that it be "small and plastic"; to this day, the Institute remains focused on its core aim of letting its members think, free of outside pressures.

To give its members the space to think, the Institute is surrounded by open green spaces, including the Institute Woods, a nature reserve that covers 240 hectares and is open to the public. Visitors are free to roam the woods and enjoy the preserved environment much as Einstein, Gödel, and others have done over the years.

The Institute itself is not open to the public, but it does have a free public lecture series that covers topics such as computability, art history, archaeology, epidemia, and anything that the members find interesting. The lecture series is interspered with regular free concerts, and all the lectures are available for free download after the event.

Practical Information

Information about the Institute for Advanced Study is available from *http://www.ias.edu/*. The Institute Woods can be accessed from the parking area of Battlefield Park (off Mercer Street).

Gödel's Incompleteness Theorem

Gödel's Incompleteness Theorem shows that given any specific system of mathematics, there are things that can be expressed but not proved within that system. For example, if you restrict yourself to doing arithmetic and algebra on whole numbers only, there are things that can't be proved without recourse to other bits of mathematics.

Gödel's first step was to convert any piece of mathematics into numbers. He imagined starting with a simple system of mathematics that had a few symbols (things like the number 0 and the + sign) and giving them each a number. Strings of symbols (such as 2 + 2 = 4) could then be turned into a single number by substituting a number for each symbol.

So his first step, in other words, was to define a formal system of mathematics and assign a number to each symbol (Table 105-1). Gödel used a system like this: 0 could be assigned the number 11, + would be 13, × would be 14, and = would be 15.

Table 105-1. Partial table of symbol numbers

Symbol	Number	Meaning
0	11	The number 0
S	12	The successor ("one more than")
+	13	Addition
×	14	Multiplication
=	15	Equals
<	16	Less than
(	17	Left parenthesis
)	18	Right paranthesis
a	19	A variable
′	20	Prime symbol

Notice that the number present in the first column is 0; any other number can be made from 0 preceded by enough applications of the S operator (e.g., 4 is SSSS0). Also, there's only one variable, a; to make more variables, a is followed by as many prime symbols (′) as necessary.

With this system of numbering, it's possible to express mathematical statements as numbers. For example, saying 2 + 2 = 4 is the same as saying SS0 + SS0 = SSSS0 which is the number 12121113121211151212121211. Any mathematical statement written using these symbols can be turned into a number.

It's also possible to take a list of statements and make a single number out of them by joining them together, with a 0 between the individual numbers representing each statement. So, for example, a = 2 followed by a' = 4 could be written 191512121101920151212121211.

These numbers are called Gödel numbers and are the key to his Incompleteness Theorem. For any statement F(a) about a variable a, there's a corresponding Gödel number that is written G(F).

Now for the big step—you may need to read the following two paragraphs twice.

The very idea of a proof or provability can be expressed using Gödel numbers. A proof is just a list of mathematical statements written in the mathematical language, the last of which is the statement that you were trying to prove. But we've seen that any list of statements can be turned into a Gödel number, so a proof is in fact just a Gödel number.

Now suppose you are trying to prove statement p. For every Gödel number g, it's possible to pose the question, "Is g a proof of p?" And since p has a Gödel number G(p), this is really a question about a relationship between two numbers: g and G(p). But this entire mathematical system is about statements involving numbers, and so whether a statement is provable turns out to be a relationship between numbers. Therefore, provability is something that the mathematical system is able to express.

Then Gödel created a statement that said, "I am not provable." That statement was actually a mathematical statement within the system of symbols defined by Gödel, and so it had a Gödel number. Such a statement is not provable; if it were, there would be a contradiction ("I am not provable" would be provable). In this way, Gödel showed that even a system with apparent consistency had theorems that could not be proved using the system itself.

106

Trinity Test Site, White Sands Missile Range, NM

33° 40′ 38.28″ N, 106° 28′ 31.44″ W

July 16, 1945

Twice a year, on the first Saturday of April and October, the Trinity Test Site, where the first nuclear bomb was exploded, is open to the public. Entry is free, there's no need to reserve, and cameras are allowed.

The Trinity Site is located inside White Sands Missile Range, New Mexico. The missile range covers over 8,000 square kilometers of New Mexico and is actively used for missile and ordnance testing. It is also where German scientists captured by the U.S., such as Wernher von Braun, worked on rocketry and launched modified V-2 missiles (renamed to the Bumper).

The range is also used today by NASA as a training area for space shuttle pilots. In 1982, the space shuttle Columbia landed there at NASA's White Sands Space Harbor.

Outside the missile range and near the main gate is the White Sands Missile Range Museum, which has an open-air missile park containing more than 50 missiles, rockets, and drones (see Chapter 108).

But the Trinity Site is the highlight of a visit, even though little remains. On July 16, 1945, the world's first nuclear explosion (Figure 106-1) occurred there; today the exact spot is marked by a small obelisk made of lava rock. The surrounding land is mildly radioactive, but safe for a short visit. The McDonald Ranch House, a small house where the bomb's final construction was done, has been restored to its 1945 state and is part of the visit.

All that's left of the tower in which the bomb was mounted are the stumps of its legs.

On the sandy floor of the test site there are pieces of trinitite, blue-green pebbles of glass that were created when the bomb exploded. Some of the trinitite is also red or black, and contains copper or iron from the vaporized bomb and tower. Most of it has been bulldozed away, but small samples have been preserved in a glass box to show the state of the sand after the explosion. A specially constructed cover opens to show the only part of the bomb crater that was not filled in.

Figure 106-1. 0.016 seconds after detonation

Also visible is a 214-ton steel container called Jumbo. The bomb that was exploded at the Trinity site contained two kinds of explosive: TNT and the plutonium nuclear bomb. The TNT was designed to explode first, inward, and compress the plutonium, which would then create the nuclear explosion. Since plutonium is highly toxic, the scientists working on the bomb worried that the TNT might explode and scatter plutonium everywhere, without the nuclear bomb going off.

So Jumbo was created to contain the entire bomb. It was strong enough to withstand the TNT explosion, and would have been vaporized had the nuclear bomb gone off. In the end, however, Jumbo was not used because the team was confident the bomb would explode, and it was left about 700 meters from the nuclear bomb. It survived the nuclear explosion and hasn't been moved since.

Also on display is the bomb casing for a Fat Man nuclear bomb, the type of bomb dropped on Nagasaki a month after the Trinity Test.

Practical Information

Information about the next open day at the Trinity Test Site is available from the website of the White Sands Missile Range at *http://www.wsmr.army.mil/wsmr.asp?pg=y&page=576*. Details of the White Sands Missile Range Museum are at *http://www.wsmr-history.org/*.

About two hours' drive from the Trinity Test Site is the Very Large Array astronomical radio observatory. Although the observatory is open daily for visitors, it offers a special guided tour twice per year on the same open days as the Trinity Test Site. See Chapter 107.

Fat Man

On August 9, 1945, the Japanese city of Nagasaki was destroyed by a bomb, nicknamed Fat Man, of the same design that had been successfully exploded in New Mexico.

The nuclear bombs exploded by the United States during the Second World War relied on bringing enough nuclear material together to create an uncontrolled chain reaction, resulting in a massive explosion.

Two materials, uranium (in the form U-235) and plutonium (in the form Pu-239), were used because they both undergo a chain reaction if enough of the material, called the critical mass, is present. To get the explosion going, a source of neutrons is needed. The neutrons hit the nuclear material, causing atoms to split apart and release more neutrons (which go on to hit other atoms and create the chain reaction) and an enormous amount of radiation and energy. It is the energy released that creates the explosion.

The Fat Man bomb used plutonium to create the explosion. The bomb's interior consisted of four concentric spheres: a small neutron source, a ball of plutonium, a layer of uranium, and a TNT exterior (see Figure 106-2).

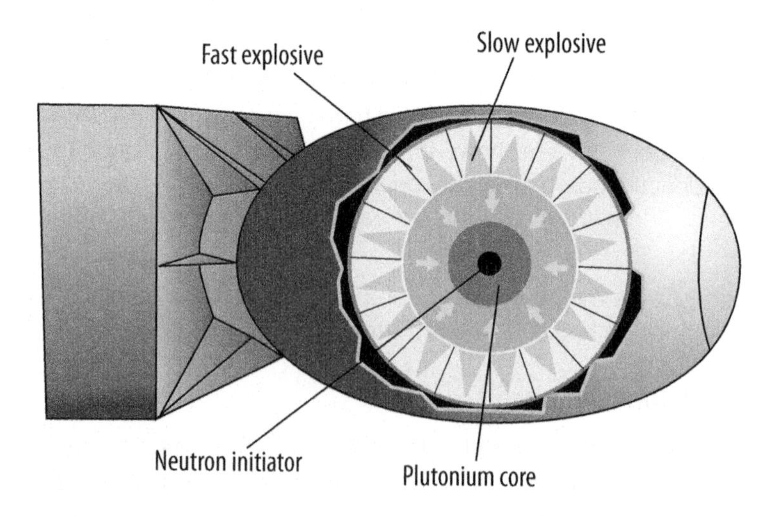

Figure 106-2. The Fat Man bomb

To set off the bomb, the TNT was exploded, which forced the rest of the bomb inward, compressing it. Once compressed, the bomb's plutonium became a critical mass, underwent a chain reaction, and exploded. To get the nuclear explosion going, the center of the bomb consisted of a neutron initiator made of polonium and beryllium.

Polonium is an alpha-emitter—it decays and gives off alpha particles (which consist of two protons and two neutrons stuck together). The alpha particles hit the beryllium, which then gives off neutrons. Those neutrons hit the plutonium to start the chain reaction.

As the chain reaction started, the plutonium heated up, causing it to expand. This expansion actually slowed down the explosion. To maximize the explosive power of the bomb, the plutonium core had to be contained while the chain reaction built up. This was achieved by wrapping the core in a layer of depleted uranium (Uranium-238), called the tamper. The tamper contained the chain reaction and reflected neutrons coming out of the core back into the plutonium, helping to increase the chain reaction.

At Nagasaki, just 6 kilograms of plutonium was enough to produce an explosion equivalent to 21,000 tons of TNT.

107

Very Large Array, Socorro, NM

34° 4′ 43.98″ N, 107° 37′ 5.49″ W

A Virtual Antenna

In western New Mexico, in the middle of an empty plain, sit 27 radio telescopes that work together to study distant galaxies, stars, quasars, and pulsars by examining their radio transmissions. The Very Large Array of dishes are mounted on railway tracks arranged in a Y shape with 21-kilometer-long branches (Figure 107-1). By mathematically combining data from all 27 radio telescopes, the array acts as if it were a single dish 36 kilometers across. (Building a 36-kilometer radio telescope would have been financially unfeasible; the Very Large Array, by contrast, cost a relatively affordable $79 million.)

Figure 107-1. The Very Large Array; courtesy of David Bales (www.davidbales.com)

The dishes in the Very Large Array are moved using a specially built transporter (which is usually part of any tour). They cycle through four major configurations, called A, B, C, and D, every 16 months. The A configuration has the dishes spread as widely apart as possible—this gives the maximum possible magnification. The D configuration has the dishes only 600 meters apart and is used to study an individual radio source in detail. The B and C configurations lie between the extremes of A and D.

Such large virtual dishes are needed because the ability of a radio telescope to distinguish details—the resolution—depends on the wavelength of the signal being listened to divided by the size of the dish. The larger the dish, the smaller the resolution possible. Because radio waves have a much longer wavelength than light waves, radio telescopes need to be much bigger. A 1-meter optical telescope has the same resolution as a radio telescope many kilometers wide.

Each dish in the Very Large Array is similar in construction to a home satellite dish—it's a parabolic reflector (see Chapter 48), but instead of having a radio receiver at the focal point of the parabola, there's a second reflector that sends the received radio signal into the middle of the dish where the actual radio receivers are located (see Figure 107-2).

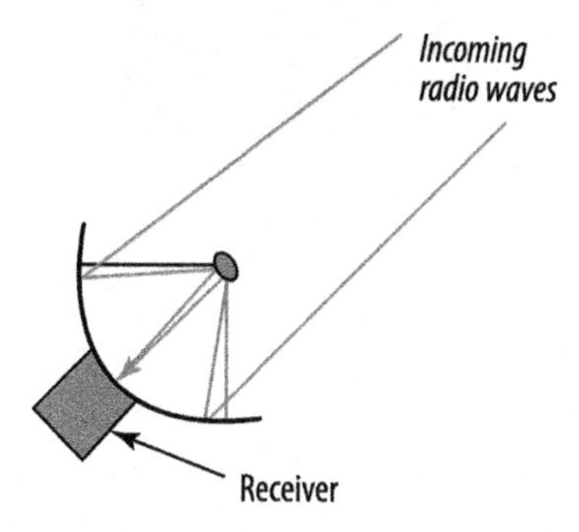

Figure 107-2. Parabolic dish with a central reflector

Data from the dishes is combined to get a radio picture of the sky. As the Earth rotates, more radio pictures are taken at different angles, building a clearer picture of the radio sources being observed.

When you arrive at the Very Large Array, the first place to go is the visitor center, where a short video introduces radio astronomy and the technique used to reconstruct a radio image from multiple dishes (interferometry). A fun experiment for children involves a pair of dishes facing each other—whisper into one dish, and the whisper is clearly heard in the other. The Very Large Array welcomes photographers, but remember to turn off your cell phone—the antennas are very sensitive and even a small phone can interfere with them.

Practical Information

Visiting information is at *http://www.vla.nrao.edu/*. The Very Large Array runs tours twice a year that coincide with the Trinity Test Site tours (see Chapter 106). It's about a two-hour drive between the two sites.

Interferometry

Real telescopes (optical or radio) suffer from a limit in their resolution caused by diffraction. A telescope observing a distant star is equivalent to a point of light passing through a small hole, and it generates a diffraction pattern called an Airy disc (Figure 107-3).

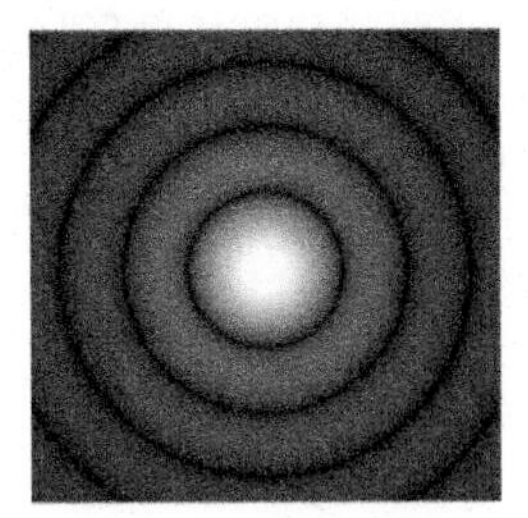

Figure 107-3. An Airy disc

The minimum resolution for the telescope is governed by the overlapping of Airy discs from the different points of light being observed. Each point will create an Airy disc—to be able to distinguish the two points, the maximum (the bright portion in the center of one Airy disc) cannot be closer than the first minimum (the innermost dark band) of the second disc.

The larger the telescope, the better the resolution. But making very large telescopes—of the size necessary to view distant stars—is either very expensive or simply impossible.

However, by using a pair of telescopes it's possible to observe a light source (a star, for example) and superimpose the two images to get a fringe pattern of dark and light bands corresponding to the interference between the same light waves as received at two different positions. A similar thing happens in the well-known double slit experiment (see Figure 107-4), where light from a lamp passes through a pair of slits and creates a fringe pattern.

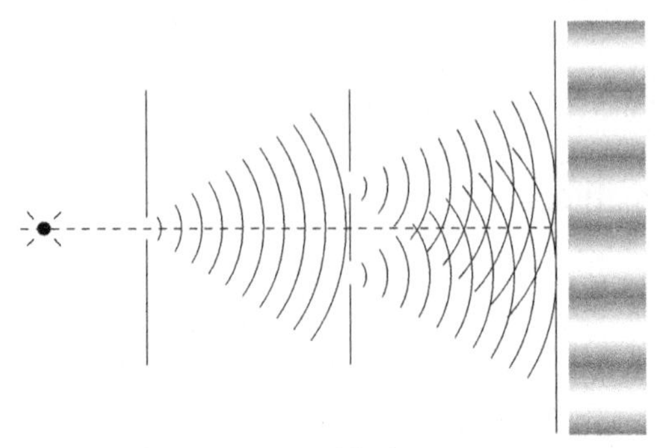

Figure 107-4. Double slit experiment

The fringe pattern can be used to reconstruct the pattern of light—for example, the star being observed—with a resolution related to the distance between the two slits (or telescopes). The spacing of the dark and light bands (and how they differ) can be used to determine the structure of the light being observed.

Radio astronomy uses the same technique—interferometry—but looking at radio waves instead of light. A pair of antennas listens to the same point in the sky, and an interference pattern is made from the observations. Because the radio waves arrive at the two antennas at slightly different times, very accurate atomic clocks are used to synchronize the signals received, by delaying one signal to match the other, and build the interference pattern (see Figure 107-5).

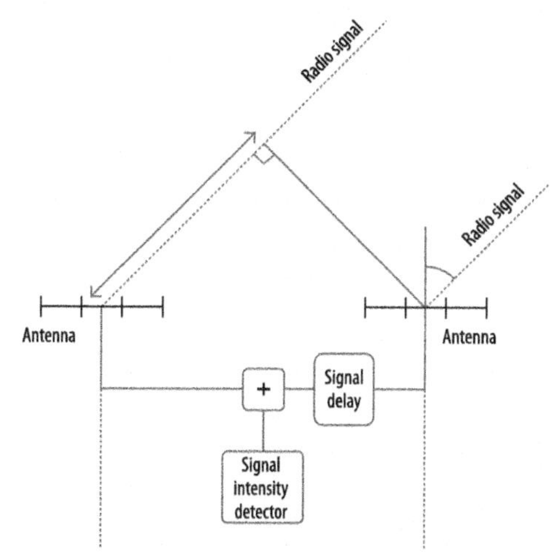

Figure 107-5. A pair of antennas

In an array of radio telescopes, each pair of telescopes contributes information about the radio signals received along the straight line joining the two telescopes. To make high-quality pictures of the sky, it's necessary to either move the antennas to build up a picture from many lines, or use many pairs of telescopes. Since the Earth also moves, it's possible to make additional observations just by observing the same radio sources as the Earth rotates.

The fringe patterns are used to determine the brightness (from a radio signal perspective) of the star being observed, and brightness measurements from each pair of telescopes are combined using a Fourier transform to build up a picture of the sky.

The Very Large Array has 27 telescopes, giving 351 possible pairs. And as the Earth moves, the telescopes can track an object, providing many more observed lines to build up a complete picture.

108

White Sands Missile Range Museum, White Sands Missile Range, NM

32° 23′ 9.06″ N, 106° 28′ 42.53″ W

Operation Paperclip

In 1944, the U.S. Army chose over 8,000 square kilometers in New Mexico to be the main site for U.S. rocket testing. The site became known as the White Sands Proving Ground, and in 1945 rocket experiments began. On July 16, 1945, the area was used for the first test of a nuclear bomb (see Chapter 106). Over in Europe, the U.S. was running Operation Paperclip to capture and offer jobs to as many valuable Nazi scientists as possible. One group, under Wernher von Braun, was transported to White Sands along with tonnes of missiles and equipment to help jump-start U.S. missile technology (see also page 71).

In 1946 the first V-2 rocket was launched at White Sands, and by the end of the year the German scientists working alongside U.S. service personnel had launched tens of V-2s across the White Sands Missile Range. The U.S. also tested versions of the V-1, renaming it the Loon.

The missile range still exists today and is in active use. Despite that, there's a museum of missile history with multiple exhibits just inside the range, including an outdoor missile park with many types of missiles on display. Notable missiles include the Loon, a newly restored V-2, a Patriot antimissile battery, Pershing I and II nuclear missiles, and a Sidewinder air-to-air missile.

There are also cruise missiles (missiles that don't follow a ballistic trajectory), including the Hound Dog (from 1960). And there are non-military missiles such as the Aerobee 170 and Aerobee Hi, which were used to carry experiments into the upper atmosphere and space to research.

Unmanned aircraft are covered by the exhibition of various drones—there's a Firebee BQM-34A rocket-propelled drone, the MQM-107D Streaker that was used until the late 1990s for target practice, the QH-50 DASH drone helicopter, and the supersonic XQ-4.

There are also displays of equipment needed for tracking missiles, such as radar antennas and high-speed camera systems.

Inside the museum are displays of missile technology including the pumps, gyroscope, and rocket motor from a V-2, a Stinger shoulder-launched missile, a special slide rule (see page 138) for doing quick rocket trajectory calculations, and the ground control equipment used for drones.

Practical Information

Details about the White Sands Missile Range Museum are available at *http://www.wsmr-history.org/*. It can easily be seen at the same time as the Trinity Test Site (Chapter 106).

The Ideal Rocket Equation

In rocketry (and especially in space travel) one value dominates calculations—the delta-v or Δv. Delta-v is the change in velocity of a rocket, and it's important in rocketry because maneuvers (such as changing orbit) are quantified by the change in velocity needed. For example, to go from low earth orbit (where the Space Shuttle and International Space Station reside) to a rendezvous with the Moon requires a delta-v of 6 kps; to go to Mars instead requires a delta-v of 6.3 kps. Getting off the Earth's surface and into low earth orbit requires a delta-v of about 8 kps.

Once the delta-v is known, then it's a question of calculating the size of rocket needed to obtain it. That's where the Ideal Rocket Equation comes in. It shows the relationship between the delta-v and the mass of fuel on board the rocket.

When a rocket engine is firing, it generates thrust by expelling a mass of burnt propellant out of its rocket nozzle. This generates thrust by Newton's Third Law (see page 172), and the thrust depends on both the speed of the expelled gases and the change in pressure at the rocket nozzle.

Since the rocket is burning fuel, its mass is constantly declining, making a calculation of its final velocity more complicated than just assuming that it undergoes constant acceleration from the thrust. The Ideal Rocket Equation incorporates the thrust and the change in mass to calculate the idealized change in velocity. Using it gives a rough estimate of the amount of fuel required for any maneuver. And it's fairly easy to derive the equation from Newton's Laws.

First, assume that a rocket has a single force, F, acting on it resulting from the engine's thrust, and that the mass of the rocket is M. A simple application of Newton's First Law shows that the thrust is related to the velocity, v, of the rocket as shown in Equation 108-1.

$$F = \frac{dMv}{dt}$$

Equation 108-1. Force acting on rocket

The thrust equation below (Equation 108-2) shows that the force is also related to the mass of propellant used per unit time, m, and the equivalent exit velocity (see page 321) of the gases leaving the rocket nozzle, v_{eq}.

$$F = \upsilon_{eq} \frac{dm}{dt}$$

Equation 108-2. Thrust equation

Putting the two equations together and eliminating dt gives the following differential equation (Equation 108-3).

$$M\, d\upsilon = \upsilon_{eq}\, dm$$

Equation 108-3. Rocket differential equation

The change in mass of the rocket, dM, is just the change in mass of the propellant, so dM = –dm (negative because the propellant is lost by ejecting it from the nozzle). Substituting that, we get the equation for the change in velocity, dv, as shown in Equation 108-4.

$$d\upsilon = -\upsilon_{eq} \frac{dM}{M}$$

Equation 108-4. Change in velocity

Integrating that equation gives you the Ideal Rocket Equation (Equation 108-5). Both sides are integrated for some time period in which the velocity changes by delta-v (which is the result of the integration of the lefthand side) and the mass changes from m_0 to m_1 (where the difference between the two is the mass of propellant used during that time).

$$\Delta\upsilon = \upsilon_{eq} \log(m_0/m_1)$$

Equation 108-5. The Ideal Rocket equation

So knowing the equivalent exit velocity for a rocket and the required change in velocity, it's possible to quickly calculate the mass of fuel needed.

109

Atomic Testing Museum, Las Vegas, NV

36° 6′ 50.98″ N, 115° 8′ 54.96″ W

History of the Nevada Test Site

From 1951 to 1992, the U.S. (with a little help from the UK) exploded 1,021 nuclear bombs above and below ground outside Las Vegas, Nevada (see Chapter 110). Many of the tests were announced in advance, and the above-ground explosions were visible from Las Vegas, where they became a tourist attraction. Today, the history of the most prolific test site in the U.S. is documented at the Atomic Testing Museum.

Since this is Las Vegas, there's a multimedia event to compete with the rest of the city's attractions—the Ground Zero Theater. The visitor is seated inside a mock concrete bunker seven miles from a nuclear explosion. The countdown begins, and the bomb explodes. A blast of air rushes into the bunker and the entire theater shakes.

However, the museum is neither a glorification of nuclear bombs, nor a glitzy tourist attraction. Instead, it's an informative and intelligent explanation of the Cold War and the chronology of nuclear testing. In addition to many photographs and films of nuclear explosions, the museum has exhibits of equipment recovered from the Nevada Test Site.

There's a very interesting exhibit concerning photography and filming of nuclear explosions, with dramatic pictures of the first microseconds of nuclear blasts. There's also a calutron: a device used for separating uranium isotopes by splitting a beam of isotopes by deflecting them using an electromagnet. Different isotopes follow different curving paths based on their weights and can then be collected.

A section on underground testing explains the technology used to bury nuclear bombs, and the after-effects of underground explosions. During the 1950s the explosions were part of popular culture, and there's a display of period pop-culture items including a book of Atomic Cocktails.

Geiger Counters

The Atomic Testing Museum has a large display of radiation-detecting equipment, including Geiger counters. A Geiger counter detects radiation indirectly by detecting ionization of a gas caused by radiation. The same technique was used by Pierre and Marie Curie in their original measurements of the strength of various radioactive sources (see Chapter 15). Visitors can use a Geiger counter to detect the amount of radioactivity from a number of common materials and objects.

A Geiger counter consists of a hollow, gas-filled glass cylinder, called a Geiger-Müller tube, that performs the detection. Inside the tube is an inert gas (such as argon). The interior of the tube is coated with a conducting material, and in the middle there's a metal needle-like electrode (see Figure 109-1).

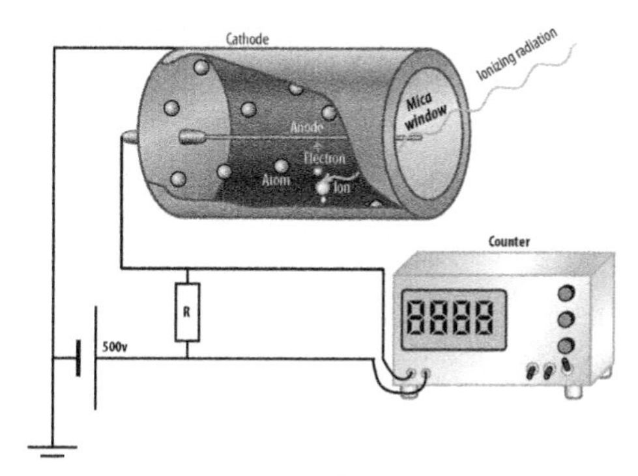

Figure 109-1. A Geiger counter

A large voltage (hundreds of volts) is applied between the needle electrode (which becomes positively charged) and the tube interior (which becomes negatively charged). No current flows.

At the end of the tube is a window, typically made of mica. The very thin mica window prevents light (to which the tube would be sensitive) from entering the tube, but lets alpha particles (helium nuclei), beta particles (high-energy electrons), and gamma radiation in.

When radiation enters the tube, it ionizes the gas, resulting in positively charged argon ions and electrons. The large voltage between the electrodes causes the ions and electrons to move rapidly toward the electrodes (the electrons to the anode and the ions to the cathode). As they move, they have enough energy from the electric field to ionize more of the gas. This avalanche of ions and electrons creates a current between the cathode and the anode.

The current is detected by the Geiger counter, which typically emits a clicking sound and counts the number of times ionization has occurred.

Peaceful uses of atomic energy are also on display, including an atomic rocket engine.

The only part of the museum that's truly a tourist trap is the shop, which sells everything from finger puppets of Curie, Einstein, Newton, and Darwin to a Miss Atomic Bomb shot glass.

Outside the museum is a weather station that confirms that Las Vegas is hot and dry, but also gives the current background radiation reading in microrems per hour.

Practical Information

The museum's details are at *http://www.atomictestingmuseum.org/*. See also Chapter 110 for information on visiting the Nevada Test Site.

110

Nevada Test Site, NV

37° 7′ 0″ N, 116° 3′ 0″ W

1,021 Explosions

At the Trinity Test Site (Chapter 106), a single nuclear bomb was tested. At the Nevada Test Site, more than 1,000 nuclear explosions were set off between 1951 and 1992. The site consists of over 3,600 square kilometers of dry lake beds and mountains, about 100 kilometers northwest of Las Vegas. Once a month, the U.S. Department of Energy provides a free, day-long tour of the Nevada Test Site's bomb craters, ground zeros, and test paraphernalia.

The tour covers around 400 kilometers of the nuclear explosion–pockmarked landscape: of the 1,021 nuclear explosions at the Nevada Test Site, only 126 occurred above ground; the rest were underground tests that left the site cratered. The largest crater of all, the Sedan Crater, is the highlight of the tour. It's almost 400 meters wide and 100 meters deep; see Figure 110-1.

Figure 110-1. The Sedan Crater; courtesy of the National Nuclear Security Administration/Nevada Site Office

Radioactive Decay and Half-Life

Radioactive elements spontaneously decay over time. Uranium-238, the most common naturally occurring isotope of uranium, decays to form thorium-234, which decays to form protactinium-234. Decay continues until the uranium has turned into lead.

The first decay occurs when uranium-238 emits an alpha particle (a pair of protons and neutrons bound together) and turns into thorium-234. This happens with a half-life of 4.47 billion years. The thorium-234 decay occurs when a beta particle (an electron) is emitted and protactinium-234 is created. That happens with a half-life of 24 days.

The next decay (also a beta particle) occurs with a half-life of 6.7 hours and results in uranium-234. Uranium-234 is relatively stable: its half-life is 246,000 years.

The half-life of a radioactive element is the amount of time it takes for half the original quantity of an element to decay. Starting with a 1-kilogram block of uranium-238, it would take 4.47 billion years for half a kilogram to decay. But starting with 1 kilogram of thorium-234, only 24 days is needed for only half a kilogram to remain. The long half-life of uranium-238 is part of the reason that it is so abundant.

Radioactive decay is an exponential process, where an element loses half its weight (or volume, or number of atoms) over a fixed period. Because half is lost with each half-life, only a small number of half-lives need to pass before the original material is completely decayed. After five half-lives, only 3% remains (see Figure 110-2).

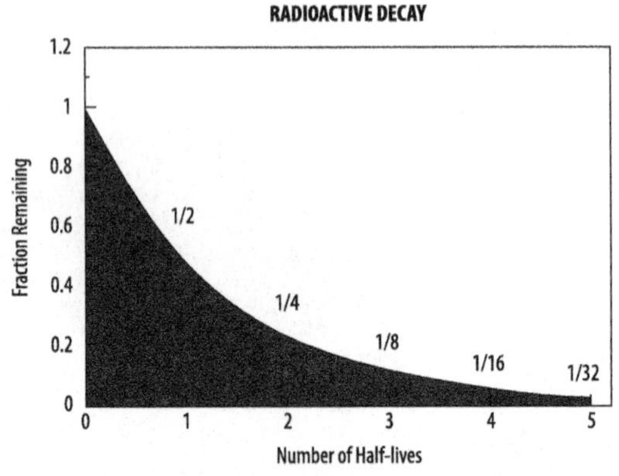

Figure 110-2. Radioactive decay

Knowing the half-lives of elements means that they can be used for dating rock samples and other ancient specimens. Radioactive dating works by counting the amount of a radioactive isotope present in an object, then counting the amount of a daughter isotope (i.e., an isotope into which it would have decayed), and comparing the two to determine the age of the object based on known half-lives.

For example, Uranium-238 eventually decays to lead-206 over about 4.47 billion years. Measuring the presence of uranium-238 and lead-206 (known as uranium-lead dating) can be used to determine age.

In dating rocks, the presence of the mineral zircon ($ZrSiO_4$) is used because, when forming, zircon is able to keep uranium as part of its crystal structure, but does not do the same for lead. This means that when a zircon crystal forms, it essentially sets an imaginary "uranium-lead clock" to time zero.

If a piece of zircon were examined and found to have equal quantities of uranium-238 and lead-206, it would mean that half the uranium had decayed. That would set the age of the zircon at the half-life of uranium-238, or 4.47 billion years. (The entire decay from uranium to lead takes about the same time as the initial uranium decay, because uranium's long half-life dominates.)

The equation used for determining the age of a sample relies on knowing the quantities of the parent isotope, P, the daughter isotope, D, and the decay constant, λ. The decay constant is ln 2 / half-life. See Equation 110-1.

$$t = \frac{1}{\lambda} \ln \left(1 + \frac{D}{P}\right)$$

Equation 110-1. Age equation

A separate measurement can be made by examining the uranium-235 decay to lead-207. That decay occurs with a half-life of 700 million years, providing two "clocks" by which to measure the age of a sample.

Sedan was created as part of Operation Plowshare, an attempt to use nuclear weapons for peaceful purposes in mining, excavating, and building harbors. Plowshare was not successful, partly because of the amount of radiation created. The Sedan explosion released one of the highest levels of radioactive material into the atmosphere of all the tests performed at the Nevada site. The tour bus stops at the Sedan Crater so that visitors can step outside (briefly, since it's still emitting a low level of radiation) and view it close up.

Another crater on the tour is the Bilby Crater, which the tour bus drives into. Bilby was created in 1963 and was the first underground test to be felt in Las Vegas. At the center of the crater are the remains of the shaft into which the nuclear bomb was inserted.

The eeriest part of the tour is what remains of the Apple II testing. To test the effects of nuclear weapons on people and property, a fake town was constructed complete with roads, houses, a school, and an electric grid. The houses were populated with dressed mannequins sitting at dinner tables. There was even food on the tables.

Not much is left of Apple II, as it was designed to be blown up, but at least one surviving house, its exterior paint completely stripped and windows blown out, is still standing (see Figure 110-3). Films of the burning and disintegrating buildings were used as part of old civil defense films.

Figure 110-3. A remaining Apple II house; courtesy of Danny Bradury

Live pigs also played a part: at the "Porker Hilton," they were exposed to explosions to test the effects of nuclear weapons on flesh and bone.

A visit to Frenchman Flat, where atmospheric nuclear tests were performed, including the very first nuclear explosion at the Nevada Test Site, rounds out the tour. This dry lake bed is scattered with material destroyed by nuclear testing, including a bank vault that survived a bomb test named Priscilla.

During the tour, all visitors must wear a radiation-measuring badge, which will be collected at the end. If you were dangerously exposed, the U.S. government will contact you.

Practical Information

Visits to the Nevada Test Site must be booked in advance, and non-U.S. citizens should expect a long wait for approval. The tour is free and can be booked from the U.S. Department of Energy's website at *http://www.nv.doe.gov/nts/*.

111

Zero G, Las Vegas, NV

Not the Vomit Comet

Zero G Corporation operates flights in a converted Boeing 727 with a padded interior that simulate weightlessness and Martian and Lunar gravity. NASA operates similar flights for astronaut training on an aircraft dubbed the "Vomit Comet." Both NASA's and Zero G's flights work on the same principle: they fly in such a way that the passengers and aircraft free-fall together. Unlike NASA's flights, Zero G's flights are open to the public, although they come with a hefty price tag.

The Zero G aircraft flies a parabolic flight path, flying upward at 45° and then into a parabolic hump where everyone becomes weightless (Figure 111-1). Once over the hump, the aircraft descends and then pulls up to start another. Weightlessness only occurs when flying the parabola and lasts for about 30 seconds. Each flight consists of 15 parabolic humps, for a total of between 7 and 8 minutes of weightlessness.

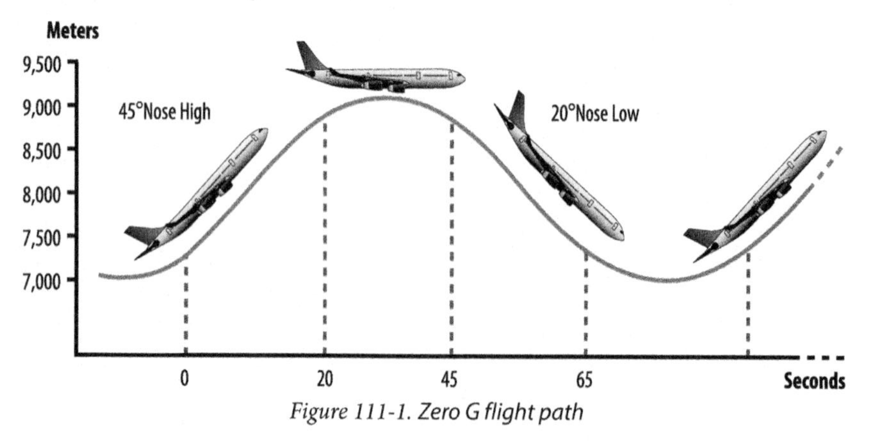

Figure 111-1. Zero G flight path

Once weightless, the passengers float inside the Boeing 727 and the pilot maneuvers the plane in an open piece of airspace between 7,000 and 10,000 meters. When the plane pulls out of the parabola, the feeling of gravity returns and increases to about 1.8g (1.8 times normal Earth gravity).

By modifying the shape of the flight path, the plane can be made to simulate weightlessness, or the weight as experienced on the Moon (the Moon's gravitational force is about 1/6 of the Earth's) with passengers walking and bouncing like the Apollo astronauts. A different shape of flight creates Martian gravity, about 1/3 of the force experienced on Earth.

Zero G has found that limiting the amount of time spent weightless—by limiting the number of total humps—minimizes motion sickness. NASA, in contrast, pushes the number of parabolic humps to over 40, and the passengers do feel motion sickness. Zero G also prescribes motion sickness medication to take before a flight.

To ease passengers into the feeling of weightlessness, the flight starts out simulating Martian gravity, then Lunar gravity, and finally "no gravity" at all.

Practical Information

Although Zero G is based in Las Vegas, Nevada, it operates flights from other airports around the U.S. Details are at *http://www.gozerog.com/*.

Weightlessness and Parabolic Flight

The name Zero G is a misnomer—there's no escaping Earth's gravity in an aircraft, or even an orbiting spacecraft—but passengers do experience the same thing real astronauts do: weightlessness.

The weight of a person is related to the gravitational force acting on his or her body. When standing on the Earth, a man who has a mass of 75 kilograms feels "weight" because the Earth's gravity acts upon his body to pull him downward, and at the same time the ground exerts an equal and opposite force upward. The force upward from the ground is felt in his body and he experiences his weight. For more on weight and mass, see Chapter 69.

When a skydiver free-falls from an aircraft, there's no ground to push upward, and he falls under the force of gravity. The only resistance to the skydiver's fall is from the air (which is minimal), and the skydiver experiences a feeling of weightlessness.

In an orbiting spacecraft, astronauts are weightless because both they and the spacecraft are free-falling together. The astronaut has no weight relative to the spacecraft, even though the Earth's gravity is acting upon them both. When a spacecraft is in orbit, it is free-falling at right angles to the Earth with sufficient velocity that it never actually falls to the ground—it keeps falling, but moves horizontally, and ends up rotating around the Earth.

In the Zero G aircraft, gravity still applies, but the parabolic flight means that the passengers and aircraft free-fall together, and relative to the aircraft the passengers have no weight (there's no force acting on them from the aircraft itself). The Zero G aircraft achieves this by flying a ballistic trajectory with just enough engine power to overcome air resistance.

If a bullet is fired from a gun at an angle, it will fly through the air with only the force of gravity acting on it (assuming that wind resistance is of little consequence). Gravity acts constantly downward on the bullet, and it travels through a parabola. If the bullet started out at ground level—the position (0, 0)—and was fired at an angle θ to the ground with velocity v, then its position at time t is $(tv \cos \theta, tv \sin \theta - 1/2\, gt^2)$, where g is the acceleration due to gravity.

An aircraft flying upward at 45° and then entering the parabola gives the maximum length of ballistic flight, just as firing a gun at 45° results in the longest distance of bullet flight.

112

Glenn H. Curtiss Museum, Hammondsport, NY

42° 23′ 49.23″ N, 77° 13′ 58.00″ W

The Lesser-Known Aviator and Fastest Man on Earth

The Wright brothers are undoubtedly the best-known aviation pioneers, and others like Blériot, Lindbergh, Earhart, Mongolfier, and Yeager are household names. But none of their names appear on pilot license number 1 issued by the Aero Club of America in 1911 (Figure 112-1). That distinction goes to Glenn Curtiss, the fastest man in the world on a motorcycle in 1907 at 219 kph, and a little-known aviation pioneer who had a very large impact on how we fly today.

Federation Aeronautique Internationale

AERO CLUB OF AMERICA

No. 1.

The above-named Club, recognized by the Federation Aeronautique Internationale, as the governing authority for the United States of America, certifies that

Glenn H. Curtiss

born 21st day of May 1878 having fulfilled all the conditions required by the Federation Aeronautique Internationale, is hereby licensed as Aviator.

Dated

President.

Secretary.

[SEAL]

Signature of Licensee:

Glenn H. Curtiss

Figure 112-1. Pilot license number 1

In 1907, Alexander Graham Bell asked Curtiss to join his Aerial Experiment Association and help him design and build aircraft (for more on Bell, see Chapter 4). Curtiss was invited because he was an expert in making lightweight internal combustion engines for his motorcycles and for dirigibles. Together, Bell and Curtiss built a number of aircraft culminating in the June Bug. On July 4, 1908, the June Bug flew over 1,500 meters to win the Scientific American trophy.

The flight was important because although the Wright brothers had been flying their aircraft since 1903, they were secretive and allowed almost no one to see their plane in flight. Curtiss's public demonstration of the June Bug caused a sensation.

In 1909 Curtiss went to France and narrowly beat Louis Blériot to the Gordon Bennett Trophy, flying at almost 75 kph. He went on to found the Curtiss Aeroplane and Motor Company.

By 1911, Curtiss had developed a viable seaplane and demonstrated it to the U.S. Navy. Eugene Ely flew a Curtiss plane from the USS *Birmingham* and later landed on the USS *Pennsylvania*, using an arrester cable to stop the plane. Curtiss himself flew a seaplane out to meet the USS *Pennsylvania*. He landed next to it—the plane was hoisted aboard, Curtiss ate lunch, and he and the plane descended into the sea and flew away. For those exploits, and for development of naval aircraft, aerial bombing, and flying boats capable of crossing the Atlantic Ocean, Curtiss is known as the "Father of Naval Aviation."

The Glenn H. Curtiss museum in Curtiss's hometown of Hammondsport, New York, covers his life and contribution to aviation. The museum has a large collection of significant aircraft, including many originals. The oldest original airplane is the 1917 Standard J-1 training aircraft that used a Curtiss engine; there's also a 1919 Curtiss Seagull flying boat. The museum has a restored 1917 Curtiss JN-4D "Jenny"—over 7,000 Jennys were made during the First World War, and aviation pioneers Lindbergh and Earhart both learned to fly in one.

The museum has a beautiful reproduction of the June Bug. The original aircraft was converted to a float plane; it accidentally sank in 1909.

There's a large collection of aircraft engines made by Curtiss and others, including a 1912 Curtiss Model "S" capable of 60 horsepower and a Pratt & Whitney R-4360. Eight of these enormous Pratt & Whitney engines powered Howard Hughes's Spruce Goose (see Chapter 117).

The collection is rounded out by motorcycles, including a reproduction of the 1907 Curtiss 8 Cylinder on which Curtiss became the "Fastest Man on Earth" (with no helmet!).

Practical Information

More details of Glenn Curtiss and his life, and complete visiting information, is available from the museum's website at *http://www.glennhcurtissmuseum.org/*.

The Aileron

All aircraft have three possible movements: yaw, pitch, and roll. The yaw is the rotation of the aircraft around a vertical axis; the pitch is the rotation of the aircraft around a left-right axis, typically through the wings; and the roll is the rotation of the aircraft around a longitudinal axis from the front to the rear of the aircraft.

Many early aircraft had yaw and pitch controls: the yaw was controlled by the rudder and the pitch by the elevators. On the Wright brothers' Wright Flyer, the elevator was mounted on the front and the rudder at the back. To control roll, the Wright brothers invented a technique called wing warping.

Wing warping involved actually bending the ends of the wings in opposite directions to roll the aircraft by changing the air flow across the wing tips. One wing tip would bend downward, increasing lift, while the other wing tip bent upward, decreasing lift, and the plane would roll. Since the Wright Flyer had flexible fabric wings, it was possible to bend the ends to roll the aircraft. By controlling roll and yaw at the same time, the Wright brothers were able to make controlled turns in their aircraft.

But wing warping only works if the wing is flexible. Modern aircraft control roll by using ailerons: movable surfaces at the ends of the wings (see Figure 112-2 for a basic diagram). The ailerons perform the same creation of lift on one wing and loss of lift on the other as the Wrights' wing warping technique, but ailerons can be attached to a fixed, solid wing.

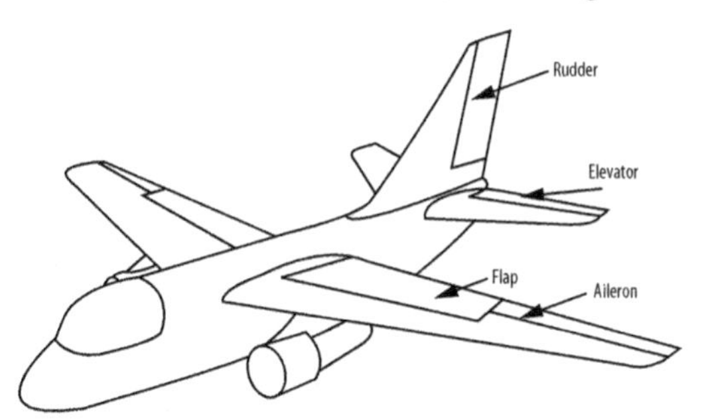

Figure 112-2. Basic flight controls

The June Bug had ailerons for roll control, and the Aerial Experiment Association was issued a patent for the invention in 1911. This led to a patent battle between the Wrights, who had patented wing warping for roll control, and Curtiss. Ultimately, the patent dispute was settled by the U.S. government because the development of aircraft had become vital to the First World War effort, and in 1929 the companies of Curtiss and the Wrights were merged.

113

John M. Mossman Lock Collection, New York, NY

40° 45′ 19.18″ N, 73° 58′ 52.74″ W

Where the Tourists Aren't

In midtown Manhattan, inside the home of the General Society of Mechanics and Tradesmen of the City of New York, is a small, barely known museum of over 370 bank and vault locks. Just two blocks from the madding crowd in Times Square, the John M. Mossman Lock Collection is a haven of peace, and a history lesson tracing locks from ancient Egypt to the 20th century.

Many of the locks on display are unique. Each was made for a particular purpose—often protecting a bank or vault. One of the locks, the Parautopic lock, was famous for being allegedly unpickable in the 19th century, and then infamous for being opened with ease by Linus Yale, Jr., who went on to help create the pin tumbler lock.

Many of the locks on display are quite beautiful—they have ornate designs and intricate gleaming mechanisms. Most are time locks, designed to protect bank vaults; they can only be opened at certain times on certain days.

The museum is a great opportunity to see locks you'll rarely see anywhere else.

Practical Information

Visiting information for the Mossman Lock Collection is available from *http://www.generalsociety.org/*. Check with the Society before visiting, as the museum does not get many visitors and you may need to coordinate with the Society's librarian to get access.

There's another excellent collection of locks in Kentucky, at the Museum of Physical Security in the Lockmasters Security Institute; details are at *http://www.lsieducation.com/museum/*.

The Pin Tumbler Lock

The most familiar lock of all is the pin tumbler lock, which many people call a Yale lock. It was invented by Linus Yale, Sr., in 1848; his son, Linus Yale, Jr., improved the design and patented it in 1861. The 1861 design is very similar to the locks still used today in homes, businesses, and padlocks.

Yale was inspired by wooden Egyptian locks that dated from about 4,000 BC and featured wooden pins that fit into grooves in a wooden key. The modern pin tumbler lock uses a very similar mechanism.

A pin tumbler lock consists of a cylinder (where the key is inserted) inside a second cylinder. The inner cylinder, called the plug, turns to open the lock. Inside the lock are five or six holes in the plug and the outer cylinder. Each hole contains a spring and a pair of metal pins (see Figure 113-1).

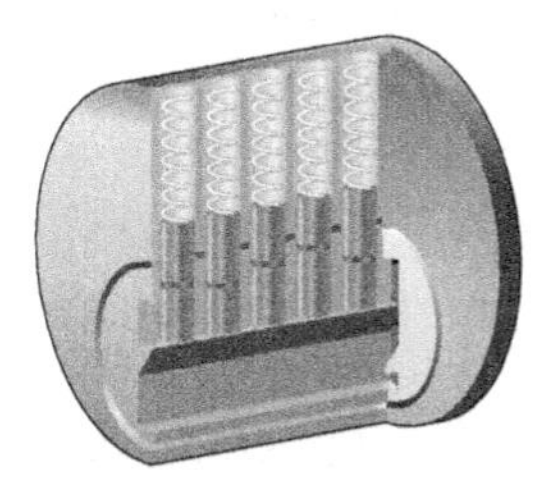

Figure 113-1. A pin tumbler lock with no key inserted

Without a key inserted, the plug cannot turn because the metal pins are forced down into the holes in the plug by the springs. When the right key is inserted, the pins are forced upward so that the tops of the bottom pins line up exactly with the point where the plug and the outer cylinder meet. The top halves of the pins are forced up into the outer cylinder.

The plug is then able to turn (Figure 113-2). If the wrong key is inserted, the pins will still be forced upward but not in the correct alignment, and the plug's movement will be blocked by the pins.

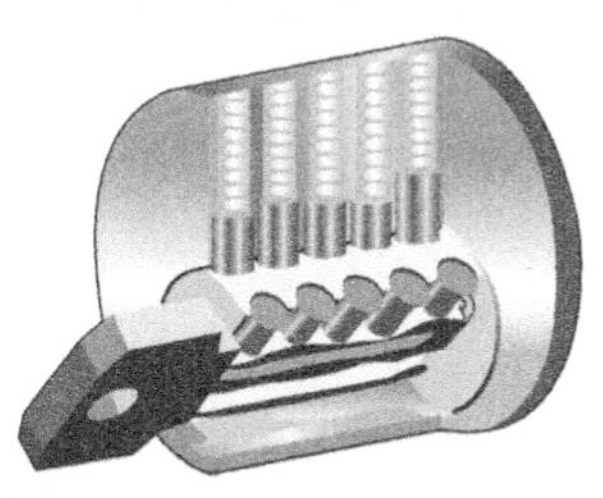

Figure 113-2. The correct key is inserted and the plug can turn

114

Sagan Planet Walk, Ithaca, NY

42° 26′ 22.53″ N, 76° 29′ 48.52″ W

The Sciencenter

The Sciencenter in Ithaca, New York, is a great science museum for children. It's full of practical demonstrations of scientific principles and hands-on exhibits without being dumbed down. A good way to reach the Sciencenter is to follow the Sagan Planet Walk, a 1.2-kilometer walk through Ithaca that starts downtown and follows a scale model of the Solar System.

But before you go, download the free MP3 tour of the Solar System that accompanies the walk. The tour starts at the Sun and works its way toward Pluto, passing the closely spaced Mercury, Venus, Earth, and Mars. After Mars, the planets become more distant; the next stop is Jupiter, then Saturn, followed by a long gap to reach Uranus. Neptune is even further, and finally at Pluto you arrive at the Sciencenter.

Each planet is represented by a concrete monolith with a picture of the appropriate planet and an explanatory panel. Stop at each planet and listen to the corresponding MP3 on your audio tour.

The Sciencenter is full of things to see and touch. The Saltonstall Animal Room has live animals such as frogs and snakes, and Connect to the Ocean has a tide pool where children can touch and interact with sea urchins, starfish, and more. For the under 4s there's the Curiosity Corner, where everything is safe for small hands.

There's also an infrared camera that projects the heat of a visitor's body onto a large plasma screen in the Mars and Stars section, where you'll also find more about the planets along the Sagan Planet Walk.

Outside, there's the Emerson Science Park, where children can clamber over a suspension bridge, a Voussoir bridge that only stays together once the keystone is in place, and a climbing frame made of common geometric shapes. The speed of sound is illustrated with a delay tube—speak into one tube and wait for it to come out the other end. Parabolic dishes (see Chapter 48) are illustrated with a pair of whisper dishes—whisper into one, and a friend hears the whisper on the other side of the park. Other exhibits show the operation of levers and pendulums.

Whether you spend two hours or half a day in the Sciencenter, it's an ideal spot for kids and science before heading off to enjoy the rest of Ithaca and Cornell University.

Practical Information

Full details of the Sciencenter and Sagan Planet Walk (including the downloadable audio tour) are available at *http://www.sciencenter.org/*.

Parsecs

It's common to speak of astronomical distances in light years (one light year being the distance light travels in one Earth year), but astronomers are more likely to talk about parsecs.

A parsec is calculated by first imagining a line connecting the Earth and the Sun and projecting a perpendicular line, starting at the Sun, into space. A triangle is completed by projecting a line from the Earth into space that meets the perpendicular at an angle of 1 arcsecond (1/3600 of a degree). The distance of the hypotenuse starting at the Earth is one parsec, which is about 3.3 light years (see Figure 114-1).

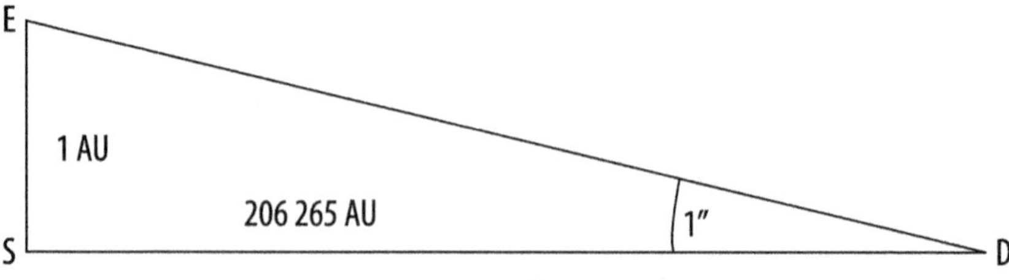

Figure 114-1. One parsec

The distance between the Earth and the Sun is called an astronomical unit (or AU), which is 149,597,870,691 meters. By basic trigonometry one parsec is calculated to be 206,265 AU or 3.09×10^{16} meters.

The name parsec comes from "parallax of one arcsecond" and refers to the stellar parallax method of determining the distance to a star. When a nearby star is viewed at different times of the year, it appears to have moved relative to the background stars because of parallax.

The parallax effect is easily seen by observing two objects at different distances. For example, if you observe a pair of trees in a park—one close to you, and one far away—and move your head relative to the closer tree, a large movement will result in a smaller apparent movement of the distant tree.

In the stellar parallax method a star is observed at two times, six months apart. By spacing the observations at six-month intervals, you ensure that the Earth will have moved as far from the initial point as possible and be on the opposite side of the Sun. Observations of the relative movement of the star from the two positions give the parallax angle, and basic trigonometry (as discussed previously) gives the distance to the star relative to the distance between the Earth and the Sun.

To get the parallax angle, astronomers measure the angle between the star being observed and a distant, fixed star (Figure 114-2). This yields two angles, one for each of the two observations, and the parallax angle is simply their sum. The distant star is assumed not to have moved relative to the Earth, making the star appear at the same angle in both observations.

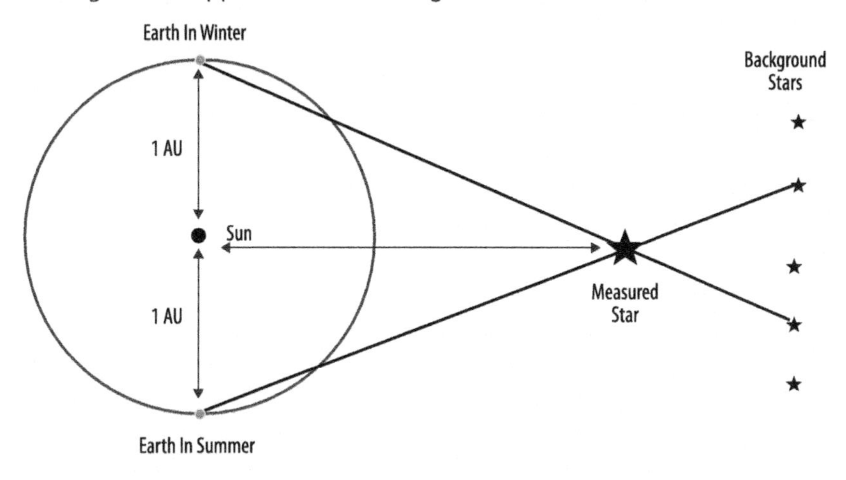

Figure 114-2. Stellar parallax

115

Early Television Museum, Hillard, OH

40° 2′ 3.67″ N, 83° 9′ 41.17″ W

What's In, Not On, the Box

There are many museums of television around the world, but the Early Television Museum is the most fun for the scientific traveler: it's dedicated to the mechanics and electronics of television, as opposed to the stars who appear on the small screen. And the museum is hands-on, with a demonstration of 1930s television technology.

The museum's collection begins with the mechanical television sets of the 1920s and 1930s, which had between 30 and 60 lines (compared with 480 visible lines on a conventional NTSC TV set). The museum's collection includes a working 1930 Baird Televisor (which originally came as a kit; see Figure 115-1), and a Davin Tri Standard built from articles that appeared in *Popular Mechanics* at the end of 1928. Many other TVs from this era are on display and in working order.

Figure 115-1. 1930 Baird televisor; courtesy of the Early Television Museum

The museum's collection continues with black-and-white electronic televisions starting in 1936. There's a large collection of U.S. televisions as well as a collection of British, European, and South American sets. Many of these are also in working order, as the Early Television Foundation (which runs the museum) works to keep history alive by restoring or repairing its exhibits.

The first color television sets on display date to the early 1950s and include an operational RCA CT-100 and a Westinghouse H840CK15: the first two color televisions made available to the public in 1954.

The transmitting side of TV is not overlooked, either. There's a collection of cameras, monitors, and test equipment, and another of TV tubes, antennas, and accessories. There's even a mobile TV transmission van dating from 1948.

And no visitor should miss the chance to appear on 1930s television. The museum has restored an RCA flying spot camera, which traces a spot of light over the visitor's face to build up an image. Stand in front of this camera, and your face will appear on a working 60-line 1930 RCA television set.

Practical Information

The museum is free, but donations are always encouraged. Information about visiting can be found at *http://www.earlytelevision.org/*. Each May, the Early Television Foundation (which runs the museum) organizes an Early Television Convention, with presentations covering television technology and an auction of TV parts.

John Logie Baird's Invention

In 1926, Scottish inventor John Logie Baird demonstrated a working television that he called the Televisor. The apparatus was crude, but it worked. The camera consisted of a rotating disc with a spiral of holes in it; light passed through the disc and onto a photocell. The resistance of the photocell changed with the amount of light falling on it, and the rotating spiral essentially scanned the image vertically, line by line. The varying current through the photocell could then be transmitted by phone or radio.

The television receiver had a similar rotating disc with a light behind it. The intensity of the light varied with the incoming television signal, so that its intensity matched the intensity of the light falling on the camera's photocell. By synchronizing the rotation of the two discs so that the same point was being scanned and projected at the same time, an image appeared on the television receiver's tiny screen.

Baird's first demonstration involved transmitting a picture of a ventriloquist's doll's head into the next room. The camera scanned 12.5 images per second with just 16 lines, and the image was faint and blurry—but for the first time, a moving image could be transmitted live.

John Logie Baird is widely credited as having been the first to demonstrate a practical television system, and he was intimately involved with the BBC's early television transmissions. But mechanical television was not a great success because the image quality was poor, and it was gradually replaced by electronic television systems invented by Kálmán Tihanyi, Vladimir Zworykin, and Philo Farnsworth.

Nevertheless, the fundamental idea of scanning an image line by line and reproducing it line by line remains intact to this day.

116

NASA Glenn Research Center, Cleveland, OH

41° 24′ 46.24″ N, 81° 51′ 44.63″ W

Little-Known NASA

Think of NASA, and Cleveland, Ohio, probably isn't the first place that comes to mind. NASA is most often associated with Cape Canaveral or Houston, Texas, and it's true that the Glenn Research Center in Cleveland isn't where rockets are fired into space or where astronauts talk to mission control. But the center is where much of the basic science and testing of space technology is done, and it's open to the public.

There's a visitor center that is particularly good for children and explains the work done at the research center. You'll find a special exhibition that pays tribute to John Glenn (the first American to orbit the Earth) with both his Mercury spacesuit from 1962 and the orange space shuttle suit he wore when he returned to space at the age of 77.

The visitor center has the Apollo Command Module that visited the Skylab 3 mission in 1973, and some pieces of moon rock returned by Apollo astronauts. The training module that astronauts on the ill-fated STS-107 Space Shuttle Columbia mission used for their preparation is also on display.

There's a fun flight simulator that lets you take the controls of all types of aircraft, from propeller planes to commercial jets. Other exhibits explain the Solar System and how NASA vehicles explore it (with a life-size model of the Mars Pathfinder rover Sojourner).

All of that is free and open 7 days a week.

But the best way to see the research center is by taking a tour. Tours are offered from April to October, on the first Saturday of each month. The tour varies depending on the accessibility of various parts of the center; a tour schedule is published on the center's website. It is vital to book well ahead.

Inside the research center, visitors get to see the Fabrication Shop, where rockets are put together. Currently this is where parts of the new Ares rocket that will replace the Space Shuttle are being manufactured.

Bernoulli's Principle

Bernoulli's principle is pretty simple: when a fluid is flowing and its speed increases, then its pressure drops. It turns out that this simple rule applies to most types of fluid flows (where the fluid is incompressible—its density cannot change) and also to flows of gases at low speeds, which makes it useful for many practical purposes. It is particularly useful in understanding the flow of air around an aircraft wing.

Bernoulli's principle is expressed by Bernoulli's equation (Equation 116-1), which links the speed of a flowing fluid, v, the pressure of the fluid, p, and its density, ρ.

$$\frac{\rho v^2}{2} + p = \text{constant}$$

Equation 116-1. Bernoulli's equation

The equation shows that the dynamic pressure (which can change) is proportional to the square of the velocity of the flow: it's the $\rho v^2/2$ part of the equation and is often referred to as simply q.

One application of the equation is the Venturi effect (see page 200), where a liquid flowing through a constriction in a tube speeds up and the pressure decreases. The difference in the pressure before entering the constriction and the pressure while it's inside is proportional to the change in flow speed. Neither flow speed needs to be actually known—just knowing the change in pressure and the areas of the two parts of the tube (before and inside the constriction) is enough to give the actual flow speed.

Bernoulli's equation is also put to good use in measuring the airspeed of a moving aircraft with a pitot tube and static port arrangement. On the outside of aircraft are small holes, called static ports, that are flush with the skin of the aircraft and measure the ambient air pressure around it. There may be more than one static port to get an average reading.

The pitot tube (which is just a small tube with a hole in the end) points out into the air and faces in the direction of flight. Through this tube, the apparent air pressure caused by the combination of the static pressure and the movement of the aircraft can be measured.

With the static port measuring a pressure of p and the pitot tube measuring a pressure of q, Bernoulli's equation can be flipped around to get the speed of the aircraft, v, from the pressure difference and air density, ρ (see Equation 116-2).

$$v = \sqrt{\frac{2(q-p)}{\rho}}$$

Equation 116-2. Velocity from pressure difference via Bernoulli's equation

Bernoulli's equation is often used to explain lift. When an aircraft moves through the air, its wing shape causes the stream of air hitting it to be deflected above and below the wing. With the correct shape, the velocity of the air above the wing is much greater than the velocity below. Bernoulli's equation shows that this difference in velocity results in a higher pressure below the wing than above it, which generates lift.

You don't need Bernoulli to explain lift, however; you can just rely on Newton's Laws. The air flowing above the wing is flowing faster than the air flowing below. From Newton's equation F = ma, it's possible to see that the lift force depends just on the mass, m, of air flowing around a wing (and in contact with it), with a velocity above, V, and below, v, measured over some time period, t (see Equation 116-3).

$$F = m(V - v)/t$$

Equation 116-3. Lift calculated from Newton's Laws

Because the air is deflected, it creates two forces on the wing (one going upward above the wing, one going downward below the wing). The net force is upward because of the greater velocity of air passing over the wing.

Visitors may also see the Zero Gravity Research Facility (which is basically a big hole in the ground), where equipment can be tested in free fall by dropping it 155 meters in a vacuum. The fall takes about five seconds, ending when the equipment falls into a well of polystyrene beads.

The facility also has wind tunnels, and on some tours it's possible to walk through the Able Silverstein Supersonic Wind Tunnel that is used to test aircraft and rocket designs up to Mach 3 and an altitude of 45 kilometers. Another wind tunnel is used for simulations of takeoffs and landings.

And then there's the Icing Research Tunnel, which can chill equipment to –34°C in winds of up to 630 kph, and cover them in ice from water sprayed into the tunnel. It is used to test the effect of severely cold weather on aircraft.

The center also has an enormous facility for understanding the acoustics of jet engines. Inside the Dome of Silence (Figure 116-1), engines can be tested and noise reduction ideas measured. The interior of the dome is entirely covered with a strange pattern of fiberglass wedges that prevents echoes.

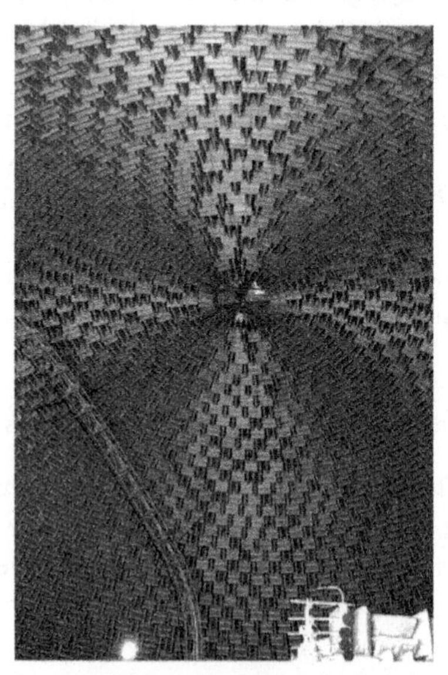

Figure 116-1. Inside the Dome of Silence; courtesy of Valerie Houghtland (val_pisces76)

Also on site is the world's largest vacuum chamber—it's 37 meters tall and 30 meters across, making it large enough to test entire spacecraft. The vacuum chamber is sometimes open for (air-filled) tours.

Practical Information

You must be a U.S. citizen to visit the NASA Glenn Research Center. Visiting information is available from *http://www.nasa.gov/centers/glenn/events/*.

117

Evergreen Aviation & Space Museum, McMinnville, OR

45° 12′ 15″ N, 123° 8′ 40″ W

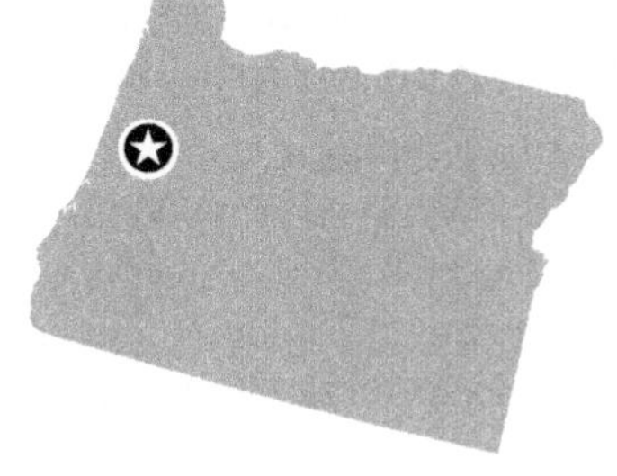

The Spruce Goose

There are many aviation museums, but only one has Howard Hughes's gigantic flying boat, the Hercules H-4 (better known as the Spruce Goose). The plane, made almost entirely from wood, flew once in 1947, and then was maintained in flight-ready condition until Hughes's death in 1976.

During the Second World War, the U.S. lost enormous numbers of ships, materials, and men crossing the Atlantic Ocean because of Nazi German U-boats. In 1942 Allied shipping losses were at their height, with more than 1,600 ships sunk. In July of that year, the U.S. government gave an $18m contract to the newly formed Hughes Kaiser Corporation to create three flying boats capable of transporting up to 750 troops high above the U-boat threat.

Because metal was needed for the war effort, Hughes chose to make the aircraft out of laminated wood (mostly birch, not spruce). The frame and ribs of the Spruce Goose were made from wood, and laminated wood sections were molded into thin sheets and glued over the frame.

Since the plane was so enormous, Hughes also invented an artificial feel system that gave the pilot the sensation of flying a small plane. The pilot only needed to apply a small force to the yoke to move the massive ailerons and rudder.

The plane was powered by eight Pratt & Whitney engines; the museum has a partially dismantled engine on display.

Hughes continued working on the project until 1947. While under investigation by the U.S. Senate for allegedly misappropriating government funds, Hughes took the H-4 out on Long Beach Harbor for taxiing tests. To the surprise of everyone, on the third test Hughes ordered the flaps to 15° and the huge plane took to the air. Hughes flew it for about one minute, about 20 meters above the water.

The Ground Effect

Some critics of the Spruce Goose claimed that it would never have flown higher than Hughes's one test flight. That claim is based on the possibility that the Spruce Goose was only flying because of the ground effect experienced by all aircraft when they are close to the ground (or sea).

When a plane is flying, its wings generate lift because the air moving across the top of the wing is at a lower pressure than the air moving underneath. At the wing tips, the high-pressure air under the wing will try to move upward to where the pressure is lower. This creates vortices at the wing tips that rotate in opposite directions (Figure 117-1).

Figure 117-1. Wing tip vortices

This leakage of air and pressure from below the wing actually reduces the lift generated while flying. Close to the ground, the wing tip vortices are compressed. This causes their negative effect on lift to be reduced, and the plane then experiences greater lift.

The ground effect—the increase in lift close to the ground due to the reduction in vortices—is most powerful below half the aircraft's wingspan. With a wingspan of 97 meters and a maximum flying height of 20 meters, the Spruce Goose would indeed have been experiencing the ground effect during its unique flight.

Whether it could have flown higher is likely to remain a mystery; the museum isn't planning to fly its star exhibit.

Today the Spruce Goose is the centerpiece of the Evergreen Aviation & Space Museum. It still holds the record for the largest wingspan—at over 97 meters it easily beats the Airbus A380's 72-meter wingspan, and the Boeing 747's 64 meters.

But the Spruce Goose isn't the only attraction at the museum. There's also an SR-71 Blackbird spy plane, a DC-3 "Dakota," a Boeing B-17 Flying Fortress, and a collection of other military aircraft, including an F-15A Eagle, a Spitfire, and a MiG17.

Practical Information

Information about the Spruce Goose and the Evergreen Aviation & Space Museum is available at *http://www.sprucegoose.org/*.

118

Joseph Priestley House, Northumberland, PA

40° 53′ 25″ N, 76° 47′ 24″ W

Oxygen, Soda Water, and More

The British clergyman and scientist Joseph Priestley is widely credited as the person who discovered oxygen. He also discovered the gases nitric oxide, nitrogen dioxide, nitrous oxide, hydrogen chloride, ammonia, sulphur dioxide, carbon monoxide, nitrogen, and silicon tetrafluoride. And he put the bubbles in soda water.

Prior to Priestley's work, three "airs" were known: air (as we know it); carbon dioxide (called, at the time, fixed air); and hydrogen. Priestley had observed carbon dioxide in a brewery in Leeds, where it settled over the fermenting beer.

Carbon dioxide is a basic byproduct of fermentation. When yeast ferments, sugar is converted into ethanol and carbon dioxide. Priestley determined that bubbling "fixed air" through water resulted in bubbly water with a pleasant taste. Soda water is simply water mixed with carbonic acid (H_2CO_3), which is a combination of water (H_2O) and carbon dioxide (CO_2). It's carbonic acid that gives soda water its characteristic "bite" when it mildly burns your tongue.

Priestley went on to develop a method of making soda water by mixing water, sulphuric acid (H_2SO_4), and chalk (calcium carbonate: $CaCO_3$). The acid reacts with the chalk to produce carbon dioxide, calcium sulphate ($CaSO_4$), and more water. The carbon dioxide is then forced through water to make carbonic acid (see Equation 118-1). Priestley described this process in his 1772 book *Directions for Impregnating Water with Fixed Air.*

$$H_2SO_4 + CaCO_3 = CaSO_4 + CO_2 + H_2O$$

Equation 118-1. Sulphuric acid and chalk reaction

But his greatest success came in 1774 with the discovery of oxygen. Priestley used a lens to focus sunlight onto mercuric oxide (HgO), heating it to over 400°C, at which point it broke down into mercury and oxygen. He noted that the gas, which he termed dephlogisticated air, made candles burn brighter, that mice lived longer in a fixed quantity of the gas than in the same quantity of air, and that breathing the gas gave a pleasant feeling.

He called oxygen "dephlogisticated" air because he believed that it was air from which the "phlogiston" had been removed. This mythical element was, at the time, thought to be the cause of flammability in materials. It was believed that when things burned, phlogiston was released; Priestley could explain that oxygen prolonged burning because more phlogiston could be released than in ordinary air (which he assumed contained a quantity of phlogiston already). He hung on to the phlogiston theory even after the Frenchman Antoine Lavoisier had demonstrated that it was not true. (Lavoisier also gave names to hydrogen and oxygen, and demonstrated that humans breathe oxygen to live.)

Priestley was an English Dissenter, a minister who broke away from the Church of England, and a political theorist who strongly supported the French Revolution. In 1791, his opinions culminated in the Birmingham Riots in which his home, church, and other buildings were burned to the ground. In 1794 Priestley emigrated to the United States of America and made his home in Northumberland, Pennsylvania, where he died in 1804.

Today, the Joseph Priestley house is a museum to his life and work. Although his original books and equipment were dispersed when he died, the house has been restored to the state in which it would have been found when he lived there, and his laboratory has reproductions of his equipment.

From time to time, Heritage Days are organized at the house, with costumed performers playing the role of the Priestley family and demonstrations of Priestley's experiments. At all other times, guided tours of the house bring to life Priestley's chemical experiments and his time in the U.S.

The house is also a site of pilgrimage for the American Chemical Society, which uses it to commemorate special moments in chemical history.

Practical Information

The Joseph Priestley House's website is at *http://www.josephpriestleyhouse.org/*.

Phlogiston, Fire, and Oxidation

The nonexistent element phlogiston can be thought of as anti-oxygen—phlogiston was supposedly released when a material was burned or oxidized (such as when iron rusts). At the time, scientists thought that when the air became saturated with phlogiston, materials would no longer burn or oxidize, and that animals could not live in phlogiston-saturated air. The opposite was the truth (as Lavoisier pointed out)—it's the presence of oxygen that makes all these things possible. Oxygen is the most common component involved in oxidation, which happens in rusting, tarnishing, burning and breathing.

There are three ways to think about oxidation: as adding oxygen, as removing hydrogen, or as losing electrons. All oxidation is accompanied by an equivalent opposite reaction called a reduction, which can in turn be thought of as losing oxygen, adding hydrogen, or gaining electrons. Together, these complementary reactions are called redox.

For example, when iron (Fe) rusts, it oxidizes to form iron oxide (Fe_2O_3). In this case, oxygen is added to iron to oxidize it (the oxidation) and the oxygen gas is completely lost (the reduction) because three oxygen molecules are used in their entirety for the oxidizing. The chemical equation is shown in Equation 118-2.

$$4\,\mathrm{Fe} + 3\,\mathrm{O_2} = 2\,\mathrm{Fe_2O_3}$$

Equation 118-2. Rusting

In a household gas-powered water heater, methane (CH_4) is burned. The carbon in the methane oxidizes by losing all four hydrogen atoms (the oxidation), and two of the oxygen atoms gain hydrogen to form water (the reduction). The chemical equation for burning methane is shown in Equation 118-3.

$$\mathrm{CH_4} + 2\,\mathrm{O_2} = \mathrm{CO_2} + 2\,\mathrm{H_2O}$$

Equation 118-3. Burning methane

The most accurate way of describing a redox reaction is in terms of electron transfer, or change in the net charge. When oxygen is added, electrons are generally transferred to the oxygen in forming the chemical bond. When oxygen is removed, there's generally a transfer of an electron from the oxygen.

For example, old camera flashes used to contain a small amount of magnesium metal (Mg) that burns brightly when ignited. When magnesium burns, it forms magnesium oxide (MgO) in a redox reaction. Here, magnesium has gained oxygen and at the same time lost electrons. The magnesium becomes Mg^{2+} (losing two electrons) and the oxygen O^{2-}. Magnesium oxide is an example of an ionic bond common in metal oxidation (such as rusting in Equation 118-2) where the metal donates electrons to the oxygen to form the bond.

Oxidation can happen without oxygen being present at all; whenever there's a transfer of electrons, there's redox. Magnesium can also react with chlorine gas (Cl_2) to form magnesium chloride ($MgCl_2$). The magnesium loses two electrons to become Mg^{2+} (oxidation), and each chlorine atom gains an electron to become Cl^- (reduction). See Equation 118-4.

$$Mg + Cl_2 = Mg^{2+} + 2Cl^-$$

Equation 118-4. Oxidation to form magnesium chloride

However the redox occurs, there's definitely no phlogiston involved.

119

Arecibo Observatory, Arecibo, Puerto Rico

18° 20′ 39″ N, 66° 45′ 10″ W

The Biggest Dish

The radio telescope at Arecibo Observatory in Puerto Rico looks like the set for a James Bond film—the 305-meter dish set into the hilly countryside near the town of Arecibo is the largest and most sensitive radio telescope ever built. Surrounded by lush vegetation and set into an ancient sinkhole, it has been listening to radio signals from the stars since 1963, with an 810-tonne platform suspended above the dish to receive signals bouncing from its surface.

The surface of the dish is made of 40,000 perforated aluminum sheets formed into a spherical shape. Since the dish itself cannot move, the receivers suspended above it can shift to listen to different parts of the sky. Hanging from the platform is a dome-like structure that can be moved so that it can focus on signals received from particular areas of the sky. Since the dish is spherical, not parabolic, its shape can be used to study different parts of the sky without moving the dish itself (see also Chapter 67 on the Sound Mirrors, which used a similar technique).

And the Arecibo Observatory can do more than just listen—the dome incorporates a 1-megawatt radar that can be used to bounce a signal off a spacecraft or planet to make measurements of their movements or surface.

Early in its history (in 1965), the dish was used to determine the length of a day on Mercury (now known to take 59 Earth days); then, in 1972, it discovered the first pulsar in a binary system (B1913+16). In 1992, the dish observed the first planets outside the Solar System, orbiting the pulsar B1257+12 (which is about 980 light years from the Sun).

The scientists who discovered B1913+16 went on to examine the orbits of the two stars involved over a period of many years (work for which they received the Nobel Prize in Physics in 1993). Observation of the orbits showed that they had changed in accordance with Einstein's theory of general relativity, which predicted the existence of gravitational waves.

In 1974, the dish transmitted a message containing 1,679 bits intended to explain to any alien civilization listening that we could count from 1 to 10, that we knew the atomic numbers of carbon, hydrogen, nitrogen, oxygen, and phosphorous (all the elements needed for DNA), and that we were made from DNA (the four nucleotides of DNA were transmitted along with a graphic of the structure of DNA). The transmission also included a figure of a human with its height measured, the current population of the Earth, a graphic of our Solar System, and a graphic of the Arecibo dish itself

Arecibo is also used by the Search for Extraterrestrial Intelligence (SETI) project, which looks for intelligent life anywhere in the Universe.

And in fact, it really was the set for a James Bond film—the dish was featured in the climax of 1995's *GoldenEye*.

If you want to play James Bond (or just a radio astronomer), you can pay a visit to the observatory, starting at the visitor center, where the work done at Arecibo is explained. There are fixed exhibits and a documentary film about radio astronomy. Then, step outside onto the observation platform to see the telescope itself.

Practical Information

The Arecibo telescope is about a 90-minute drive from San Juan airport. Information about visiting can be found at *http://www.naic.edu/*.

SETI and the Wow! Signal

The SETI project works by listening to radio signals received by radio telescopes (in particular the Arecibo dish) and looking for narrowband radio signals (i.e., radio signals that only occupy a small part of the radio spectrum, just like a normal Earth radio). Since SETI doesn't know the frequency on which extraterrestrials might be transmitting (or when), it is forced to sample enormous numbers of radio signals, break them up into narrow bands of frequencies, and examine each for potential communications.

Since the 1960s, SETI has been using radio telescopes and antennas around the world to look for narrowband signals. Since 1999, anyone with a computer has been able to assist by running the SETI@home software. This software uses spare computer cycles (such as when the screen saver is running) to search for narrowband signals in chunks of received radio signals sent to home computers over the Internet.

On August 15, 1977, SETI detected what could have been a genuine radio signal coming from an alien source. The radio transmission, sometimes referred to as 6EQUJ5, is commonly called the Wow! signal after the handwritten note Dr. Jerry R. Ehman inscribed on the computer printout after spotting the anomaly (see Figure 119-1).

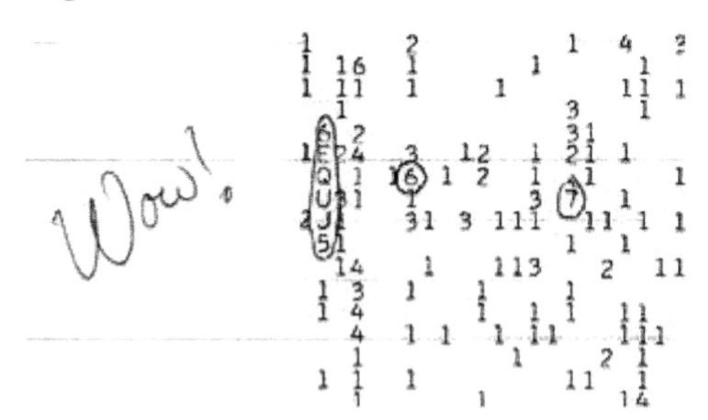

Figure 119-1. The Wow! signal

The printout showed the radio signals received by the Big Ear radio telescope at Ohio State University, with numbers and letters representing the intensity of the signals. In Figure 119-1, a space indicates that no signal was received; numbers 1 through 9 indicates increasing intensities. To represent numbers starting from 10, letters were substituted (with A meaning 10, B meaning 11, etc.).

Each column of the printout represented a 10 kHz band of the received signal. An intense signal received on a single narrow band would seem to indicate an alien transmission on that particular frequency. Most of the time, the printout was filled with 1s and spaces, indicating nothing more than background noise from the cosmos.

But the Wow! signal was something else—a clear strong signal on one frequency that increased in intensity and then faded away. Each row on the printout represented 12 seconds on listening. Plotting the seconds before and after the signal shows how it grew, faded, and stood out from the noise (see Figure 119-2).

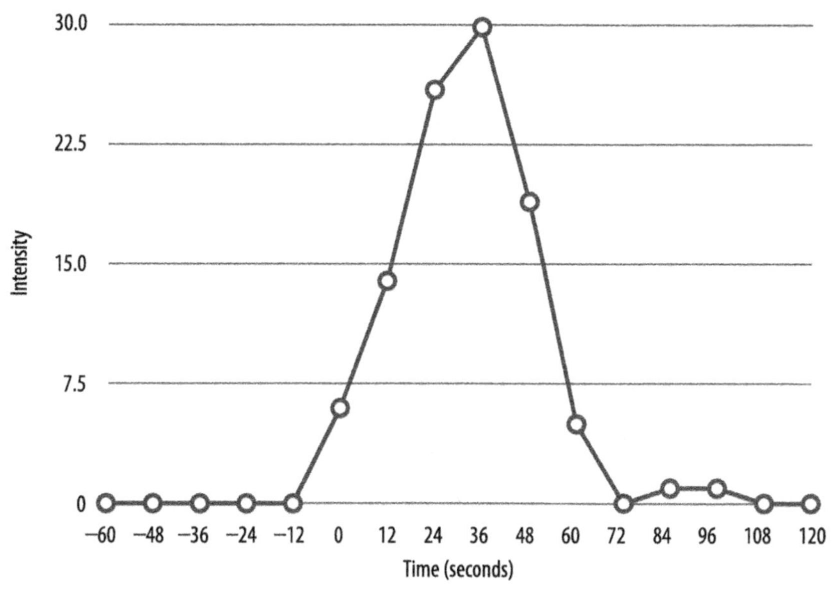

Figure 119-2. Plot of the Wow! signal

The signal lasted for a full 72 seconds—the most tantalizing part of the entire Wow! signal story. Since the Big Ear telescope was not moving, it swept the sky due to the rotation of the Earth. If a signal was being received from a distant, alien radio source, it would have appeared faintly on the printout, increased in intensity, and then faded away over a period of…72 seconds (due to the speed of the Earth's rotation).

Unfortunately, the signal was never detected again. Even worse, Big Ear actually had two antennas that would have listened to the same spot of sky three minutes apart, but the Wow! signal was only heard once.

120

X-10 Graphite Reactor, Oak Ridge, TN

35° 55′ 34.85″ N, 84° 18′ 59.27″ W

The World's First Viable Nuclear Reactor

The Oak Ridge National Laboratory was created in 1943, at the same time the town of Oak Ridge, Tennessee, was built to house the laboratory's workers and their families. The laboratory was originally created as part of the Manhattan Project to build the first nuclear bomb, and its role was the production of the uranium and plutonium needed for the bomb.

To produce the plutonium (see page 376), the laboratory first needed to extract the fissile Uranium-235 isotope and then place it in a nuclear reactor. So not only did the uranium extraction plant have to be created, but the first significant nuclear reactor as well.

The only working nuclear reactor in 1943 was a small proof-of-concept, created by Enrico Fermi (see Chapter 97) and known as the Chicago Pile (it had been built under Stagg Field stadium in Chicago). But the Chicago Pile was much too small for real use, and so the X-10 Graphite Reactor was built at Oak Ridge. It operated from 1943 to 1963 and is now open to the public.

The reactor consists of a cube of graphite with edges over 7 meters long and encased in concrete 2 meters thick. Cut into the graphite are 1,248 diamond-shaped slots, into which aluminum cylinders containing a mixture of Uranium-235 and Uranium-238 were inserted. The entire reactor could contain almost 50 tonnes of fuel. The fuel cylinders were inserted by hand (Figure 120-1) from the east side of the reactor; when they were used up, they were pushed through to the western side using long rods and fell out into a deep pool of water. There they lay until the radiation had died down and they could be collected for further processing.

Since the Uranium-235 is fissile, it decays to give off neutrons that go on to hit other Uranium-235 atoms and split them apart, leading to a chain reaction (see page 376). In the X-10 reactor, the fast neutrons from the reaction are slowed by the graphite blocks to become "thermal neutrons"; these have the right energy level to be absorbed by Uranium-235 to keep the reaction going.

Figure 120-1. Inserting the fuel cylinders

To control the speed of the reaction, the reactor had vertically mounted rods made of boron and steel. The rods could be inserted into holes in the reactor to slow (or even stop) the chain reaction. Another set of boron-and-steel rods passed through the reactor horizontally to control the reaction. The boron absorbed the neutrons, preventing (or slowing down) the chain reaction.

The X-10 reactor was the test bed for many nuclear technologies and proved the viability of nuclear reactors. In addition to producing plutonium for the first atomic bombs, it was later used to create radioisotopes for medical use.

Visitors can view the reactor control room and the east side (where fuel was loaded) on a guided tour of the laboratory. Tours start at the American Museum of Science and Energy, which tells the story of the Oak Ridge National Laboratory and the Manhattan Project. The museum also presents nuclear power in the wider context of energy production.

Practical Information

The American Museum of Science and Energy is open to all; its website is at *http://www.amse.org/*. Tours of the Oak Ridge National Laboratory start at the museum and are held daily; however, due to the sensitive nature of the laboratory's work, tours are only open to U.S. citizens with advance bookings and identification. Details are available from *http://www.ornl.gov/ornlhome/visiting.shtml*.

Uranium Enrichment

Uranium-235 is fissile and will undergo a chain reaction, but it is not abundant naturally. Naturally occurring uranium consists of three isotopes comprising 99.284% uranium-238, 0.711% uranium-235, and 0.005% uranium-234. To obtain enough uranium-235 for a chain reaction, it's necessary to separate it out. Typical nuclear reactors used a mixture of uranium-238 and uranium-235 (see page 376 for reaction details), and the process of increasing the percentage of uranium-235 present in a sample of natural uranium is called enrichment.

The basic process is as follows (don't try this at home!).

Uranium ore (which is mined around the world) contains a high percentage of uranium oxide (U_3O_8). The uranium oxide is dissolved in nitric acid (HNO_3), which creates uranyl nitrate ($UO_2(NO_3)_2$) plus water. The uranyl nitrate is heated to create UO_3. The UO_3 is then reacted with hydrogen gas in a kiln and undergoes a reduction reaction to create UO_2 and water.

UO_2 is then mixed with hydrogen fluoride (HF) to obtain more water and uranium tetrafluoride (UF_4). Finally, the UF_4 is reacted with fluorine gas (F_2) to create uranium hexafluoride (UF_6), which is gaseous when above 56°C.

The uranium hexafluoride is the chemical that enters the enrichment process. A common method of separating the uranium hexafluoride that contains uranium-235 from uranium-238 is by exploiting the difference in weight between the two isotopes in a centrifuge.

To do this, the uranium hexafluoride gas is placed inside a centrifuge (Figure 120-2). The centrifuge rotates at high speed, and the uranium-238 flies out toward the sides of the cylinder. The lighter uranium-235 stays in the middle of the centrifuge, and can then be sucked out. By cascading centrifuges one after another, the percentage of uranium-235 can be increased to a useful level.

Figure 120-2. Nuclear centrifuges in Piketon, Ohio; courtesy of the U.S. Department of Energy

Finally, uranium oxide is typically used for fuel in nuclear reactors and has to be extracted from the uranium hexafluoride. This can be done in a number of ways. The uranium hexafluoride (UF_6) will react with hydrogen gas to produce uranium tetrafluoride (UF_4) and hydrogen fluoride. The UF_4 can then be reacted with steam to produce uranium oxide (UO_2) and more hydrogen fluoride.

121

Kryptos Sculpture, Langley, VA

38° 57′ 6.50″ N, 77° 8′ 44.00″ W

An Unbroken Code

Unless you work for the CIA (or the CIA decides that you must visit their headquarters), you'll have to make do with a virtual visit of the Kryptos sculpture on the CIA website. The sculpture (Figure 121-1), installed in 1990, is next to the new CIA HQ building and contains four encrypted messages. Three of the messages have been decrypted by the CIA and by members of the public. One section, however, remains a mystery to this day, and is believed not to have been cracked even by employees of the CIA and NSA.

Figure 121-1. The Kryptos sculpture

The sculpture is made of petrified wood, slate, and quartz, but its main part is a large S-shaped piece of copper with almost 2,000 letters cut into it. This copper panel consists of two distinct parts: a Vigenère tableau and the encrypted messages.

The Vigenère tableau is a table of letters used for encrypting and decrypting messages using the Vigenère cipher (see sidebar). When decrypted, the first three parts read as follows (the spelling errors are in the original text):

1. BETWEEN SUBTLE SHADING AND THE ABSENCE OF LIGHT LIES THE NUANCE OF IQLUSION
2. IT WAS TOTALLY INVISIBLE HOWS THAT POSSIBLE ? THEY USED THE EARTHS MAGNETIC FIELD X THE INFORMATION WAS GATHERED AND TRANSMITTED UNDERGRUUND TO AN UNKNOWN LOCATION X DOES LANGLEY KNOW ABOUT THIS ? THEY SHOULD ITS BURIED OUT THERE SOMEWHERE X WHO KNOWS THE EXACT LOCATION ? ONLY W_W THIS WAS HIS LAST MESSAGE X THIRTY EIGHT DEGREES FIFTY SEVEN MINUTES SIX POINT FIVE SECONDS NORTH SEVENTY SEVEN DEGREES EIGHT MINUTES FORTY FOUR SECONDS WEST X LAYER TWO
3. SLOWLY DESPARATLY SLOWLY THE REMAINS OF PASSAGE DEBRIS THAT ENCUMBERED THE LOWER PART OF THE DOORWAY WAS REMOVED WITH TREMBLING HANDS I MADE A TINY BREACH IN THE UPPER LEFT HAND CORNER AND THEN WIDENING THE HOLE A LITTLE I INSERTED THE CANDLE AND PEERED IN THE HOT AIR ESCAPING FROM THE CHAMBER CAUSED THE FLAME TO FLICKER BUT PRESENTLY DETAILS OF THE ROOM WITHIN EMERGED FROM THE MIST X CAN YOU SEE ANYTHING Q

The first two parts were encrypted using the Vigenère cipher; the third part—the description of the opening of Tutankhamun's tomb in 1922—is a transposition cipher (where the letters of the original message are shuffled). As for the unsolved fourth message, all that's known are the 97 letters of encrypted text: ?OBKRUOXOGHULBSOLIFBBWFLRVQQPRNGKSSOTWTQSJQSSEKZZWATJKLUDIAWINFBNYPVTTMZFPKWGDKZXTJCDIGKUHUAUEKCAR.

The second part of the message refers to a latitude and longitude, 38° 57′6.5″ N, 77° 8′44″ W. This is about 45 meters from the sculpture, tucked in the garden in the center of the CIA's HQ building. No one appears to know what's there.

James Sanborn, the artist who made Kryptos, has created a number of other sculptures with secret messages on them. His Cyrillic Projector is on display at the University of North Carolina in Charlotte. At night it is illuminated, and the Cyrillic letters are projected onto the floor and walls surrounding it. Its code was broken in 2003—the message consists of secret KGB documents, in Russian.

Sanborn's 1997 sculpture Antipodes contains the text of both the Cyrillic Projector and Kryptos, and can be viewed at the Hirshhorn Museum and Sculpture Garden in Washington, DC.

The Vigenère Cipher

The simplest ciphers just substitute one letter for another. The simplest of all, the Caesar cipher, just shifts the letters by a fixed distance in the alphabet. For example, a Caesar cipher shift of three would turn A into D, B into E, etc. Using that cipher, the message WE ATTACK AT DAWN would be enciphered as ZH DWWDFN DW GDZQ.

The problem with Caesar ciphers is that they are easily broken: it doesn't take too long to try all 26 possible alphabets, and looking at the frequency of letters and comparing with the frequency of English letters can reveal the alphabet even more quickly. If E is always encrypted as H, then lots of Hs indicate that a shift of three was used.

A variant of the Caesar cipher, called a polyalphabetic cipher because it uses more than one alphabet for encryption, is the Vigenère cipher. Blaise de Vigenère was a 16th-century French diplomat and cryptographer. The Vigenère cipher is named after him, although it had originally been described by an Italian cryptographer.

The Vigenère cipher works by choosing a word as a key and using that word to pick a different alphabet for each letter of the message to be encrypted. For example, to encrypt WE ATTACK AT DAWN using the Vigenère cipher with the key BELASO, the message is written out with the key (repeated as necessary) above it:

```
BELASOBELASOBE
WEATTACKATDAWN
```

The encryption continues using a Vigenère tableau (see Figure 121-2). To encrypt each letter of the message, the corresponding letter from the key is used to find a row in the tableau. Then, the letter from the message is looked up on the top row to find a column, and the corresponding letter at the intersection is written down. For example, the letter W is encrypted as X by looking at the row that starts with B, finding W on the top row, and then by finding the letter at the intersection.

The complete encryption of WE ATTACK AT DAWN results in a scrambled message with the same alphabet reused every six letters (because the key has six letters):

```
BELASOBELASOBE
WEATTACKATDAWN
XILTLODOLTVOXR
```

	A	B	C	D	E	F	G	H	I	J	K	L	M	N	O	P	Q	R	S	T	U	V	W	X	Y	Z
A	A	B	C	D	E	F	G	H	I	J	K	L	M	N	O	P	Q	R	S	T	U	V	W	X	Y	Z
B	B	C	D	E	F	G	H	I	J	K	L	M	N	O	P	Q	R	S	T	U	V	W	X	Y	Z	A
C	C	D	E	F	G	H	I	J	K	L	M	N	O	P	Q	R	S	T	U	V	W	X	Y	Z	A	B
D	D	E	F	G	H	I	J	K	L	M	N	O	P	Q	R	S	T	U	V	W	X	Y	Z	A	B	C
E	E	F	G	H	I	J	K	L	M	N	O	P	Q	R	S	T	U	V	W	X	Y	Z	A	B	C	D
F	F	G	H	I	J	K	L	M	N	O	P	Q	R	S	T	U	V	W	X	Y	Z	A	B	C	D	E
G	G	H	I	J	K	L	M	N	O	P	Q	R	S	T	U	V	W	X	Y	Z	A	B	C	D	E	F
H	H	I	J	K	L	M	N	O	P	Q	R	S	T	U	V	W	X	Y	Z	A	B	C	D	E	F	G
I	I	J	K	L	M	N	O	P	Q	R	S	T	U	V	W	X	Y	Z	A	B	C	D	E	F	G	H
J	J	K	L	M	N	O	P	Q	R	S	T	U	V	W	X	Y	Z	A	B	C	D	E	F	G	H	I
K	K	L	M	N	O	P	Q	R	S	T	U	V	W	X	Y	Z	A	B	C	D	E	F	G	H	I	J
L	L	M	N	O	P	Q	R	S	T	U	V	W	X	Y	Z	A	B	C	D	E	F	G	H	I	J	K
M	M	N	O	P	Q	R	S	T	U	V	W	X	Y	Z	A	B	C	D	E	F	G	H	I	J	K	L
N	N	O	P	Q	R	S	T	U	V	W	X	Y	Z	A	B	C	D	E	F	G	H	I	J	K	L	M
O	O	P	Q	R	S	T	U	V	W	X	Y	Z	A	B	C	D	E	F	G	H	I	J	K	L	M	N
P	P	Q	R	S	T	U	V	W	X	Y	Z	A	B	C	D	E	F	G	H	I	J	K	L	M	N	O
Q	Q	R	S	T	U	V	W	X	Y	Z	A	B	C	D	E	F	G	H	I	J	K	L	M	N	O	P
R	R	S	T	U	V	W	X	Y	Z	A	B	C	D	E	F	G	H	I	J	K	L	M	N	O	P	Q
S	S	T	U	V	W	X	Y	Z	A	B	C	D	E	F	G	H	I	J	K	L	M	N	O	P	Q	R
T	T	U	V	W	X	Y	Z	A	B	C	D	E	F	G	H	I	J	K	L	M	N	O	P	Q	R	S
U	U	V	W	X	Y	Z	A	B	C	D	E	F	G	H	I	J	K	L	M	N	O	P	Q	R	S	T
V	V	W	X	Y	Z	A	B	C	D	E	F	G	H	I	J	K	L	M	N	O	P	Q	R	S	T	U
W	W	X	Y	Z	A	B	C	D	E	F	G	H	I	J	K	L	M	N	O	P	Q	R	S	T	U	V
X	X	Y	Z	A	B	C	D	E	F	G	H	I	J	K	L	M	N	O	P	Q	R	S	T	U	V	W
Y	Y	Z	A	B	C	D	E	F	G	H	I	J	K	L	M	N	O	P	Q	R	S	T	U	V	W	X
Z	Z	A	B	C	D	E	F	G	H	I	J	K	L	M	N	O	P	Q	R	S	T	U	V	W	X	Y

Figure 121-2. Vigenère tableau

Because a different alphabet is used for each letter, repeating every six letters, breaking the cipher is difficult. Looking at letter frequencies doesn't help because a single letter may be encrypted differently each time it appears. In the example, A has been encrypted as both L and O.

The Vigenère cipher was considered for some time to be unbreakable, but in the 19th century both Friedrich Kasiski and Charles Babbage (see Chapter 77), discovered that it could be broken by looking at repeated sequences of letters in the encrypted message.

In the example, the letters AT (in ATTACK and AT) happen to line up with LA in the key. That means they are both encrypted in the same way, as LT. And those LTs are six letters apart, which reveals the length of the key used. Once you know the length of the key, you can split the message up—in this case into six separate blocks, one corresponding to each letter in the key—and attack it just like a Caesar cipher. Not only that, but the repeated letters are likely to come from a common English word such as THE, giving more clues about the key used.

The first two parts of Kryptos use a slightly modified Vigenère cipher, where the alphabet used to build up the tableau is reordered by moving the letters K R Y P T O S to the start of the alphabet.

Practical Information

The virtual Kryptos tour can be found at *https://www.cia.gov/about-cia/virtual-tour/kryptos/*. In addition, detailed information about the sculpture and the cryptography used to create it is available from Elonka Dunin's website at *http://www.elonka.com/kryptos/*.

122

Shot Tower Historical State Park, Austinville, VA

36° 52′ 8.52″ N, 80° 52′ 14.27″ W

The Science of the Simple

Some of the simplest things are full of interesting science. Take making lead shot. In 1782 a British man named William Watts patented a method of making lead shot that involved pouring molten lead through a copper sieve and letting it fall a long distance before being cooled in a pool of water.

He built a tower in Bristol, England, and went to work. His tower is long gone, but others around the world from Melbourne to Virginia are still standing, testaments to a really simple technology with lots of science behind it.

Of course, Watts was less concerned with the science than with the result. But if he'd asked, "How does this work?" the answers would have covered surface tension, intramolecular forces, and terminal velocity. But before getting into that, visit the shot tower still standing in the Shot Tower Historical State Park in Virginia.

That shot tower was built in 1807 and has an unusual design—the tower is almost 23 meters tall, with a hole cut into the ground that extends the distance the lead shot could drop by another 23 meters. At the bottom of the shaft was a kettle filled with water, which was accessed by a horizontal shaft that led from the bank of the nearby New River (also a handy water supply for refilling the kettle).

The tower stands in the state park, and for a small fee you can climb up to the top to see where workers poured molten lead to make shot. While you are up there, you can reflect upon the following questions and their answers.

Why use a copper sieve? This one's easy: the melting point of copper is 1084°C, and the melting point of lead is 327°C. Molten lead won't melt a copper sieve. Tungsten has an even higher melting point (3422°C), but copper is far more abundant.

Why does this work at all? When the molten lead passes through the sieve, it turns into thin streams like water from a tap. As the lead falls, it turns into droplets, and these droplets become spherical because of surface tension. Surface tension is caused by the intramolecular forces between the molecules of lead.

The forces pull evenly in all directions except at the surface of the drop, where they can only pull along the surface. This even pulling creates a sphere.

Why does the stream of lead break up into drops in the first place? Once again, surface tension plays a role. The British scientist Lord Rayleigh and Belgian scientist Joseph Plateau determined in the late 1800s that a circular column of falling water will break up into drops if its length exceeds its circumference. As the water falls, any imperfections that prevent it from being a perfect cylinder become exaggerated by surface tension.

Rayleigh and Plateau showed that over time, the imperfections result in a wave forming on the surface of the falling column of water, which grows until the column splits into drops. This is called the Rayleigh-Plateau instability, and because of it water (and lead) breaks up into similar-sized drops corresponding to the wavelength.

How fast is the lead shot going when it hits the water? When anything (lead shot, sky divers, etc.) falls through air, it eventually reaches a constant speed and stops accelerating. Gravity is still acting on the falling body, but this is balanced by the drag from the air.

Bodies falling through air experience an upward force caused by drag, which is proportional to the square of their velocity. The force is defined by the equation $F = -1/2\ C\rho Av^2$. The force is negative because it is acting in the opposite direction to the movement of the body (i.e., upward) and depends on the velocity, v, the cross-sectional area, A; the drag coefficient, C; and the density of air, ρ.

The downward force on the lead shot is from gravity, and is equal to mg (g is acceleration due to gravity, and m the mass of the lead shot). The lead shot stops accelerating when $mg = -1/2\ C\rho Av^2$, which can be rearranged to get the terminal velocity (see Equation 122-1).

$$v = \sqrt{\frac{2mg}{\rho AC}}$$

Equation 122-1. Terminal velocity

So for a #0 buck shot with a diameter of 7.62 millimeters and a weight of 2.62 grams, and a drag coefficient of 0.4 (the value for a rough sphere) and air density of 1.2 kg/m^3 (an average value for air density at sea level at 20°C), the terminal velocity is 48 m/s (or about 172 kph).

Why don't the balls of lead become deformed by drag? When the molten lead is falling through the air, surface tension causes it to form a sphere, but there's a force acting upward on the bottom of the sphere caused by drag. Lead has a very high surface tension (around 430 dynes/cm—a dyne is 10 micronewtons), which is enough to overcome the pressure caused by drag.

But water has a surface tension of just 73 dynes/cm, so what is the shape of raindrops? See the sidebar for the answer.

Practical Information

Visiting information for the Shot Tower Historical State Park is available from the Virginia Department of Conservation and Recreation website at *http://www.dcr.virginia.gov/state_parks/shottowr.shtml*.

The Shape of Raindrops

Despite popular notions, raindrops are not shaped like tears, with a round bottom and a pointed top. However, most raindrops are not spherical either.

Raindrops are formed by the same surface tension that makes lead shot spherical, but the force of air on the bottom of the raindrop distorts the sphere because the surface tension isn't large enough to overcome the pressure.

The shape of a raindrop depends on its size. There's been plenty of research on raindrop shapes by using high-speed photography in nature and in wind tunnels, and theoretical research has been performed to understand the effect of drag on rain.

Raindrops split into three major shapes: spheres, hamburger buns, and parachutes. Raindrops below about 1 millimeter in size remain spheres as they fall. Above 4.5 millimeters, the pressure is so high that the raindrop turns into a little parachute, with a thin film of water billowing upward under the pressure of air on the raindrop's underside.

In between, the spherical raindrop becomes flattened on the bottom by the pressure of falling through air, and remains spherical on top where the pressure is lower. It ends up resembling a hamburger bun or a small bean. See Figure 122-1.

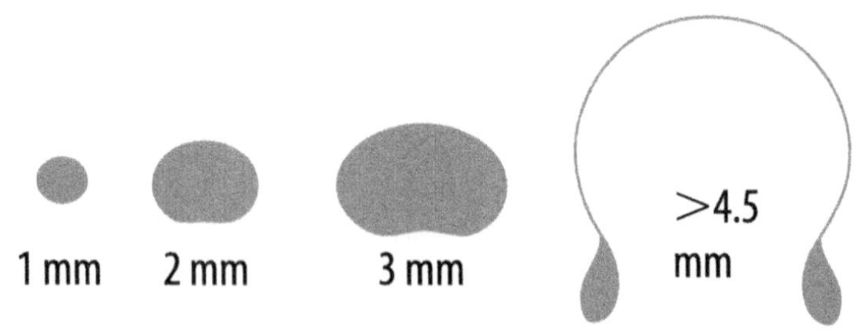

Figure 122-1. Raindrop shapes

123

American Museum of Radio and Electricity, Bellingham, WA

48° 45′ 5.25″ N, 122° 28′ 49.39″ W

CQD

Tune your car radio to KMRE-LP 102.3 FM as you drive into Bellingham, Washington, and you might think you've passed through a time warp—KMRE-LP plays recordings from the media collection of the American Museum of Radio and Electricity. You'll find yourself listening to radio programs from the 1920s to the 1950s.

Far from being a pile of dusty old equipment only of interest to true radio enthusiasts, the museum is a small and delightfully presented collection of radio equipment that tells the story of radio from early experiments with electricity through the Golden Age of Radio, before television displaced the radio as the focal point of most living rooms (see Figure 123-1). To set the mood for visitors, the museum has a mockup of an 18th-century laboratory (as might have been used by Benjamin Franklin) and a full-scale copy of the *Titanic* radio room, including a Marconi radio set from another White Star Line ship. The last message sent from the *Titanic*'s radio room was CQD in Morse code—CQ meant that the message was of interest to anyone receiving, and the D indicated distress.

Figure 123-1. Part of the collection: vacuum tubes; courtesy of Christopher Bellevie

Beat Frequencies and the Theremin

In 1919, Russian inventor Lev Sergeivich Termen (who went by the name Léon Theremin in the West) created the Theremin musical instrument after having worked on proximity sensors. The Theremin has two controls (in the form of antennas), one for pitch and one for volume, neither of which is touched by the musician. But both controls use the hands of the musician as one half of a capacitor in an LC resonant circuit (for more on LC circuits, see page 232).

To generate variable pitch, a Theremin contains a pair of oscillators, generating radio frequency signals. One oscillator creates a fixed frequency; the other is variable. The variable oscillator works by attaching an antenna to an LC circuit. When the musician's hand approaches the antenna, a small capacitance is created from the antenna to the ground through the musician. This human capacitance changes the capacitance in the LC oscillator, thus changing its frequency. When the musician moves his hand, he varies the capacitance and hence the pitch.

Both oscillators are working with radio frequency, so to get an audible frequency the two signals are heterodyned (which is just a fancy way of saying combined). In a Theremin the two frequencies are subtracted to create a beat frequency. The beat frequency is equal to the difference between the two frequencies. By choosing the fixed frequency and the range of the variable frequencies, the Theremin is able to produce a wide range of sounds from small changes in capacitance.

Figure 123-2 shows two sine waves, as might be produced by the two Theremin oscillators, interfering with each other to produce a new wave with a different frequency.

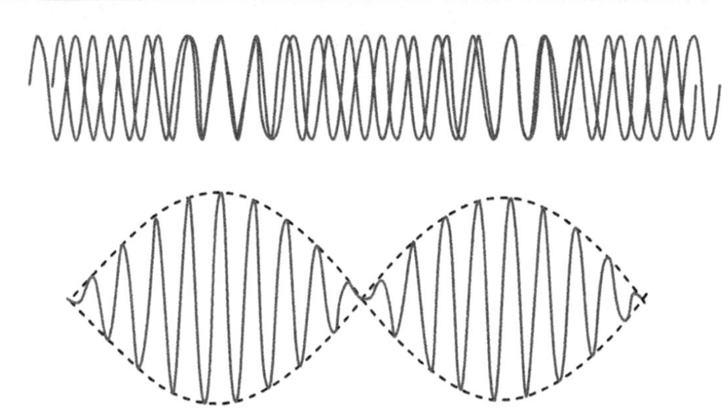

Figure 123-2. Two waves combine to create a beat frequency

The Theremin uses a similar process to control volume.

Heterodyning radio signals is very common—household AM and FM radios use the same principle. When you tune into KMRE-LP 102.3 FM, the radio superimposes a signal at 113 MHz to translate the received signal to 10.7 MHz. The rest of the radio works with the 10.7 MHz signal to produce the sound heard from the loudspeakers.

This has the advantage that almost the entire radio is designed just to work with 10.7 MHz signals; only the first piece of tuning apparatus, called the front end, needs to tune across the FM dial.

There's also a mockup of a 1930s living room with a working 1936 Zenith radio set. Visitors can listen to classic radio and news broadcasts from the 1930s, including the *Lone Ranger* and *Green Hornet*. And there's a working vintage telephone that visitors can use to call the museum front desk with any questions.

Naturally, the museum has a large exhibition of radio equipment, and it's worth booking a tour to get a complete understanding of the equipment on display.

The most recent display is a hands-on exhibit explaining static electricity. The Static Electricity Learning Center is an ideal stop for children, and contains spark machines for generating static electricity, Leyden jars (an early type of battery that could be recharged), and electroscopes.

Probably the most fun thing to do at the museum is play the 1929 Theremin. The Theremin was one of the first electronic instruments, and is played without being touched at all. The eerie Theremin sound was made famous by films like *Spellbound* and *The Day the Earth Stood Still*.

There's also a fine surviving example of Giuseppe Zamboni's 18th-century Perpetual Motion Machine—actually a pendulum that swings between the terminals of a battery. Zamboni built one that ran until well after his death.

If you are serious about radio, there's too much here to see in a day, but for a quick introduction, the museum's explanatory posters can be digested in a couple of enjoyable hours.

Practical Information

Information about the American Museum of Radio and Electricity can be found at *http://www.amre.us*; tours should be booked in advance by calling the museum. The KMRE-LP radio station can be heard over the Internet via SHOUTcast: *http://www.shoutcast.com/*.

124

Grand Coulee Dam, Coulee Dam, WA

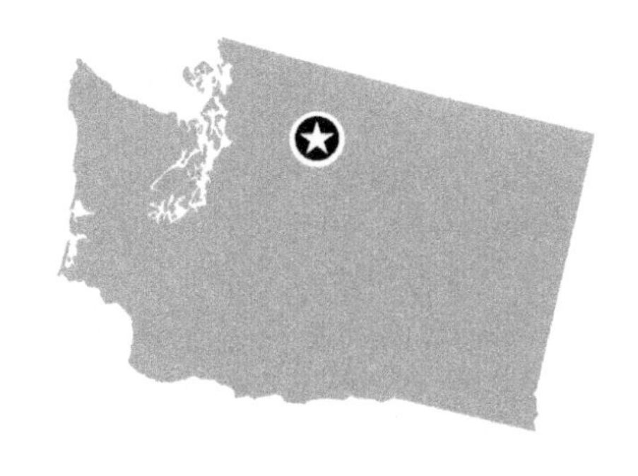

47° 57′ 24″ N, 118° 59′ 0″ W

The Largest Hydroelectric Plant in North America

The Grand Coulee Dam is one of the largest concrete structures in the world, and is the largest hydroelectric station in North America, capable of generating 6,800 megawatts of power. It has a free visitor center, and free tours in a glass elevator inside one of the power plants.

The dam was constructed between 1933 and 1941. It's 168 meters high and 1.6 kilometers wide. It consists of over 9 million cubic meters of concrete. In addition to power generation, the dam is used to control flooding on the Columbia River, and provides water for irrigation projects. The lake it creates, Lake Roosevelt, is used for recreational boating and fishing.

The visitor center explains the construction of the dam, how it is used, and the process of generating and transmitting electricity (for more on electricity generation, see Chapter 30). The visitor center also has some hands-on exhibits: you can try out a pneumatic drill and generate electrical power.

But the highlight of a visit to the dam is the tour inside the Third Powerplant. The power plant was added to the dam starting in 1966 and houses six generators—three are capable of providing 600 megawatts each, and the other three 800 megawatts. Any one of the generators is capable of producing enough power for a city the size of Seattle.

The power plant tour involves descending the face of the dam in a glass elevator and viewing the generators themselves. You can see the tops of the generators, along with a massive crane, one of the largest in the world, that is used for servicing the generator components.

The turbines cannot be seen, because they are immersed in water, but the huge stainless steel turbine shaft linking the turbine to the generator is visible rotating under the force of water. The sound and sight of the shaft brings to life the incredible power of the turbine below.

Francis Turbine

Another Grand Coulee Dam superlative is that it has the largest Francis turbines in the world. These are the turbines that drive the generators in the Third Powerplant.

The Francis turbine (see Figure 124-1 and Figure 124-2) was invented by British/American engineer James Francis in 1849. Today, the turbines are widely used for power generation (the largest hydroelectric plant in the world, the Three Gorges Dam in China, also uses Francis turbines).

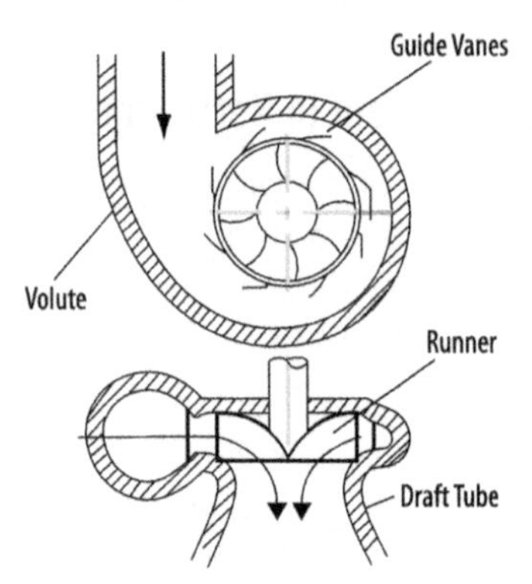

Figure 124-1. Francis turbine schematic

The turbine is mounted horizontally, and water from the dam enters it via a spiraling tube. The worm-like tube ensures that the incoming water enters all parts of the turbine uniformly. The water enters from the outside of the turbine, moves toward the center, and eventually exits downward, perpendicular to the spiral tube and parallel to the turbine shaft.

Unlike in a simple water wheel, in a Francis turbine it is not only the velocity of the water entering the turbine that turns the shaft; changes in water pressure also help turn it. The space between the turbine blades creates a nozzle that lowers the pressure of the water passing through. When water enters the turbine, its velocity pushes the turbine blade; as the water exits and the pressure lowers, its velocity increases. The increase in velocity creates an equal and opposite reaction force (see Chapter 69 for more on Newton's Laws) that further turns the turbine shaft.

Figure 124-2. Installing a Francis turbine at the Grand Coulee Dam

Since Francis turbines use both water velocity and pressure to turn the shaft, they are highly efficient (achieving up to 90% efficiency).

The generators consist of a moving magnet called the rotor that is powered by the turbine, and a fixed stator, which is the coil of wire in which the electrical current is induced (for more on electrical induction, see Chapter 75). The generators in the Grand Coulee Dam use electromagnets in the rotor powered by DC to create a very powerful moving magnetic field.

When outside, it's possible to see the transformers that take power from the generators, increase the voltage, and pass the power on to the grid.

Between Memorial Day and the end of September, it's worth sticking around until after dark—there's a free laser show projected onto the dam itself.

Practical Information

Information about the Grand Coulee Dam, including the visitor center, the power plant tour, and the laser light show, is available from the U.S. Bureau of Reclamation website at *http://www.usbr.gov/pn/grandcoulee/gcvc/*.

125

Reber Radio Telescope, Green Bank, WV

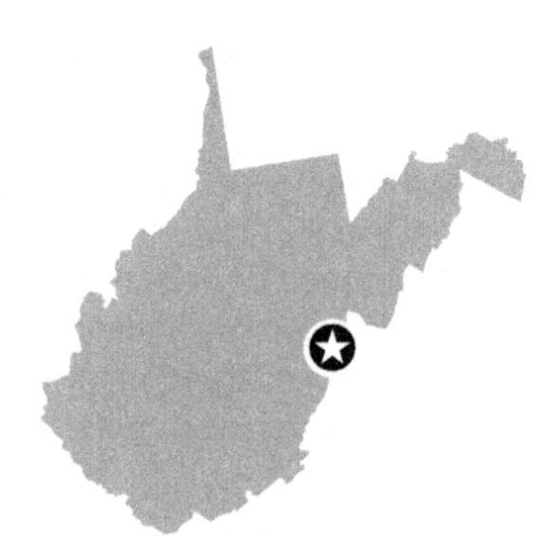

38° 25′ 48.61″ N, 79° 49′ 4.45″ W

The Homemade Telescope

In 1937, an amateur astronomer and radio ham named Grote Reber hand-built his own radio telescope in his backyard in Wheaton, Illinois. The almost 10-meter parabolic dish was the only operating radio telescope in the world until after the end of the Second World War. Using the telescope, he drew radio maps of the sky and discovered that some apparently dark areas (with no stars) nevertheless were producing radio signals. Between 1937 and 1948, Reber and his home-built telescope revolutionized astronomy, and his telescope marked the birth of radio astronomy.

In 1948 Reber sold the telescope to the U.S. National Bureau of Standards, and it began a countrywide journey—first being moved from Illinois to Virginia, then to Colorado, and finally to the National Radio Astronomy Observatory (NRAO) in Green Bank, West Virginia, where it sits today. It is still in working order.

Reber didn't have enough money to build a fully steerable telescope—it took half a year's salary just to get the telescope built—and so his backyard dish could only rotate in one direction, relying on the motion of the Earth to scan in the other direction. When the telescope was bought by the NRAO, it was mounted on a rotating platform to make it fully steerable; Reber helped supervise the work before moving to Tasmania to continue radio astronomy.

The NRAO has many other radio telescopes, including part of the horn antenna used by Harold Ewen and Edward Purcell to detect the 21-centimeter hydrogen line (see sidebar).

Happily, the general public is welcome at the observatory. The recently constructed Green Bank Science Center has many exhibits, mostly aimed at schoolchildren. For adults, there's a bus tour of the observatory that takes in the historic telescopes and the Robert C. Byrd Green Bank Telescope, the largest fully steerable radio telescope in the world.

21-Centimeter Hydrogen Line

About 240,000 years after the Big Bang, hydrogen and helium atoms began to form. Initially they were ionized, but after a period of cooling, the ionized atoms captured free electrons and became neutral. When this occurred, the universe became transparent and photons were free to travel, resulting in the cosmic background microwave radiation (see page 413). However, the hydrogen present also emitted radiation because of hyperfine splitting.

A neutral hydrogen atom has a single proton and a single electron. In 1913, the physicist Niels Bohr introduced a model of the hydrogen atom that explained why the spectrum of light (and other electromagnetic radiation) from excited hydrogen atoms consisted of specific lines of color. By applying thousands of volts to a glass tube containing hydrogen gas, it's possible to obtain a spectrum of light from the gas (see Figure 125-1).

Figure 125-1. Hydrogen emission lines

Bohr's model showed that the radius of the electron's orbit around the proton in a hydrogen atom was constrained to certain fixed positions (called quanta), and that when the electron changed orbit it either absorbed energy from a photon, or emitted one (see Figure 125-2).

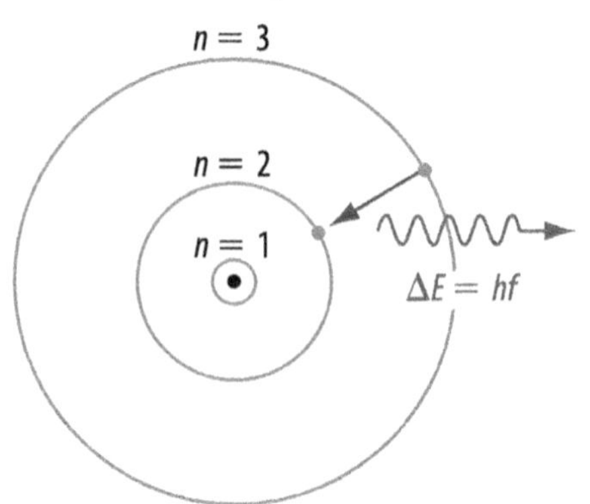

Figure 125-2. Bohr's model of hydrogen

Bohr also suggested that the energy level associated with the electron depended on the orbit. The closer the orbit, the lower the energy, and when jumping from a far-to-near orbit, energy (in the form of a photon) is emitted. Bohr's rule was that the frequency of the emitted photon, f, is determined by Planck's constant and the change in energy level, ΔE: $f = \Delta E / h$.

Bohr's model was approximately correct, but was only useful for hydrogen. It was subsequently refined by quantum mechanics, which helped explain an anomaly in the emitted spectrum of hydrogen.

When at the lowest possible energy level, with the electron closest to the proton, there are actually two separate energy levels for the atom. These occur because not only does the electron orbit the proton (actually, "orbiting" is just an approximation, but close enough), both the electron and proton have a spin. The spin can be thought of as the particle rotating about its own axis.

This rotation leads to a magnetic field created by the proton and the electron. When the magnetic fields are in the same direction (i.e., both norths are in the same direction), the atom has a slightly higher energy than when they are opposed.

When the electron spin changes direction, there is a change in energy. If the energy decreases, then a photon is emitted, and the emitted wavelength depends on the difference in the energy level between the two possible electron spins—essentially as Bohr had predicted. This is called hyperfine splitting of the emission spectrum (because a single line of spectrum is actually made up of more than one closely spaced line), and in hydrogen the emitted photon has a wavelength of 21 centimeters.

This transition between spins occurs with a very, very small probability. However, the universe is filled with neutral hydrogen, and hence microwave radiation at 21 centimeters is detectable and is evidence of the Big Bang. It's also useful to astronomers because the radiation can be used to map the presence of hydrogen throughout the universe, which can aid in the understanding of the structure of the universe after the Big Bang.

The NRAO is in a wonderfully quiet location, partly because of its remoteness in the Potomac Highlands, and partly because most electronic equipment is banned—no cell phones, no digital cameras, no MP3 players. Electronic items might interfere with the very sensitive radio receivers used for listening to the sky.

Practical Information

Find details about visiting the NRAO at Green Bank at *http://www.gb.nrao.edu/*.

126

The Greenbrier, White Sulphur Springs, WV

37° 47′ 7.44″ N, 80° 18′ 29.88″ W

The VIP Bunker

About 400 kilometers southwest of Washington, DC, is one of the most luxurious and exclusive resorts in the U.S.—The Greenbrier. Set in over 2,600 hectares of the Allegheny mountains in West Virginia, with everything from golf to fly-fishing to an onsite clinic at hand, The Greenbrier is the kind of resort that welcomes kings, presidents, and other dignitaries. But The Greenbrier has a secret—it was also the location chosen to house the entire U.S. Congress in the event of nuclear war (Figure 126-1).

Figure 126-1. Inside The Greenbrier bunker; courtesy of The Greenbrier

Radiation Protection

Nuclear explosions produce an enormous fireball, an electromagnetic pulse capable of knocking out electrical items (see page 196), hurricane-force winds, and enormous changes in air pressure. They also produce large amounts of radiation in four main forms: neutrons, alpha particles, beta radiation, and gamma rays.

The biggest health risk from the radiation comes from the neutrons. It's the neutrons that actually keep the nuclear explosion going (see page 420) by colliding with uranium (or plutonium) atoms and causing them to split apart and release energy (and more neutrons). Some of the neutrons hit other types of atoms (such as those in the air or ground) and cause them to become radioactive. These newly radioactive atoms get sucked up into the explosion and then come down as radioactive fallout.

The fallout is one of the main reasons to build a bunker underground, especially a bunker like The Greenbrier's, which was not designed to survive a direct hit by a nuclear weapon. The fallout also makes filtered air and clean water necessary.

Gamma rays are a form of electromagnetic radiation with a very short wavelength, and they are very difficult to stop. They have high energy and will easily enter the body where they can cause damage to DNA. This DNA damage can directly cause cancers, and can also create hereditary diseases. Protection against gamma radiation requires many centimeters of lead or meters of concrete. Gamma rays can also travel great distances through the air.

Luckily, the other two types of radiation are much easier to deflect. Beta radiation (which is just high-energy electrons) can be stopped by even a thin sheet of metal. It can enter the body through the skin and can cause cancers. Beta radiation is also used in cancer treatments because its effect is fairly localized—a source of beta radiation can be used to kill cancerous cells without affecting the rest of the body. Alpha radiation (which is just helium nuclei) can be stopped by the skin or even a sheet of paper (see Figure 126-2).

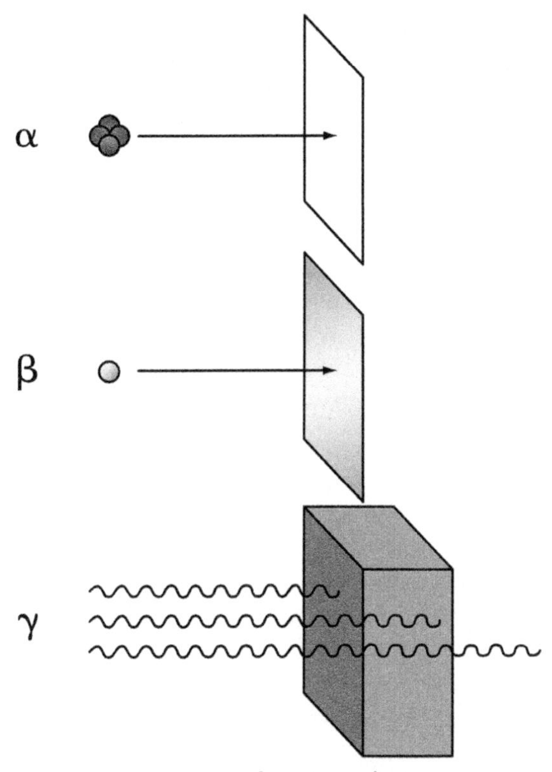

Figure 126-2. Stopping radiation

But alpha and beta radiation are dangerous if they are produced inside the body. If a source of either is breathed in or swallowed, the radiation emitted inside the body can cause cancer and other internal damage. For this reason, the radioactive gas radon is very dangerous. Radon is present in the Earth, and can enter homes through the floor in certain parts of the world. As radon decays (and eventually becomes lead), it turns into a number of other elements. Of these the most dangerous is Polonium-210, which decays, releasing alpha radiation.

In 2006, the Russian exile Alexander Litvinenko died after drinking a cup of tea containing polonium-210. Polonium-210 is a particularly effective poison because it is hard to detect: it gives off very little gamma radiation, making it safe to carry around and unlikely to be picked up by radiation detectors. However, once inside the body, it causes damage as alpha particles attack soft tissues.

In 1958, The Greenbrier began building a new wing with space for its clinic. This wing apparently needed enormously deep foundations, because the construction company building it excavated far into the earth. In fact, they were making a hole big enough to construct a 10,500-square-meter bunker where 1,100 people could sleep, eat, and work. And beneath The Greenbrier's own clinic was a fully functioning secret clinic to keep the lawmakers healthy.

The bunker stayed operational until 1992, when its secret was revealed in a newspaper article. For 30 years, a fake company had operated audio-visual services for The Greenbrier while maintaining the bunker in a state ready to receive 1,100 occupants with four hours' notice. On arrival at the bunker, members of Congress would have stripped naked, showered to remove contamination, and been given fresh, drab clothing to wear.

Once inside, the bunker could be sealed off behind massive blast doors and survive on its own power (from diesel generators), water (from storage tanks), and air. The bunker had rooms for both houses of Congress to meet, a television studio for communication with what was left of the outside world, and a spartan dormitory. Until its closure, staff kept nameplates updated on every bunk bed, so that each member of Congress would know exactly where he or she would be sleeping. Only the leadership of the House and Senate had private rooms. Everyone would eat together in a large canteen capable of seating 400, and would subsist on an enormous supply of freeze-dried food kept stocked inside the bunker.

Part of the bunker was used by the hotel for meetings—little did the VIP guests know that they were sitting in rooms that could be hermetically sealed in seconds and that were designed for meetings of the displaced government of the U.S. In its 30-year existence, the bunker became operational just once—during the Cuban Missile Crisis, the bunker was readied to receive occupants, and historical documents were transported to The Greenbrier for storage in the bunker's extensive vault.

The bunker is now open for tours, which take in the entire site. Visitors see everything from the power plant, clinic, and kitchen to the dormitories and meeting rooms. There are even stacks of supplies waiting for use. The eeriest part is the incinerator, which stands ready to deal with any member of Congress who didn't survive their visit.

Given the cost of a room, it's lucky that you don't have to stay at The Greenbrier to take the tour. As it stands, it will still set you back a bit—at the time of this writing, it's $30 a head.

Practical Information

The Greenbrier's website is at *http://www.greenbrier.com/*. Information about touring the bunker is at *http://www.greenbrier.com/site/bunker.aspx*.

127

Magicicada Brood X, East Coast, U.S.

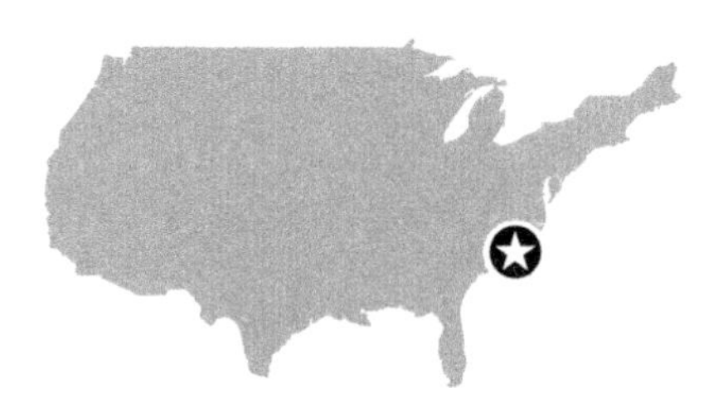

2021, 2038, 2055, ...

Here's a sight you'll have to wait until 2021 to see: the emergence of the Magicicada periodical cicada along the eastern coast of the U.S. And if you miss it in 2021, then your next chance won't be until 2038. Of course, if you're squeamish, you might want to avoid the emergence of the entire cicada population altogether—a cicada is a large, flying insect, and even though it doesn't sting or bite, it has a nasty habit of treating humans as a suitable landing place.

Cicadas are found all over the world, and most come out of hiding each year. But the Magicicada of North America comes out of hiding only once every 13 or 17 years, depending on the specific species.

Since 1907, scientists have been tracking 13 broods of Magicicada, numbered using Roman numerals I through XIII. Brood X (sometimes called the Great Eastern Brood) is the largest of all and last appeared in May 2004, blanketing Delaware, Georgia, Indiana, Kentucky, Maryland, Michigan, North Carolina, New Jersey, New York, Ohio, Pennsylvania, Tennessee, Virginia, and West Virginia (see Figure 127-1).

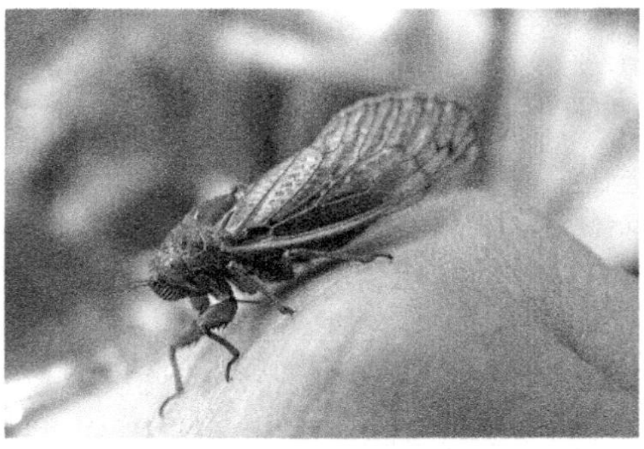

Figure 127-1. A Magicicada from 2004; courtesy of Terry Spears/TERSAN Photography

If you're in the area in 2021, then you'll only have a few weeks to see this remarkable insect. The Magicicada in Brood X spends almost its entire life underground, feeding off the roots of trees. Then, after 17 years in the soil, they all emerge at the same moment in the spring.

The broods are enormous, with up to 370 insects per square meter. As the cicadas leave their hiding places, the sky becomes filled and darkened by insects in search of a mate. The cicada can only mate once it has left the ground, and in order to attract a mate it makes a clicking sound that can reach 120 dB. It is one of the loudest insects of all.

Once the cicada has mated, the female lands on a tree, cuts into the bark, and lays her eggs. After the eggs hatch, they fall to the ground, burrow into the roots of the tree to a depth of about 20 centimeters, and start their 17-year wait. When they finally emerge, they are about 3 centimeters long with a black body and red eyes. The cicada climbs its tree, sheds its outer skin (leaving the abandoned crispy shell), and flies away. Within weeks, the cicada will have mated and died.

It's common to mistake cicadas for locusts, but despite their huge numbers, cicadas do little damage to the plants and trees they land on.

A good base to experience the next emergence of Brood X is Washington, DC. Since the U.S. capital is filled with museums (see Chapter 92), you can bide your time until one of the three Magicicada species in Brood X makes its appearance.

Once the cicadas are out (any TV or radio station in the area will keep you apprised), head west out of the city up the Potomac River, following George Washington Memorial Parkway. You should already be able to hear the cicadas chirping along this leafy drive. Exit at Chain Bridge Road/Dolley Madison Boulevard and drive past the entrance to the CIA in Langley, Virginia. Turn off onto Route 193 and follow it to Draneville District Park (also known as Scotts Run Nature Preserve).

The park is on the Virginia/Maryland border and will be teeming with cicadas. They won't bite, but if you stand still, they are sure to land on you.

Other broods (also on 13- or 17-year cycles) appear in Iowa, Kansas, Missouri, Oklahoma, Wisconsin, and Michigan. The Great Southern Brood (Brood XIX) is a 13-year Magicicada and is due to appear from Arkansas to North Carolina in 2011. This is the first opportunity to see a periodical cicada if you can't wait for the explosion of Brood X in 2021.

Practical Information

Visit Cicada Mania for everything you need to know about cicadas (including whether they are safe to eat): *http://www.cicadamania.com/cicadas/*. You can find more information, as well as detailed maps of cicada sightings, at *http://www.magicicada.org/*.

Prime Numbers

If you are of a mathematical mind, you'll have instantly noticed that the Magicicada cycles are the prime numbers 13 and 17. Prime numbers are not divisible by any other numbers, and are important in many branches of mathematics (including cryptography). The mathematical properties of 13 and 17 seem to be important to Magicicada.

One theory is that Magicicada has a prime-numbered cycle to help ward off predators. Suppose that the cicada instead appeared every 12 years: any cicada predator that appeared every 1, 2, 3, 4, 6, or 12 years would be able to attack the entire cicada population (see Figure 127-2).

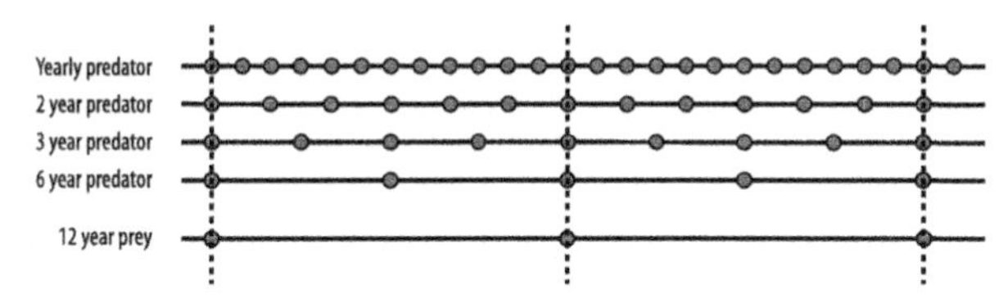

Figure 127-2. Preying on insects with a 12-year cycle

Those numbers are not picked at random, obviously: they are the factors of 12. Since a prime number has no factors, a predator would have to exactly match the cycle of the cicada. If it does not, then it will not synchronize with the cicada's appearance very often.

For example, a predator that appears every 2 years would only synchronize with a 13-year Magicicada every 26 years (see Figure 127-3).

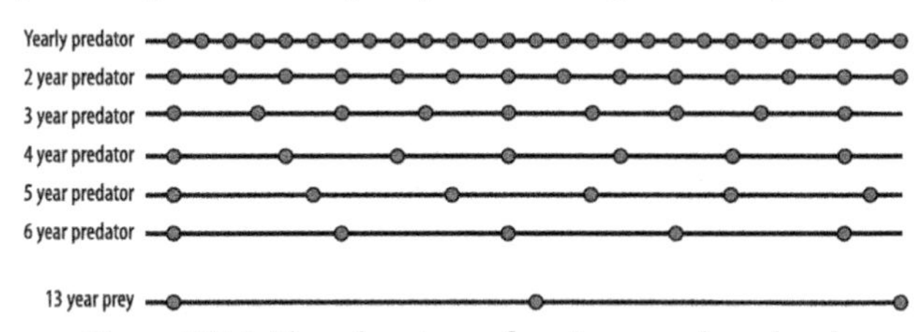

Figure 127-3. The advantage of a prime-numbered cycle

Another theory is that the long cycle is intended to prolong the life of the cicada if there's an unusually cold winter. By staying underground longer, the likelihood that they'll need to emerge on a particularly cold year is reduced.

Yet another advantage of a prime-numbered cycle is that different species of cicada are unlikely to emerge during the same year. Periodic cicadas with cycles of 13 years and 17 years will meet only once every 221 years. That reduces the chances of breeding between species, which would likely mess up the prime-numbered cycle of the offspring.

128

Magnetic North Pole

82° 42′ 0″ N, 114° 24′ 0″ W

Somewhere in Canada

Not only is the Magnetic North Pole difficult to get to, but it also moves over 10 kilometers per year, and it even moves in a rough oval as much as 80 kilometers each day. But it is roughly the spot to which compasses point when determining the direction "north." Until GPS completely displaces compasses, for the purposes of navigation it's important to know the location of the Magnetic North Pole and the lines of the Earth's magnetic field (Figure 128-1).

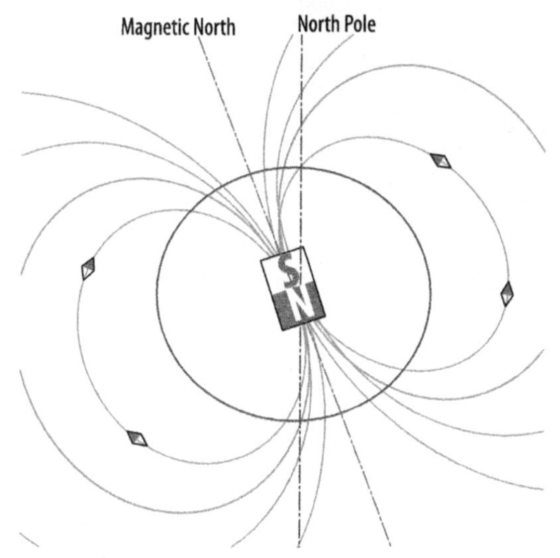

Figure 128-1. The North Pole and Magnetic North Pole

The Earth rotates on its axis, at 23.5° to the perpendicular to the Earth's orbit around the Sun. The axis passes through the Geographic North and South Poles. The location of the geographic poles sets the lines of longitude; latitude is measured from the Equator, which sits halfway between the poles. The axis and the poles are the basic reference points for navigation.

The Earth's Magnetic Field

Although the Earth's magnetic field may appear to come from a permanent bar magnet, it does not. Even though the center of the Earth contains iron, it is not permanently magnetized because it's too hot. At a temperature called the Curie point, ferromagnetic materials lose their magnetism; for iron that happens at 768°C.

The Earth consists of four major layers: from outside to inside, these are the crust (on which we live), the mantle, the outer core, and the inner core (Figure 128-2). The inner core is believed to be mostly composed of iron at over 5,000°C—well above the Curie point, preventing it from being a permanent magnet.

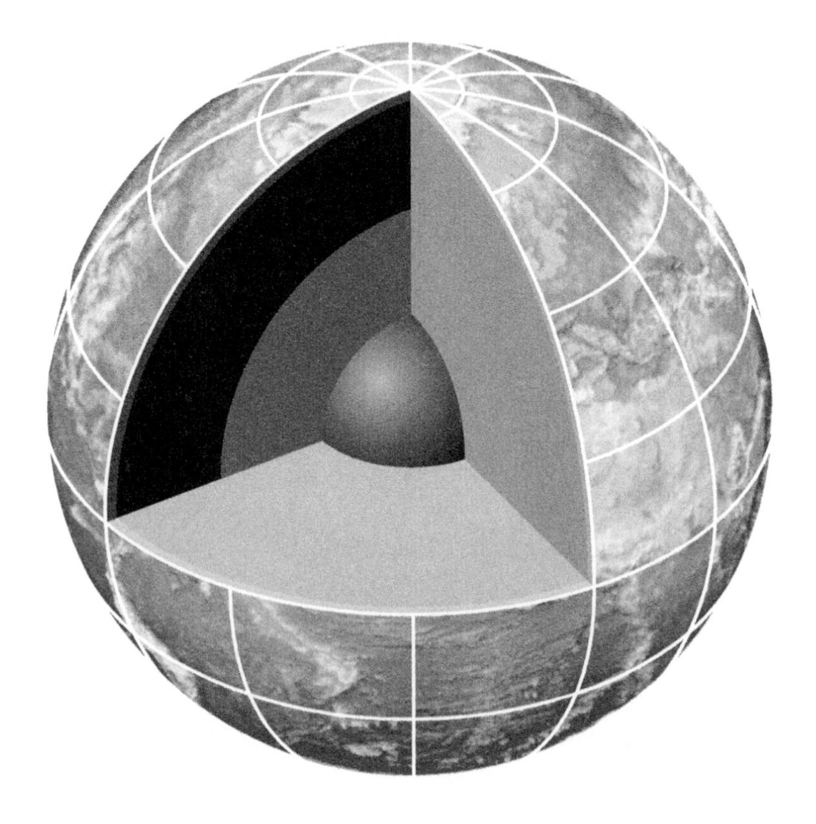

Figure 128-2. Earth's layers

Surrounding the inner core is the outer core. The outer core is believed to be a liquid, hot alloy of nickel and iron. It's the outer core that is thought to be the source of Earth's magnetic field, and the motion of the outer core's liquid explains the changing nature of the Earth's magnetic field.

The liquid outer core apparently experiences a dynamo effect: as the liquid passes through the Earth's magnetic field, a current is induced in the liquid metal; this current creates a magnetic field, which in turn induces a current in the moving liquid. In this way, the magnetic field is self-sustaining.

The outer core liquid is moving because of convection currents generated by the intensely hot inner core, and it moves relative to the surface of the Earth because of the Coriolis effect. The Coriolis effect is an apparent error in Newton's laws of motion that occurs when the vantage point isn't fixed.

For example, if a cannon ball is launched a long distance, directly northward along a meridian line, it will not land on the meridian line. This happens because the Earth rotates while the ball is in the air, and because the speed of the rotation depends on the latitude. These two factors mean that the ball falls to the east of the meridian line because its landing point is moving much more slowly than its launch point.

From the perspective of an Earthbound viewer, an unknown (and nonexistent) force appears to have deflected the ball from its straight path. This is called the Coriolis effect, and the nonexistent force is the Coriolis force.

The overall effect of these forces is that the liquid appears to create dynamos along a roughly north-south axis. These dynamos are responsible for the Earth's magnetic field.

Unfortunately, compasses do not point to the Geographic North Pole, but instead follow the Earth's magnetic field and point approximately toward the ever-moving Magnetic North Pole. To compensate for this problem when using a compass, it is necessary to know the declination. At any point on the Earth, the declination is the angle between a line drawn to the Geographic North Pole and the direction in which the compass is indicating north.

The Earth's magnetic field can be approximated by imagining that the Earth contains a large bar magnet whose south pole points toward the Magnetic North Pole and whose north pole points toward the Magnetic South Pole. (In reality, the Magnetic North and South Poles are not diametrically opposite each other, and the Earth's magnetic field is not even uniform across the Earth's surface.)

The imaginary bar magnet is titled approximately 11° degrees from the Earth's rotational axis. The south pole of this imaginary magnet points to a spot called the Geomagnetic North Pole, which is located near Greenland at 79.74° N, 71.78° W. But only about 90% of the Earth's magnetic field can be approximated by a bar magnet—the other 10% causes the Earth's magnetic field to vary by location and so the Magnetic North Pole and the imaginary Geomagnetic North Pole do not coincide.

Currently, the Magnetic North Pole is located in the far north of Canada. In 2001, it was found to be at 81.3°N, 110.8°W. When navigating north with a compass, a traveler will follow the lines of the Earth's magnetic field, eventually arriving at the Magnetic North Pole, but not by the most direct route because of variations in the field. For this reason, the compass only approximately points to the Magnetic North Pole, and declination varies from place to place.

It is possible to reach both the Magnetic North Pole and the Geographic North Pole by starting in the far north of Canada at Resolute Bay in Nunavut. The Qausuittuq Inn at Resolute Bay has been the starting point for many such expeditions (ask there for directions!).

The truly hardy can enter the biannual Polar Race from Resolute Bay to the Magnetic North Pole.

Practical Information

The Qausuittuq Inn's website is *http://www.resolutebay.com/*. The Polar Race can be found at *http://polarrace.com/*.

Acknowledgments

Many anonymous people picked up telephones at museums, universities, tourist offices, town halls, and companies around the world and answered my questions. Others answered my unsolicited emails asking for press kits, detailed information, photographs, and technical help. Amongst them were a number of professors to whom I'm grateful for their expertise and willingness to help me understand some of the technical topics that were new to me.

Those whose names I know are: Carolyn Rule, Denise Chamberlain, Cathy Avent, Gian Piero Siroli, Cristina Bueti, Keith Matthew, Floyd B. Hanson, Junya Ishihara, Naoki Iimura, Tony Evans, Rory Cook, Emilie Collin, Nathalie Slinckx, Richard Scarth, Matthew Trainer, Jason McFall, Dreas van Donselaar, Caitlin Hawke, Caroline Seats, Mark Frank, Daniel Gambis, Kevin Brown, Cathy Liberatore, Pauline Shepheard, Kate Cook, Dianne Knippel, Patrick Weidman, Andrew Croxton, Phil Green, David Gordon, Lynn Swann, Howard Perlman, Ken Beard, Caroline Durbin, Judy Strebel, Derek Ingram, Lynda Sather, Michael Friendly, Jack Kirby, Catherine Masteau, Susanne Giehring, Manni Heumann, Michaela Jarkovska, Kai Hampel, Robert Hulse, Katreena Dare, Diane Dodd, Tom Vine, Isabel Lara, Katharine St. Paul, and Emily McLeish.

Two people, in particular, encouraged me to write (despite the fact that I was an apparently illiterate computer-nerd): Jeff Sanders, my doctoral supervisor, and Bill O'Neill, who let me write freelance for *The Guardian*.

I'm particularly grateful to the Flickr users who freely gave their photographs of places and objects around the world. They are acknowledged alongside each of their photographs (where requested, I have included the user's Flickr name so that you can view their photostreams). The many editors of Wikipedia helped greatly, especially with citations linking directly to original material; I was able to make a small contribution by correcting the few errors I found as I researched this book.

The HyperPhysics website from Georgia State University is a wonderful resource for anyone interested in physics. It has clear explanations and simple diagrams that elucidate difficult topics (***http://hyperphysics.phy-astr.gsu.edu/hbase/HFrame.html***). NASA's Jet Propulsion Laboratory website explains clearly the physics of flight and rocketry (***http://www.jpl.nasa.gov/***).

The Nobel Foundation's excellent website (***http://nobelprize.org/***) is a superb resource, as it contains PDF versions of Nobel Prize presentation speeches, and lectures given by Nobel laureates. Many original scientific papers are still under copyright and must be licensed from journals before reading them, which made for some painful experiences as I attempted to read original papers to ensure accuracy. In contrast, the Nobel Foundation makes access to the scientific lectures given by its laureates simple and free.

I owe an unrepayable debt to my parents, for having recognized before I could walk that I was fascinated not by plants or animals or stories but by machines (and in particular clocks). I only hope that I can "pay it forward" by recognizing the particular desires of my own children.

Although I wrote this book, it didn't end up in your hands by my efforts alone. I'm very grateful to the large team of people at O'Reilly who contributed by designing, copyediting, illustrating, proofreading, editing, managing production, and countless other tasks that were hidden from me for my own good. I would particularly like to thank the people I dealt with most frequently for their patience, good humor, and effort to put this book in your hands: Julie Steele, Rachel Monaghan, and Emily Quill. Thanks also to technical reviewers Howard Dierking, Cameron Freer, Bryan Lincoln, and Phil Tavernier.

Finally, *merci* to my wife, who read my drafts despite being only vaguely interested in science and technology, who put up with me working odd and long hours, and who gave me much-needed encouragement.

Index

B

C

D

E

H

I

J

K

L

M

N

O

P

Q

R

S

T

U

V

W

X

Y

Z

About the Author

John Graham-Cumming is a wandering programmer who's lived in the UK, California, New York, and France. Along the way he's worked for a succession of technology start-ups, written the award-winning open source POPFile email program, and churned out articles for publications such as *The Guardian* newspaper, *Dr. Dobb's Journal*, and *Linux Magazine*. His previous effort writing a book was the obscure and self-published computer manual *GNU Make Unleashed*, which saturated its target market of 100 readers. Because he has a doctorate in computer security, he's deeply suspicious of people who insist on being called Doctor, but doesn't mind if you refer to him as a geek. He is the proud owner of a three-letter domain name where he hosts his website: *http://www.jgc.org*.

Colophon

The text font is Adobe Myriad Birka; the heading and cover font is House Industries' Chalet Comprime.

CPSIA information can be obtained at www.ICGtesting.com
Printed in the USA
BVOW06s1101230715

409930BV00042B/178/P